Research on Plant Genomics and Breeding 2023

Research on Plant Genomics and Breeding 2023

Guest Editors

Zhiyong Li
Jian Zhang

MDPI

Basel • Beijing • Wuhan • Barcelona • Belgrade • Novi Sad • Cluj • Manchester

Guest Editors

Zhiyong Li
State Key Lab of Rice Biology and Breeding
China National Rice Research Institute
Hangzhou
China

Jian Zhang
State Key Lab of Rice Biology and Breeding
China National Rice Research Institute
Hangzhou
China

Editorial Office
MDPI AG
Grosspeteranlage 5
4052 Basel, Switzerland

This is a reprint of the Special Issue, published open access by the journal *Plants* (ISSN 2223-7747), freely accessible at: www.mdpi.com/journal/plants/special_issues/T0429MTBKK.

For citation purposes, cite each article independently as indicated on the article page online and using the guide below:

Lastname, A.A.; Lastname, B.B. Article Title. *Journal Name* **Year**, *Volume Number*, Page Range.

ISBN 978-3-7258-3044-2 (Hbk)
ISBN 978-3-7258-3043-5 (PDF)
https://doi.org/10.3390/books978-3-7258-3043-5

Contents

plants

Editorial

Research on Plant Genomics and Breeding: 2023

Zhihui Chen [1], Xiaohong Tong [2], Jian Zhang [2], Jie Huang [2,*] and Zhiyong Li [2,*]

1 Tea Research Institute of Fujian Academy of Agricultural Sciences, Fuzhou 350013, China; chenzhihui75@sina.com
2 State Key Laboratory of Rice Biology and Breeding, China National Rice Research Institute, Hangzhou 311400, China; tongxiaohong@caas.cn (X.T.); zhangjian@caas.cn (J.Z.)
* Correspondence: huangjie@caas.cn (J.H.); lizhiyong@caas.cn (Z.L.)

Over the past two decades, the rapid progress made in plant breeding has been significantly driven by the integration of knowledge in the fields of plant genomics and genetics, and by the application of state-of-the-art biotechnologies. Furthermore, collaborative initiatives within community-based projects have yielded substantial contributions to these advancements. This article aims to furnish a comprehensive overview of recent publications in *Plants*, with a specific emphasis on a diverse array of molecular biology research methodologies associated with plant genomics and breeding.

1. Genetic Basis Underpinning Plant Trait

Amalova and colleagues from the Institute of Plant Biology and Biotechnology, Kazakhstan, attempted to identify Quantitative Trait Loci (QTLs) associated with wheat adaptation traits using a Nested Association Mapping (NAM) population [1]. The authors evaluated plant adaptation-related traits in two contrasting regions, including the heading date, the seed maturation date, the plant height, the peduncle length, and the yield of 290 recombinant inbred lines (RILs) of the NAM population. The integration of findings from 10,448 polymorphic SNP (single-nucleotide polymorphism) markers in the genome-wide association study (GWAS) led to the identification of 74 QTLs associated with four agronomic traits under study. Notably, 16 of these QTLs had been consistently identified in previous studies. To date, this study represents the inaugural analogous investigation in terms of applying NAM to cereals in Kazakhstan. It yielded a valuable data source for the identification of novel genes associated with the adaptation of wheat plants.

Anther morphology, particularly anther length, is genetically controlled by QTLs. Researchers have detected QTLs for anther length in different populations, but no candidate genes have been identified despite the critical role of this parameter in determining hybrid rice seed production. A research team from the China National Rice Research Institute reported the fine mapping of *qAL5.2* for anther size using backcross inbred lines (BILs) [2]. Using gene chip technology, the authors identified six *qALs*, including *qAL5.1* and *qAL5.2*, which explained 35.73% and 25.1% of the phenotypic variance in the population, respectively. *ORF8* encoded a lipase belonging to the alpha/beta-hydrolase (ABH) family. It is a likely candidate for *qAL5.2* and is possibly involved in regulating alpha-linolenic acid metabolism, which controls anther length.

Similarly, Yin et al. (2024) analyzed the genetic basis of yield traits in rice and validated a novel QTL for grain width and weight. Nine yield-related traits were investigated using an F2 population containing 190 plants [3]. The authors discovered that grain yield per plant was positively correlated with six yield traits, except for grain length and width. The highest correlation coefficient—0.98—was observed with regard to the number of filled grains per plant. Simultaneously, a linkage map covering the whole genome was constructed by genotyping each plant with the selected polymorphic markers using the F_2 population. All told, 36 QTLs for the yield traits were detected on nine chromosomes, and the phenotypic variation was explained by a single QTL that ranged from 6.19% to 36.01%. The identification of a major QTL for grain width and weight, denoted as *qGW2-1*,

check for updates

Citation: Chen, Z.; Tong, X.; Zhang, J.; Huang, J.; Li, Z. Research on Plant Genomics and Breeding: 2023. *Plants* **2024**, *13*, 2998. https://doi.org/10.3390/plants13212998

Received: 10 October 2024
Revised: 23 October 2024
Accepted: 25 October 2024
Published: 26 October 2024

was refined to an interval of approximately 2.94 megabases. Upon comparative analysis of QTL locations, it was observed that no previously reported QTLs for grain width (GW) and thousand grain weight (TGW) shared an overlapping interval with *qGW2-1*. This suggests that *qGW2-1* represents a novel QTL, with substantial potential for application in molecular design breeding strategies for rice.

Kenaf (*Hibiscus cannabinus* L.), originating from Africa, is an essential crop for fiber and other industrial materials, but its genetics have yet to be explored. A recent publication in *Plants* reported the GWAS analysis of seven agronomic traits in Kenaf: days to flowering, plant height, fresh weight, dry weight, flower color, stem color, and leaf shape [4]. In a cohort of 96 germplasm accessions, encompassing 55 gamma irradiation-derived mutant lines, an analysis using genotyping by sequencing (GBS) revealed the presence of 49,241 single-nucleotide polymorphisms (SNPs). Of these, 49 SNPs are associated with important agronomic traits, such as days to flowering, plant height, fresh weight, dry weight, flower color, stem color, and leaf shape. Except for fresh and dry weights, no single-nucleotide polymorphisms (SNPs) were detected in the genic region. Most SNPs were identified within the exon of the respective genic regions. The identified genes exhibit a homology of 45% to 96% with plants of the Malvaceae and Betulaceae families and are known to play a role in plant growth and development through various pathways. These findings are anticipated to enhance selective breeding programs for Kenaf, thereby improving overall plant quality.

As integral constituents of the SCF complexes, F-box proteins play pivotal roles in responding to both abiotic and biotic stresses in plants. The wheat F-box protein, TaFBA1, has been recognized for its reactivity to abiotic stress in tobacco. Nonetheless, its precise biological functions within wheat itself are yet to be fully elucidated. Using transgenic wheat with enhanced and suppressed expression of *TaFBA1*, Li et al. (2024) found that *TaFBA1* is a positive regulator of drought tolerance [5]. Compared to WT and RNAi lines, OE plants showed a higher survival rate and fresh weight and a more robust antioxidant capacity, as well as the maintenance of stomatal openings. *TaFBA1* is implicated in the regulation of genes associated with antioxidants, fatty acid, and lipid metabolism, as well as with cellulose synthesis. Consequently, this study yields a valuable gene reservoir for use enhancing abiotic stress tolerance in crops through genetic interventions.

The proteins responsible for reading DNA methylation can specifically adhere to methylated CpG dinucleotides via methyl-CpG-binding domains (MBDs) or SET- and ring-finger-associated domains. Presently, the *Arabidopsis thaliana* genome has revealed the existence of at least thirteen AtMBD proteins, although the functions of certain members remain ambiguous. A recent publication in *Plants* presented the functional characterization of *AtMBD3* [6]. The *AtMBD3* gene exhibits high expression levels in pollen and seeds, displaying a preference for binding to methylated CG and CHG and to unmethylated DNA sequences. Consistent with its murine ortholog, *mbd3* mutants demonstrated reduced mature pollen production and experienced early- or late-stage embryo abortion, likely stemming from asymmetric divisions. Transcriptome analysis of the seeds revealed the differential expression of numerous key transcription factors involved in embryo development. Moreover, in yeast and in vivo, a pollen-specific protein, PBL6, was discovered to interact with AtMBD3. Hence, this study offers significant functional insights into AtMBD3, contributing to a more comprehensive understanding of the AtMBD family.

2. Gene Function Studies Using RNA Molecules

RNA interference (RNAi) has been found to trigger genemic epigenetic modifications, effectively regulating gene expression in plants. RNAi constitutes a regulatory and defensive mechanism, pivotal in the growth, development, and management of plant responses to pathogens and abiotic stresses. In the context of spray-induced gene silencing (SIGS), the force known as exogenous RNA interference, complementary to target plant genes, reduces the expression of said target genes. In their study, Suprun et al. (2024) examined the influence of specific exogenous dsRNAs on the promoter, coding sequence (CDS), and intron regions of the *SlTRY* tomato gene, which encodes an R3-type MYB repressor of anthocyanin biosynthesis [7]. Their findings indicate that achieving efficient gene silencing in plants necessitates the targeting of a protein-coding region with dsRNA of at least 50 base pairs in length, exemplified by dsRNA-Prom2. Interestingly, dsRNA-Intron treatment increased the expression of the *SlTRY* gene compared to the control under normal growth conditions. The suppression of the *SlTRY* gene increased the concentration of anthocyanins and led to the simultaneous elevation of other compounds within the phenylpropanoid pathway. This research underscores the efficacy of dsRNA as a potent instrument for expeditiously probing gene functionality in plant science.

As mentioned above, the application of double-stranded RNAs (dsRNAs) to plant surfaces has been greatly beneficial in plant biotechnology and gene function studies. However, the comprehensive impact of dsRNA, particularly on the intricate network of microRNAs (miRNAs), remains unclear. Nityagovsky and colleagues tested the effects of the exogenous application of chalcone synthase (*CHS*)-encoding dsRNA in terms of altering the miRNA transcriptome in the rosette leaves of Arabidopsis thaliana [8]. Compared to the non-specific dsRNA control, a two-day treatment of *AtCHS*-dsRNA induced alterations in the expression pattern of 59 miRNAs. These miRNAs can be categorized into 17 major gene ontology (GO) groups, notably including the aromatic compound biosynthesis pathway associated with *CHS* activity. qRT-PCR results verified that the transcription of the predicted targets of DEG miRNAs changed accordingly, implying a negative correlation between the expression of miRNAs and their predicted targets. Thus, exogenous plant gene-specific dsRNAs induce substantial changes in the plant miRNA composition, ultimately affecting the expression of a wide range of genes. These findings have profound implications for our understanding of the effects of exogenously induced RNA interference, which can have broader effects beyond targeted mRNA degradation. Through the lens of miRNA profile, this work revealed that gene-specific dsRNAs induce specific and massive cellular responses at the epigenetic level after exogenous application.

Short interfering RNAs (siRNAs) are converted by dsRNAs with specialized enzymes, known as dicer-like enzymes, and are integrated into RNA-induced silencing complexes to effectively inhibit gene expression in plants. The work of Kiselev et al. (2024) found that simultaneous foliar application of gene-specific dsRNAs can decrease gene expression in response to activating anthocyanin accumulation in *Arabidopsis thaliana* [9]. They discovered that a combination of external dsRNAs or siRNAs, administered at two different doses, could silence the expression of the target gene, leading to a more effective increase in anthocyanin content than using a single dsRNA treatment.

3. Using OMICs to Understand Key Biological Process in Plants

The presence of a "beany flavor" in soybeans represents an unfavorable characteristic that diminishes their market value. Dongfudou 3 is a highly coveted soybean variety because it lacks this undesirable trait. However, the absence of the genome sequence of Dongfudou 3 has hindered the investigation into the genes responsible for this beany flavor. Duan and colleagues, originating from China, conducted a genomic survey of a distinct soybean variety utilizing next-generation sequencing technology [10]. The survey revealed an estimated genome size of approximately 1.07 Gb, with 72.50% of this comprising repetitive sequences. The 916.00 Mb draft genome sequence was successfully anchored onto 20 chromosomes, housing 46,446 genes and 77,391 transcripts. Notably, employing the principles of collinearity and sequence similarity, three genes, namely, *GmLox1*, *GmLox2*, and *GmLox3*, were identified and mapped within the Dongfudou 3 genome. Mutations in these genes significantly reduced transcript abundance and enzyme activity, potentially contributing to the absence of a beany flavor in Dongfudou 3. Utilizing third-generation sequencing, this study not only facilitates the formulation of advanced strategies for constructing a high-quality genome of Dongfudou 3, but also provides a preliminary comprehension of specific characteristics of Dongfudou 3 through the assembled draft genome.

Plant-derived bioactive compounds, specifically glucosinolates (GSLs), play a crucial role in human health. A research team from Korea conducted a comprehensive analysis of 17 glucosinolate compounds in leaf samples obtained from 65 pak choy germplasm accessions using an acuity ultra-high-performance liquid chromatography (UHPLC) analytical system [11]. The variation in GSL compounds led to the categorization of the pak choy germplasm accessions into three distinct groups. This classification has promising prospects for use in future breeding endeavors seeking to enhance the glucosinolate content within the crop.

The biosynthesis of anthocyanin, an antioxidant with functional food properties for humans, is significantly influenced by light availability. Danpreedanan et al. (2023) conducted a study to examine the impact of shading on the yield and anthocyanin content of purple rice [12]. The findings revealed that increased shading resulted in higher anthocyanin content in the grain pericarp while concurrently reducing yield. The four tested rice varieties displayed varying responses to shading treatment in terms of grain yield, leaf chlorophyll content, and anthocyanin content. Phenotypical variation was observed to be independent of leaf colors. Although it has already been established that the *OsDFR* gene leads to a red grain color, the expression level of the *OsDFR* gene showed a weak correlation with anthocyanin content under shading stress, suggesting the involvement of other genes in anthocyanin biosynthesis when light intensity is low. This study indicates that certain varieties of purple rice can thrive and produce satisfactory yields under low-light conditions. These findings could potentially be valuable for the commercial cultivation of purple rice when employed in an optimal shading technique to enhance the antioxidant content of purple rice grains.

In addition to synthesizing metabolites, light intensity can significantly affect fruit development. A study by Gao et al. (2023) found that in watermelon, low-light stress led to a decrease in fruit size and the content of soluble sugar and amino acids compared to the control [13]. Fruits at 0–15 days after pollination were very sensitive to shading in terms of fruit expansion. The authors conducted transcriptome analysis at four stages and identified 8837 differentially expressed genes (DEGs), including 55 DEGs related to oxidation reduction, secondary metabolites, carbohydrate and amino acid metabolism, and transcriptional regulation. This study lays the groundwork for further research into the functions of low-light-stress-responsive genes and the molecular mechanisms underlying the effects of low-light stress on watermelon fruit expansion.

4. Conclusions and Perspectives

Plant genomics and breeding play crucial roles in terms of addressing the global challenges of food security, climate change, and sustainable agriculture. By studying the genetic makeup of plants, scientists can identify genes responsible for desirable traits such as high yield, disease resistance, and tolerance to environmental stresses. In summary, this Editorial thoroughly reviewed the 13 research articles that were recently published in the Special Issue of *Plants* entitled *"Research on plant genomics and breeding 2023"*. These studies represent the latest progress in the genetic basis exploration of major agronomic traits using traditional genetic mapping or modern genotype-to-phenotype association technologies and cover various plant species like rice, wheat, Kenaf, and *Arabidopsis*. Employing genomics, transcriptomics, and metabolomics technologies, scientists gained deep insights into the biological puzzles of fruit development and beneficial metabolite biosynthesis. Additionally, researchers investigated the intricacy of SIGC in terms of target sites, dsRNA interaction, and effects on miRNA profiles. Overall, integrating plant genomics and breeding offers promising opportunities to enhance crop productivity, nutritional quality, and resilience, ultimately contributing to global food security and environmental sustainability.

Author Contributions: All the authors participated in the editing of this Research Topic. Z.C. wrote the draft, and all the other authors provided suggestive comments on the editorial. All authors have read and agreed to the published version of the manuscript.

Funding: This project was funded by the Zhejiang Provincial Natural Science Foundation of China (LY22C130002), the National Natural Science Foundation of China (32071986), the Natural Science Foundation of Fujian Province (2021J01490), and Basic Scientific Research Funds of Public Welfare Scientific Research Institutes of Fujian Province (2024R1076).

Data Availability Statement: No new data were created or analyzed in this study. Data sharing is not applicable to this article.

Conflicts of Interest: The authors declare that the research was conducted in the absence of any commercial or financial relationships that could be construed as a potential conflict of interest.

References

1. Amalova, A.; Babkenov, A.; Philp, C.; Griffiths, S.; Abugalieva, S.; Turuspekov, Y. Identification of Quantitative Trait Loci Associated with Plant Adaptation Traits Using Nested Association Mapping Population. *Plants* **2024**, *13*, 2623. [CrossRef] [PubMed]
2. Liu, X.; Yu, Z.; Tong, X.; Chang, L.; Huang, J.; Wang, Y.; Ying, J.; Li, X.; Ni, S.; Zhang, J. Fine Mapping of qAL5.2 Controlling Anther Length in Oryza sativa. *Plants* **2024**, *13*, 1130. [CrossRef] [PubMed]
3. Yin, M.; Tong, X.; Yang, J.; Cheng, Y.; Zhou, P.; Li, G.; Wang, Y.; Ying, J. Dissecting the Genetic Basis of Yield Traits and Validation of a Novel Quantitative Trait Locus for Grain Width and Weight in Rice. *Plants* **2024**, *13*, 770. [CrossRef] [PubMed]
4. Kim, W.J.; Yang, B.; Lee, Y.J.; Kim, J.H.; Kim, S.H.; Ahn, J.W.; Kang, S.Y.; Kim, S.H.; Ryu, J. Genome-Wide Association Study for Agronomic Traits in Gamma-Ray-Derived Mutant Kenaf (*Hibiscus cannabinus* L.). *Plants* **2024**, *13*, 249. [CrossRef] [PubMed]
5. Li, Q.; Zhao, X.; Wu, J.; Shou, H.; Wang, W. The F-Box Protein TaFBA1 Positively Regulates Drought Resistance and Yield Traits in Wheat. *Plants* **2024**, *13*, 2588. [CrossRef] [PubMed]
6. Shu, J.; Yin, X.; Liu, Y.; Mi, Y.; Zhang, B.; Zhang, A.; Guo, H.; Dong, J. MBD3 Regulates Male Germ Cell Division and Sperm Fertility in Arabidopsis thaliana. *Plants* **2023**, *12*, 2654. [CrossRef] [PubMed]
7. Suprun, A.R.; Manyakhin, A.Y.; Trubetskaya, E.V.; Kiselev, K.V.; Dubrovina, A.S. Regulation of Anthocyanin Accumulation in Tomato *Solanum lycopersicum* L. by Exogenous Synthetic dsRNA Targeting Different Regions of SlTRY Gene. *Plants* **2024**, *13*, 2489. [CrossRef] [PubMed]
8. Nityagovsky, N.N.; Kiselev, K.V.; Suprun, A.R.; Dubrovina, A.S. Impact of Exogenous dsRNA on miRNA Composition in Arabidopsis thaliana. *Plants* **2024**, *13*, 2335. [CrossRef] [PubMed]
9. Kiselev, K.V.; Suprun, A.R.; Aleynova, O.A.; Ogneva, Z.V.; Dubrovina, A.S. Simultaneous Application of Several Exogenous dsRNAs for the Regulation of Anthocyanin Biosynthesis in *Arabidopsis thaliana*. *Plants* **2024**, *13*, 541. [CrossRef] [PubMed]
10. Duan, Y.; Li, Y.; Zhang, J.; Song, Y.; Jiang, Y.; Tong, X.; Bi, Y.; Wang, S.; Wang, S. Genome Survey and Chromosome-Level Draft Genome Assembly of Glycine max var. Dongfudou 3: Insights into Genome Characteristics and Protein Deficiencies. *Plants* **2023**, *12*, 2994. [CrossRef] [PubMed]
11. Kim, S.H.; Ochar, K.; Hwang, A.; Lee, Y.J.; Kang, H.J. Variability of Glucosinolates in Pak Choy (*Brassica rapa* subsp. chinensis) Germplasm. *Plants* **2023**, *13*, 9. [CrossRef] [PubMed]
12. Danpreedanan, N.; Yamuangmorn, S.; Jamjod, S.; Prom, U.T.C.; Pusadee, T. Genotypic Variation of Purple Rice in Response to Shading in Yield, Anthocyanin Content, and Gene Expression. *Plants* **2023**, *12*, 2582. [CrossRef] [PubMed]
13. Gao, W.; She, F.; Sun, Y.; Han, B.; Wang, X.; Xu, G. Transcriptome Analysis Reveals the Genes Related to Water-Melon Fruit Expansion under Low-Light Stress. *Plants* **2023**, *12*, 935. [CrossRef] [PubMed]

plants

MDPI

Article

Identification of Quantitative Trait Loci Associated with Plant Adaptation Traits Using Nested Association Mapping Population

Akerke Amalova [1], Adylkhan Babkenov [2], Charlie Philp [3], Simon Griffiths [3], Saule Abugalieva [1,4] and Yerlan Turuspekov [1,4,*]

1 Institute of Plant Biology and Biotechnology, Almaty 050040, Kazakhstan; akerke.amalova@gmail.com (A.A.); absaule@yahoo.com (S.A.)
2 Alexandr Barayev Scientific-Production Center for Grain Farming, Shortandy 021600, Kazakhstan; babkenov64@mail.ru
3 John Innes Centre, Norwich NR4 7UH, UK; charlie.philp@jic.ac.uk (C.P.); simon.griffiths@jic.ac.uk (S.G.)
4 Faculty of Biology and Biotechnology, Al-Farabi Kazakh National University, Almaty 050040, Kazakhstan
* Correspondence: yerlant@yahoo.com

Abstract: This study evaluated 290 recombinant inbred lines (RILs) of the nested association mapping (NAM) population from the UK. The population derived from 24 families, where a common parent was "Paragon," one of the UK's spring wheat cultivar standards. All genotypes were tested in two regions of Kazakhstan at the Kazakh Research Institute of Agriculture and Plant Industry (KRIAPI, Almaty region, Southeast Kazakhstan, 2019–2022 years) and Alexandr Barayev Scientific-Production Center for Grain Farming (SPCGF, Shortandy, Akmola region, Northern Kazakhstan, 2019–2022 years). The studied traits consisted of plant adaptation-related traits, including heading date (HD, days), seed maturation date (SMD, days), plant height (PH, cm), and peduncle length (PL, cm). In addition, the yield per m^2 was analyzed in both regions. Based on a field evaluation of the population in northern and southeastern Kazakhstan and using 10,448 polymorphic SNP (single-nucleotide polymorphism) markers, the genome-wide association study (GWAS) allowed for detecting 74 QTLs in four studied agronomic traits (HD, SMD, PH, and PL). The literature survey suggested that 16 of the 74 QTLs identified in our study had also been detected in previous QTL mapping studies and GWASs for all studied traits. The results will be used for further studies related to the adaptation and productivity of wheat in breeding projects for higher grain productivity.

Keywords: bread wheat; nested association mapping; genome-wide association studies; plant adaptation-related traits

Citation: Amalova, A.; Babkenov, A.; Philp, C.; Griffiths, S.; Abugalieva, S.; Turuspekov, Y. Identification of Quantitative Trait Loci Associated with Plant Adaptation Traits Using Nested Association Mapping Population. *Plants* **2024**, *13*, 2623. https://doi.org/10.3390/plants13182623

Academic Editors: Jian Zhang and Zhiyong Li

Received: 8 July 2024
Revised: 10 September 2024
Accepted: 18 September 2024
Published: 20 September 2024

1. Introduction

Bread wheat is one of the most important agricultural commodities in the world market [1]. In the 2023/2024 year, the global production volume of wheat amounted to almost 785 million tons [2]. In order to continue providing the world's population with enough wheat in 2050, its yield should be increased by 60% [3]. Therefore, constant productivity and quality improvement are essential for wheat breeding [4]. Wheat cultivation worldwide requires cultivars to adapt to various environmental and climatic conditions. This adaptability is achieved through variations in phenology and traits related to plant architecture [5]. Key phenological and agronomic traits such as heading/flowering time, plant height, and seed maturity time are crucial for adaptation and maximizing yield potential and stability. Identifying the genes underlying this variation and understanding how they interact and perform in different environments are crucial for improving wheat adaptability and optimizing yield potential [6].

Phenology plays a crucial role in crop adaptation to specific environments [7]. The perception of the genetic control of phenological traits is essential for breeders to develop cultivars better suited to their local environments. The major genes affecting wheat adaptation include those associated with phenology and plant architecture, such as vernalization (*Vrn*) [8–10], photoperiod (*Ppd*) [11–13], earliness per se (*Eps*) [14–16], and reduced height (*Rht*), in addition to other minor-effect loci. The interaction of these genes determines a genotype's adaptation to specific environments [13]. Therefore, developing new competitive high-yielding cultivars and adapting them to different environments are key priorities in wheat breeding projects.

Plant height is a complex trait influenced by various genetic and environmental factors. One of the key genetic factors affecting plant height is the presence of genes related to gibberellin biosynthesis and response, commonly known as the "Green Revolution" genes or reduced height (*Rht*) genes [17]. *Rht* genes are associated with the semi-dwarf phenotype observed in many modern cereal varieties. Semi-dwarf plants have shorter stems compared to their wild-type counterparts, which helps reduce lodging (stem bending or breaking) and allows for a more efficient allocation of resources to reproductive structures, ultimately increasing grain yield [18]. In wheat, the *Rht* genes were first identified in the 1960s. The two main *Rht* genes in wheat are *Rht-B1* and *Rht-D1*, located on the short arms of chromosomes 4B and 4D, respectively. Mutations in these genes reduce stem elongation and contribute to the semi-dwarf phenotype [19].

Plant adaptation and related traits are complex traits controlled by multiple genes [20]. The study of complex quantitative traits in cereals used two common methods: quantitative trait loci (QTL) [21,22] and mapping genome-wide association studies (GWASs) [23–30]. Particularly, with the availability of large-scale genomic resources, the GWAS has emerged as an addition to QTL mapping for complex traits [31]. The GWAS analyzes genetically diverse lines that harbor numerous historical and hereditary recombination events. Additionally, utilizing diverse germplasm in GWASs can potentially capture superior alleles overlooked by conventional breeding practices [31].

One of the ways to combine the strength of biparental and association mapping is to employ nested association mapping (NAM) populations [32,33]. NAM populations offer several advantages, including high allelic diversity, high mapping resolution, and low sensitivity to population structure. NAM population is usually developed by using many related progeny within multiple biparental mapping populations, which are developed by selecting a diverse set of founder lines and crossing them to a common reference parent [34]. Founder lines are carefully selected for genetic diversity, allowing them to encompass a wide range of genetic backgrounds. Consequently, these lines may include exotic germplasm, wild relatives, and landraces. The resulting F1 progeny undergo at least four generations of selfing to produce recombinant inbred lines (RILs), whose genomes are mosaics of the parental genomes [34]. This process of shuffling the parental genomes breaks down population structure, introduces recent recombinations, and creates new allele combinations [35]. Consequently, it aids in detecting small effects of QTL and rare alleles from specific parents [35].

Multiparent populations have advantages over biparental as they produce additional recombination breakpoints and increase QTL detection's allelic diversity and power [36]. Currently, NAM and multiparent advanced generation inter-cross (MAGIC) showed their high potential, for instance, in studies in wheat [37], barley [38,39], durum [40], rice [41], maize [34], sorghum [42], soybean [43], etc. Additionally, QTL studies of traits such as the grain quality of wheat [44,45], grain protein content [46], yield and its components [47], stay-green [48], nitrogen-deficiency tolerance [49], drought tolerance [50], and disease resistance [51,52] in wheat were mainly based on the use of an NAM population. Also, considering genotype–environment interaction patterns suggests a strong influence of the growth environment on the detection of QTLs for plant adaptation [53,54]. As environmental conditions may greatly impact the timing of the heading date and seed maturation, they may also significantly alter yield [55,56]. In Kazakhstan, analogous studies using NAM for cereals, including for wheat, have not been conducted so far. The present study aims to identify quantitative trait nucleotides (QTNs) associated with plant adaptation traits: heading date, seed maturation date, plant height, and peduncle length in the NAM population in the northern and southeastern regions of Kazakhstan through the GWAS.

2. Materials and Methods

2.1. Plant Materials

The nested association mapping (NAM) population consisted of 290 spring wheat recombinant inbred lines (RILs) derived from 24 families using a single-seed descent method in greenhouse conditions by the John Innes Centre (Norwich, UK). Paragon, a standard UK spring wheat cultivar, was the common parent utilized in the NAM population. The spring wheat NAM panel comprises twenty-four accessions selected as second parental lines, which include (1) 19 landraces sourced from the A.E. Watkins collection, (2) 2 lines from CIMMYT Core Germplasm (CIMCOG), and (3) 2 cultivars: Baj and Wylakatchem (Table 1) [57,58].

Table 1. The list of accessions used as parental lines for the developed nested association mapping population.

Cultivars	Origin (Countries)	Mapping Population	Number of RIL
Watkins34	India (Asia)	Paragon × Watkins34	8
Watkins141	China (Asia)	Paragon × Watkins141	10
Watkins216	Morocco (North Africa)	Paragon × Watkins216	10
Watkins223	Burma (Asia)	Paragon × Watkins223	11
Watkins254	Morocco (North Africa)	Paragon × Watkins254	13
Watkins264	Canary Islands (Western Europe)	Paragon × Watkins264	13
Watkins273	Spain (Western Europe)	Paragon × Watkins273	14
Watkins291	Cyprus (Western Europe)	Paragon × Watkins291	14
Watkins292	Cyprus (Western Europe)	Paragon × Watkins292	11
Watkins299	Türkiye (Middle East)	Paragon × Watkins299	11
Watkins349	Bulgaria (Eastern Europe)	Paragon × Watkins349	12
Watkins396	Portugal (Western Europe)	Paragon × Watkins396	10
Watkins397	Portugal (Western Europe)	Paragon × Watkins397	13
Watkins398	Palestine (Middle East)	Paragon × Watkins398	9
Watkins420	India (Asia)	Paragon × Watkins420	12
Watkins546	Spain (Western Europe)	Paragon × Watkins546	13
Watkins566	Greece (Western Europe)	Paragon × Watkins566	12
Watkins685	Spain (Western Europe)	Paragon × Watkins685	12
Watkins811	Tunisia (North Africa)	Paragon × Watkins811	13
BAJ		Paragon × BAJ	15
CIMCOG 47	Mexico	Paragon × CIMCOG 47	16
CIMCOG 49	Mexico	Paragon × CIMCOG 49	16
Wylakatchem	Australia	Paragon × Wylakatchem	15
PFAU	Mexico	Paragon × PFAU	7

2.2. Evaluation of the Nested Association Mapping Population for Variation in Studied Traits

The studied plants of the NAM population were tested in the field of two regions of Kazakhstan: (1) at the Kazakh Research Institute of Agriculture and Plant Industry (KRIAPI, Almaty region, Southeast Kazakhstan, 2019–2022 years) and (2) Alexandr Barayev

Scientific-Production Center for Grain Farming (SPCGF, Shortandy, Akmola region, Northern Kazakhstan, 2019–2022 years). All genotypes and two local standards (check cultivars) "Kazakhstanskaya 4" in KRIAPI and "Astana" in SPCGF were planted in both locations with two replications in a one-meter plot using a randomized complete block design. The distance between rows was 15 cm, with a 5 cm distance between plants [59]. The table presented in Table 2 displays the meteorological conditions recorded during the trials. The studied traits consisted of plant adaptation-related traits, including heading date (HD, days), seed maturation date (SMD, days), plant height (PH, cm), and peduncle length (PL, cm). HD was recorded as the number of days from emergence to the day when half of the spikes appeared in 50% of the plants. SMD was measured as the number of days between heading days and maturation days. PH was measured at harvest maturity, when the maximum height was achieved, from the ground level to the top of the spikes (excluding awns). PL was measured as the length of the first peduncle. Each one-meter plot consisted of seven rows, and three randomly selected plants per row were analyzed for PH and PL. We studied 21 plants per each of 290 genotypes per replication. A similar approach was taken for the second replication. The mean for two replications was calculated using averages in each replication. In addition, the yield per m^2 (YM2, g/m^2) was analyzed in both regions.

Table 2. Location, environment, and weather data for the two study regions in Kazakhstan.

Site/Region	**KRIAPI (Almaty Region, Southeast Kazakhstan)**				**SPCGF (Akmola Region, Northern Kazakhstan)**		
Latitude/Longitude	43°21′/76°53′				51°40′/71°00′		
Soil type	Light chestnut (humus 2.0–2.5%)				Southern carbonate chernozem (humus 3.6%)		
Conditions	Rainfed				Rainfed		
Year	2019	2020	2021	2022	2020	2021	2022
Annual rainfall, mm	299	279	183	250	426	112	117
Mean temperature, °C	19.8	19.8	21.8	22.2	19.2	18.0	18.4
Max temperature, °C	27.0	24.2	27.4	26.5	20.7	20.4	21.1
Min temperature, °C	12.9	14.2	12.4	16.7	17.6	14.7	15.7

Note: KRIAPI—Kazakh Research Institute of Agriculture and Plant Industry; SPCGF—Alexandr Barayev Scientific-Production Center for Grain Farming.

2.3. Genotyping, Population Structure, and Genome-Wide Association Studies

The studied collection was genotyped using the Axiom Wheat Breeder's Genotyping Array with 35K single-nucleotide polymorphisms (SNPs) [60]. In total, 10,448 polymorphic SNP markers were used in the GWAS after filtering missing data (≥50%) and the minor allele frequency (MAF) ≥ 5%. [61]. Estimation of the linkage disequilibrium (LD) for each chromosome in the 290 RILs of the NAM population was performed in TASSEL version 5.0, and it was estimated and visualized at r^2 = 0.1 using the R packages version 4.3.0. The association mapping was conducted using a multivariate linear mixed model (MLMM) in the Genome Association and Prediction Integrated Tool (GAPIT version 3) [62]. MLMM was selected for its balanced integration of fixed and random effects and its ability to detect multiple loci while effectively accounting for population structure and kinship, making it particularly suitable for analyzing the complex traits targeted in this study. The population structure was analyzed using a model-based clustering method using STRUCTURE version 2.3.4 [63,64]. Manhattan plots and SNP density plots were generated using the rMVP package (https://cran.r-project.org/web/packages/rMVP/index.html, accessed on 30 May 2024) [65]. The BLAST tool available on Ensembl Plants for the reference genome of *T. aestivum* (https://plants.ensembl.org/Triticum_aestivum/Tools/Blast, accessed on 25 June 2024) [66] was used to identify the protein-coding genes overlapping with the identified significant QTLs. The analysis of variance (ANOVA), principal component analysis (PCA), and correlation analysis were performed using Rstudio software version 4.3.0 (POSIT, Boston, MA, USA) [67]. The broad-sense heritability index (h_2^b), indicating the proportion of phenotypic variation due to genetic factors, was calculated based on the ANOVA results according to Genievskaya et al. [68].

3. Results

3.1. Phenotypic Variation of 290 RIL NAM Population for Studied Traits

The phenotypic assessment of 290 RILs in 24 NAM families was analyzed in seven environments (year-by-location) at the KRIAPI (2019–2022) and SPCGF (2020–2022) (Tables S1 and 3). The phenotypic variability of four traits between the two regions, including the mean HD, ranged from 42.22 ± 0.23 days at the SPCGF to 58.05 ± 0.19 days at the KRIAPI (Figure 1). The mean PH valued from 49.03 ± 0.38 cm at SPCGF to 79.14 ± 0.67 cm at the KRIAPI (Figure 1), which showed that the mean PH was 30 cm taller in the Almaty region (KRIAPI, south). The average value of YM2 ranged from 253.69 ± 4.03 g/m^2 (KRIAPI) to 370.90 ± 5.52 g/m^2 (SPCGF). In general, the *t*-test suggested that the average values of HD, SMD, PH, and PL in the two contrasted regions (KRIAPI and SPCGF) were significantly different ($p < 0.0001$).

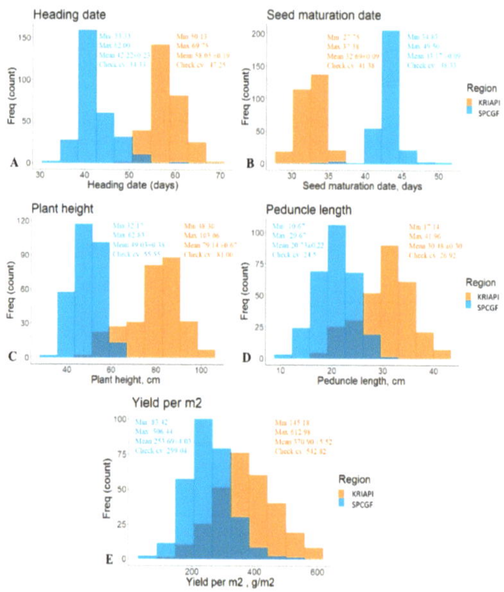

Figure 1. The distributions of four traits: heading date (**A**), seed maturation date (**B**), plant height (**C**), peduncle length (**D**), and yield per m^2 (**E**) averaged data in the nested association mapping population in the two regions.

The field performance of PH showed that in the two regions, the mean of the NAM population was shorter than that of the local standard (check cultivars) "Astana" and "Kazakhstanskaya 4" (Figure 1). In contrast, the average HD was from 8 (north) to 11 (southeast) days longer than that of the check cultivars (local standard). Similarly, the average SMD was from 5 (southeast) to 5 (north) days longer than those of the check cultivars (Figure 1). The assessment of the mean YM2 revealed that the yield at the southeastern station was 2.4 times higher than that at the southern station. A total of 11 and 60 accessions showed higher values than check cultivars at the KRIAPI and SPCGF, respectively.

In general, the *t*-test suggested that the average values of HD, SMD, PH, and PL in the two contrasted regions were significantly different ($p < 0.0001$) (Table S2).

Table 3. Phenotypic distribution of trait nested association mapping (NAM) of 24 families in two regions.

NAM Population/Region	HD, days		SMD, days		PH, cm		PL, cm		YM2_g/m^2	
	KRIAPI	SPCGF	KRIAPI	SPCGF	KRIAPI	SPCGF	KRIAPI	SPCGF	KRIAPI	SPCGF
Paragon × Watkins34	53.4 ± 0.65	37.7 ± 1.00	34.5 ± 0.58	43.9 ± 0.54	76.3 ± 1.20	42.3 ± 0.99	30.8 ± 1.23	18.1 ± 0.50	421.5 ± 30.49	188.0 ± 16.28
Paragon × Watkins141	62.7± 0.76	47.2 ± 1.33	31.6 ± 0.46	42.5 ± 0.78	86.2 ± 1.26	54.0 ± 1.96	30.4 ± 0.80	21.0 ± 0.99	299.9 ± 17.67	271.7 ± 17.70
Paragon × Watkins216	59.2 ± 0.48	43.7 ± 1.18	32.6 ± 0.26	42.5 ± 0.62	86.2 ± 1.21	52.9 ± 1.25	33.0 ± 0.91	22.1 ± 0.71	262.0 ± 12.93	206.3 ± 9.66
Paragon × Watkins223	55.1 ± 0.77	37.7 ± 0.80	34.4 ± 0.39	43.8 ± 0.61	83.1 ± 1.20	45.6 ± 0.93	32.4 ± 0.84	19.6 ± 0.73	449.4 ± 33.14	223.9 ± 12.56
Paragon × Watkins254	61.5 ± 0.83	44.7 ± 1.13	31.6 ± 0.53	43.0 ± 0.30	81.1 ± 2.26	53.1 ± 1.80	30.2 ± 1.46	22.3 ± 0.99	332.8 ± 26.15	234.0 ± 11.45
Paragon × Watkins264	57.4 ± 0.53	42.1 ± 0.86	33.4 ± 0.35	43.3 ± 0.23	87.7 ± 1.84	52.6 ± 1.00	36.8 ± 0.89	23.2 ± 0.73	389.7 ± 20.47	264.6 ± 23.04
Paragon × Watkins273	57.0± 0.67	42.9± 1.20	33.4 ± 0.24	44.4 ± 0.47	86.7 ± 1.80	51.1 ± 1.34	34.4 ± 0.91	22.1 ± 0.60	305.4 ± 28.33	228.6 ± 26.03
Paragon × Watkins291	59.0 ± 0.82	42.0 ± 0.81	32.1 ± 0.42	42.9 ± 0.18	84.5 ± 1.96	50.6 ± 1.42	29.8 ± 0.74	20.7 ± 0.74	392.1 ± 22.60	237.9 ± 13.54
Paragon × Watkins292	59.0 ± 0.83	45.2 ± 1.42	31.6 ± 0.36	42.0 ± 0.75	81.4 ± 1.20	49.2 ± 1.73	31.0 ± 1.24	19.8 ± 1.09	375.1 ± 15.81	177.2 ± 17.76
Paragon × Watkins299	57.8 ± 0.56	39.6 ± 0.83	32.6 ± 0.44	44.2 ± 0.24	82.1 ± 1.43	46.0 ± 1.26	33.3 ± 0.76	20.2 ± 0.63	423.6 ± 34.68	240.3 ± 12.00
Paragon × Watkins349	61.0 ± 0.51	42.0 ± 0.59	32.5 ± 0.32	43.4 ± 0.15	87.0 ± 1.85	53.5 ± 1.27	32.7 ± 0.48	24.0 ± 0.61	439.1 ± 27.96	288.8 ± 20.49
Paragon × Watkins396	63.2 ± 1.01	44.7 ± 0.96	31.4 ± 0.56	43.3 ± 0.27	79.2 ± 1.98	55.1 ± 1.54	31.3 ± 1.17	24.2 ± 0.88	331.6 ± 26.65	310.2 ± 19.02
Paragon × Watkins397	59.0 ± 0.83	41.8 ± 0.83	32.8 ± 0.52	43.7 ± 0.21	90.1 ± 1.55	52.7 ± 1.29	35.1 ± 0.99	23.5 ± 0.42	387.8 ± 23.74	247.4 ± 17.50
Paragon × Watkins398	58.2 ± 0.47	42.7 ± 1.04	33.3 ± 0.27	42.5 ± 0.49	79.7 ± 1.75	49.5 ± 1.67	31.4 ± 1.60	21.1 ± 0.71	391.3 ± 24.11	241.8 ± 16.33
Paragon × Watkins420	59.5 ± 0.73	43.9 ± 1.94	32.9 ± 0.53	43.4 ± 0.61	80.8 ± 1.30	52.0 ± 1.49	30.8 ± 1.14	22.5 ± 0.77	392.9 ± 28.24	289.8 ± 14.00
Paragon × Watkins546	59.7 ± 0.87	42.3 ± 1.00	34.2 ± 0.46	42.5 ± 0.56	87.7 ± 1.75	55.4 ± 0.78	33.5 ± 0.96	24.6 ± 0.71	433.2 ± 20.52	293.9 ± 12.24
Paragon × Watkins566	60.4 ± 0.49	45.5 ± 1.01	32.9 ± 0.32	42.7 ± 0.54	89.2 ± 2.09	53.9 ± 0.69	31.5 ± 0.73	22.4 ± 0.67	359.5 ± 13.15	252.5 ± 15.94
Paragon × Watkins685	57.8 ± 0.52	43.5 ± 1.23	31.9 ± 0.38	41.8 ± 0.74	80.8 ± 1.50	50.2 ± 0.86	31.8 ± 0.44	22.3 ± 0.60	347.0 ± 9.45	202.3 ± 21.86
Paragon × Watkins811	56.4 ± 0.34	43.6 ± 0.88	32.0 ± 0.18	43.0 ± 0.21	85.6 ± 2.27	49.7 ± 1.28	33.6 ± 1.38	21.5 ± 0.79	407.5 ± 19.53	245.4 ± 13.51
Paragon × BAJ	56.0 ± 0.59	41.4 ± 0.71	32.2 ± 0.25	43.0 ± 0.20	60.5 ± 1.37	41.1 ± 1.00	21.7 ± 0.43	14.5 ± 0.50	361.5 ± 18.76	290.1 ± 21.00
Paragon × CIMCOG 47	54.7 ± 0.34	40.1 ± 0.47	32.7 ± 0.23	42.9 ± 0.29	61.6 ± 1.53	40.9 ± 1.05	25.0 ± 0.77	17.0 ± 0.56	403.3 ± 19.31	297.6 ± 14.18
Paragon × CIMCOG 49	55.9 ± 0.55	40.4 ± 0.53	32.9 ± 0.37	44.0 ± 0.21	65.4 ± 1.66	45.7 ± 1.15	26.1 ± 0.83	18.9 ± 0.74	376.8 ± 24.98	237.7 ± 19.08
Paragon × PFAU	52.8 ± 0.69	38.1 ± 0.82	33.5 ± 0.49	44.3 ± 0.42	59.3 ± 2.61	38.3 ± 2.01	21.8 ± 1.48	15.2 ± 1.32	208.5 ± 20.90	269.9 ± 28.27
Paragon × Wylakatchem	56.5 ± 0.45	40.0 ± 0.39	32.4 ± 0.26	43.2 ± 0.18	64.1 ± 1.91	41.8 ± 0.84	25.4 ± 0.96	17.1 ± 0.68	328.5 ± 23.61	298.7 ± 13.23

Note: HD, days—heading date; SMD, days—seed maturation date; PH, cm—plant height; PL, days—peduncle length; YM2, g/m^2—yield per m^2; KRIAPI—Kazakh Research Institute of Agriculture and Plant Industry; SPCGF—Alexandr Barayev Scientific-Production Center for Grain Farming.

ANOVA was performed using field data collected from 280 RILs across two locations: KRIAPI (2021–2022) and SPCGF (2020). The ANOVA showed a highly significant difference between the two factors (genotype, environment) for all four studied phenotypic traits. The index of heritability (h_b^2) was analyzed for all traits (Table 4), and the highest h_b^2 value was noted for HD (29.8%).

Table 4. Analysis of variance (ANOVA) results for studied traits of nested association mapping population grown in Kazakhstan.

Traits	Factor	Df	Sum Sq	Mean Sq	F-Value	h_b^2
HD, days	Genotype (G)	279	30,899	111	14.59 ***	29.8%
	Environment (E)	2	47,631	23,815	3138.26 ***	
	G:E	558	18,932	34	4.47 ***	
	Residuals	840	6375	8		
SMD, days	Genotype (G)	279	3380	12	1.97 ***	2.2%
	Environment (E)	2	136,499	68,250	11,093.20 ***	
	G:E	558	5366	10	1.56 ***	
	Residuals	840	5168	6		
PH, cm	Genotype (G)	279	114,177	409	11.96 ***	13.9%
	Environment (E)	2	624,924	312,462	9128.80 ***	
	G:E	558	52,838	95	2.77 ***	
	Residuals	840	28,752	34		
PL, cm	Genotype (G)	279	27,909	100	5.09 ***	10.0%
	Environment (E)	2	213,112	106,556	5421.09 ***	
	G:E	558	20,949	38	1.91 ***	
	Residuals	840	16,511	20		
YM2, g/m^2	Genotype (G)	279	8,638,103	30,961	5.79 ***	14.9%
	Environment (E)	2	30,095,833	15,047,916	2815.83 ***	
	G:E	558	14,653,053	26,260	4.91 ***	
	Residuals	840	4,488,995	5344		

Note: *p*-values are provided with significance level indicated by the asterisks; *** $p < 0.001$; HD, days—heading date; SMD, days—seed maturation date; PH, cm—plant height; PL, days—peduncle length; YM2, g/m^2—yield per m^2.

Pearson's correlation of the average phenotypic values in the two regions suggested different associations. In the southeast region (KRIAPI), the YM2 was negatively correlated with DH and positively correlated with SMD, PH, and PL (Table 5). In the northern region (SPCGF), none of the average data over three years (2020–2022) correlated with YM2. However, when the correlation was analyzed for each year, it was visible that YM2 positively correlated with PH in two out of three years of data (Table 5). At the same time, YM2 was negatively correlated with HD in 2020 and positively correlated in 2022 (Table 5). This controversial correlation was most probably affected by annual rainfall in these years (Table 2).

The PCA for the studied traits showed a relationship of 290 RILs of the NAM population using PC1 and PC2, which explain 36.6% and 21.6% of the total variation, respectively. The results of the PCA of traits are similar to those of the Pearson correlation analysis (Table 5). A similar negative correlation was also noted between HD and SMD, with arrows pointing in different directions (Figure 2). The same trend of negative correlation in terms of yield components was revealed between YM2 and HD at KRIAPI. It is obvious from Figure 2 that the correlation indices were positive for two traits (PH, PL).

Table 5. Pearson's correlation index by years (2020–2022) of five studied traits in spring wheat nested association mapping population grown in the southeast and north of Kazakhstan.

	KRIAPI					**SPCGF**			
					2020				
	SMD	PH	PL	YM2		SMD	PH	PL	YM2
HD	−0.44 ***	0.15 *	−0.12 *	0.02 ns	HD	−0.06 ns	0.39 ***	0.16 **	−0.30 ***
SMD		0.31 ***	0.28 ***	0.21 ***	SMD		0.02 ns	0.01 ns	−0.03 ns
PH			0.75 ***	0.32 ***	PH			0.68 ***	−0.09 ns
PL				0.22 ***	PL				0.01 ns
					2021				
	SMD	PH	PL	YM2		SMD	PH	PL	YM2
HD	−0.70 ***	0.12 *	0.02 ns	−0.46 ***	HD	−0.59 ***	0.48 ***	0.21 ***	0.05 ns
SMD		0.03 ns	0.11 ns	0.23 ***	SMD		−0.22 ***	−0.05 ns	−0.10 ns
PH			0.80 ***	0.30 ***	PH			0.76 ***	0.29 ***
PL				0.26 ***	PL				0.24 ***
					2022				
	SMD	PH	PL	YM2		SMD	PH	PL	YM2
HD	−0.60 ***	0.35 ***	−0.02 ns	−0.31 ***	HD	−0.79 ***	0.22 ***	0.13 *	0.13 *
SMD		−0.21 ***	−0.01 ns	0.24 ***	SMD		−0.19 **	−0.13 *	−0.13 *
PH			0.60 ***	0.02 ns	PH			0.72 ***	0.23 ***
PL				0.27 ***	PL				0.12 *
					mean				
	SMD	PH	PL	YM2		SMD	PH	PL	YM2
HD	−0.59 ***	0.33 ***	−0.06 ns	−0.09 ns	HD	−0.79 ***	0.22 ***	0.13 ***	0.13 ns
SMD		0.10 ns	0.32 ***	0.23 ***	SMD		−0.19 *	−0.13 ns	−0.13 ns
PH			0.76 ***	0.25 ***	PH			0.72 ***	0.23 ns
PL				0.27 ***	PL				0.12 ns

Note: *p*-values are provided with significance level indicated by the asterisks; * $p < 0.05$, ** $p < 0.01$, *** $p < 0.001$; ns—not significant; HD—heading date (days); SMD—seed maturation date (days); PH—plant height (cm); PL—peduncle length (cm); YM2—yield per m^2 (g/m^2). KRIAPI—Kazakh Research Institute of Agriculture and Plant Industry; SPCGF—Alexandr Barayev Scientific-Production Center for Grain Farming.

3.2. SNP Genotyping and Population Structure of the NAM Population

The GWAS analysis was conducted using 10,448 polymorphic SNP markers, of which 40% were mapped to the A genome, 48% to the B genome, and 12% to the D genome (Figure 3A). The minimum number of SNPs (73) was assigned to chromosome 4D, while the maximum number of SNPs was assigned to chromosome 5B (915). Homoeologous group 2 chromosomes contained the largest number of markers, at 1790 SNP markers, having a subgenome A of 573 SNPs, subgenome B of 877 SNPs, and subgenome D of 340 SNPs. Homoeologous group 4 chromosomes had the smallest number, with only 921 markers, with a subgenome A of 438 SNPs, subgenome B of 410 SNPs, and subgenome D of 73 SNPs (Figure 3A). The smallest size was found in chromosome 6D (461 Mb), and the longest was found in chromosome 3B (829 Mb).

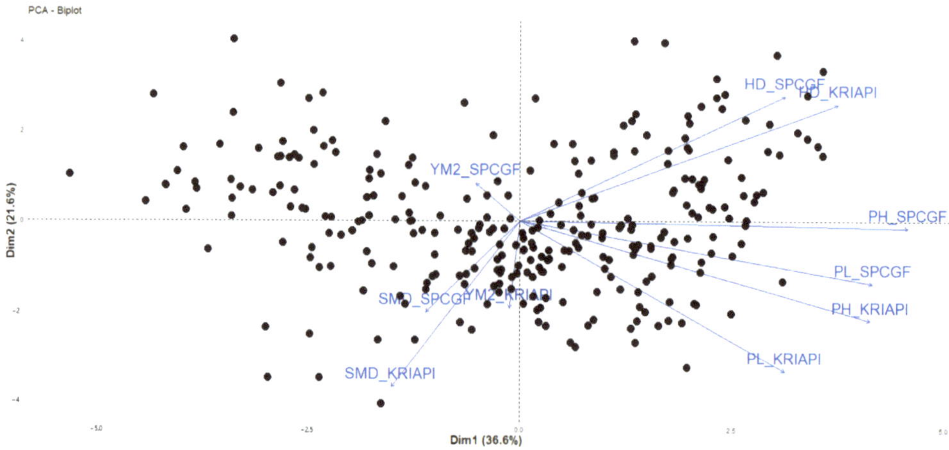

Figure 2. Principal component analysis for studied traits in two regions. Note: RILs—back color, directions of traits—blue color. HD, days—heading date; SMD, days—seed maturation date; PH, cm—plant height; PL, days—peduncle length; YM2, g/m^2—yield per m^2; KRIAPI—Kazakh Research Institute of Agriculture and Plant Industry; SPCGF—Alexandr Barayev Scientific-Production Center for Grain Farming.

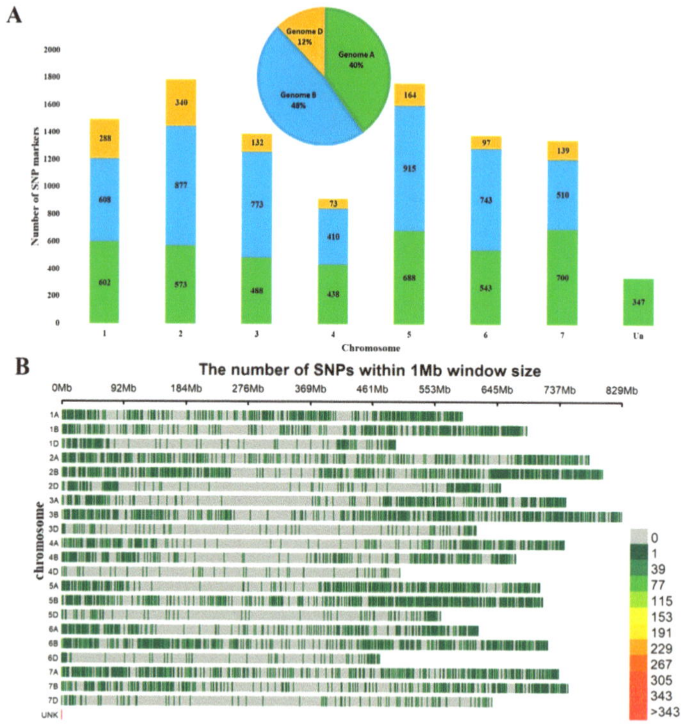

Figure 3. Distribution of 10,448 single-nucleotide polymorphism (SNP) markers mapped to the wheat genome: (**A**) the distribution of SNPs on seven chromosomal groups and the numbers of mapped SNPs on subgenomes A, B, and D; (**B**) the density plot of SNP markers in wheat chromosomes.

The results of the population structure NAM population and STRUCTURE Harvester analyses suggested that K = 3 was the optimal number of clusters for studying 290 RILs (Figure S1). The estimated r^2 values for all pairs of linked SNP loci were used to assess the extent of LD decay in this study. As expected, the r^2 value decreased as the physical distance between markers increased (Figure S2).

3.3. Identification of Marker–Trait Associations for Studied Traits

In two regions, the GWAS for four studied traits led to the detection of 74 significant quantitative trait nucleotides (QTNs) for two or more environments (Tables 6 and S3). Specifically, 28 and 11 QTLs were identified only at the KRIAPI and SPCGF. A comparison of the GWAS results from both regions revealed that 35 QTLs were notably significant in both regions (Table 6). The majority of QTLs were localized on the chromosomes of genome B (30), followed by genomes A (29) and D (8). In the studied traits, the number of identified QTLs varied from 12 for PL to 26 for HD (Table 6).

Table 6. Summary of identified marker–trait associations in the NAM population spring wheat based on field performance in the two locations.

Trait	Identified QTL	KRIAPI	SPCGF	Both Regions
Heading date (HD, days)	26	8	5	13
Seed maturation date (SMD, days)	22	8	2	12
Plant height (PH, cm)	14	7	2	5
Peduncle length (PL, cm)	12	5	2	5
Total	74	28	11	35

Note: KRIAPI—Kazakh Research Institute of Agriculture and Plant Industry; SPCGF—Alexandr Barayev Scientific-Production Center for Grain Farming.

The GWAS detected 74 significant QTLs, 26 for HD, 22 for SMD, 14 for PH, and 12 for PH (Table 7).

Table 7. The list of QTLs for four studied traits identified using 290 RILs of the NAM population in the two regions.

QTLs	SNP	Chr.	Pos., bp	*p*-Value	Regions
QHD.ta.NAM.ipbb-1A.1	*AX-94561041*	1A	41,901,010	6.36×10^{-4}	both
QHD.ta.NAM.ipbb-1A.2	*AX-94768074*	1A	474,699,818	4.45×10^{-5}	KPIAPI
QHD.ta.NAM.ipbb-1B	*AX-94592638*	1B	678,266,710	2.33×10^{-5}	KPIAPI
QHD.ta.NAM.ipbb-2A.1	*AX-95255993*	2A	31,811,157	5.39×10^{-4}	SPCGF
QHD.ta.NAM.ipbb-2A.2	*AX-95098442*	2A	43,299,265	3.05×10^{-4}	both
QHD.ta.NAM.ipbb-2A.3	*AX-94665800*	2A	603,549,569	2.54×10^{-4}	both
QHD.ta.NAM.ipbb-2A.4	*AX-94504542*	2A	729,298,590	4.50×10^{-4}	KPIAPI
QHD.ta.NAM.ipbb-2B.1	*AX-94681430*	2B	18,941,804	1.05×10^{-9}	SPCGF
QHD.ta.NAM.ipbb-2B.2	*AX-94393895*	2B	788,664,980	1.32×10^{-4}	both
QHD.ta.NAM.ipbb-3A	*AX-94701190*	3A	719,763,842	1.38×10^{-6}	both
QHD.ta.NAM.ipbb-3B	*AX-95249280*	3B	571,763,709	7.37×10^{-4}	KPIAPI
QHD.ta.NAM.ipbb-3D	*AX-94713011*	3D	484,808,321	6.69×10^{-4}	both
QHD.ta.NAM.ipbb-4A	*AX-95633345*	4A	707,039,327	1.82×10^{-4}	KPIAPI

Table 7. *Cont.*

QTLs	SNP	Chr.	Pos., bp	*p*-Value	Regions
QHD.ta.NAM.ipbb-5A.1	*AX-94603117*	5A	476,763,775	1.19×10^{-4}	both
QHD.ta.NAM.ipbb-5A.2	*AX-94654737*	5A	588,761,524	3.56×10^{-20}	both
QHD.ta.NAM.ipbb-5A.3	*AX-94725943*	5A	673,709,691	4.11×10^{-5}	both
QHD.ta.NAM.ipbb-5B.1	*AX-95256298*	5B	460,267,677	3.01×10^{-4}	SPCGF
QHD.ta.NAM.ipbb-5B.2	*AX-94386712*	5B	591,836,342	7.67×10^{-6}	both
QHD.ta.NAM.ipbb-5D	*AX-95122517*	5D	462,988,586	3.04×10^{-6}	KPIAPI
QHD.ta.NAM.ipbb-6A	*AX-94943644*	6A	140,607,311	9.31×10^{-4}	SPCGF
QHD.ta.NAM.ipbb-6B	*AX-94570953*	6B	658,818,818	1.23×10^{-5}	both
QHD.ta.NAM.ipbb-6D	*AX-94562028*	6D	468,842,171	9.40×10^{-5}	SPCGF
QHD.ta.NAM.ipbb-7A	*AX-94755544*	7A	127,676,409	1.90×10^{-4}	KPIAPI
QHD.ta.NAM.ipbb-7B.1	*AX-94810990*	7B	9,702,461	7.28×10^{-6}	both
QHD.ta.NAM.ipbb-7B.2	*AX-94684729*	7B	676,144,642	2.37×10^{-4}	both
QHD.ta.NAM.ipbb-UNK	*AX-95256830*	UNK	30,120	4.45×10^{-4}	KPIAPI
QSMD.ta.NAM.ipbb-1A.1	*AX-94500759*	1A	128,626,137	1.00×10^{-6}	KPIAPI
QSMD.ta.NAM.ipbb-1A.2	*AX-94964616*	1A	517,415,353	2.36×10^{-7}	both
QSMD.ta.NAM.ipbb-1B.1	*AX-95208428*	1B	478,053,661	1.07×10^{-4}	both
QSMD.ta.NAM.ipbb-1B.2	*AX-94610095*	1B	587,823,781	1.17×10^{-4}	both
QSMD.ta.NAM.ipbb-1D	*AX-94636030*	1D	53,381,669	2.10×10^{-4}	both
QSMD.ta.NAM.ipbb-2A	*AX-95099971*	2A	94,003,182	1.07×10^{-7}	KPIAPI
QSMD.ta.NAM.ipbb-3A.1	*AX-94605747*	3A	54,939,425	4.85×10^{-4}	SPCGF
QSMD.ta.NAM.ipbb-3A.2	*AX-94866541*	3A	568,383,306	1.65×10^{-5}	both
QSMD.ta.NAM.ipbb-3B.1	*AX-94808751*	3B	431,589,634	5.62×10^{-5}	SPCGF
QSMD.ta.NAM.ipbb-3B.2	*AX-94483125*	3B	781,461,038	4.12×10^{-5}	both
QSMD.ta.NAM.ipbb-4A	*AX-94542577*	4A	614,111,171	1.93×10^{-4}	both
QSMD.ta.NAM.ipbb-5A.1	*AX-95235821*	5A	8,237,880	5.80×10^{-7}	KPIAPI
QSMD.ta.NAM.ipbb-5A.2	*AX-94690257*	5A	706,429,847	1.44×10^{-6}	KPIAPI
QSMD.ta.NAM.ipbb-5B.1	*AX-94817648*	5B	25,666,462	8.85×10^{-5}	KPIAPI
QSMD.ta.NAM.ipbb-5B.2	*AX-94890794*	5B	566,685,969	3.23×10^{-5}	both
QSMD.ta.NAM.ipbb-6B	*AX-94609735*	6B	−1	3.04×10^{-4}	both
QSMD.ta.NAM.ipbb-7B	*AX-94510416*	7B	707,698,825	1.49×10^{-11}	both
QSMD.ta.NAM.ipbb-7D.1	*AX-94696494*	7D	−1	1.37×10^{-5}	both
QSMD.ta.NAM.ipbb-7D.2	*AX-94747939*	7D	58,869,306	5.28×10^{-5}	KPIAPI
QSMD.ta.NAM.ipbb-UNK.1	*AX-94597695*	UNK	9,920	3.04×10^{-4}	KPIAPI
QSMD.ta.NAM.ipbb-UNK.2	*AX-94779279*	UNK	19,750	4.70×10^{-8}	KPIAPI
QSMD.ta.NAM.ipbb-UNK.3	*AX-95254671*	UNK	30,050	3.59×10^{-13}	both
QPH.ta.NAM.ipbb-1A	*AX-95104178*	1A	340,249,943	6.52×10^{-5}	KPIAPI
QPH.ta.NAM.ipbb-2B.1	*AX-94818538*	2B	−1	3.14×10^{-4}	SPCGF
QPH.ta.NAM.ipbb-2B.2	*AX-95150897*	2B	115,839,405	3.19×10^{-4}	SPCGF

Table 7. *Cont.*

QTLs	SNP	Chr.	Pos., bp	*p*-Value	Regions
QPH.ta.NAM.ipbb-2D	*AX-94705599*	2D	577,454,929	3.20×10^{-5}	KPIAPI
QPH.ta.NAM.ipbb-3A	*AX-95083017*	3A	699,419,434	4.80×10^{-4}	KPIAPI
QPH.ta.NAM.ipbb-3B	*AX-95208494*	3B	661,827,596	4.15×10^{-4}	both
QPH.ta.NAM.ipbb-4B	*AX-95630372*	4B	169,935,701	3.50×10^{-4}	both
QPH.ta.NAM.ipbb-5B.1	*AX-94541915*	5B	−1	5.91×10^{-5}	KPIAPI
QPH.ta.NAM.ipbb-5B.2	*AX-94392836*	5B	679,687,601	3.16×10^{-5}	both
QPH.ta.NAM.ipbb-6A.1	*AX-94783460*	6A	127,189,675	1.38×10^{-4}	both
QPH.ta.NAM.ipbb-6A.2	*AX-94575241*	6A	573,496,900	1.36×10^{-4}	KPIAPI
QPH.ta.NAM.ipbb-7A	*AX-94492491*	7A	581,848,865	1.84×10^{-5}	both
QPH.ta.NAM.ipbb-7B	*AX-94439304*	7B	334,455,703	2.15×10^{-4}	KPIAPI
QPH.ta.NAM.ipbb-UNK	*AX-94659909*	UNK	31,410	1.18×10^{-4}	KPIAPI
QPL.ta.NAM.ipbb-1B	*AX-95022601*	1B	106,765,751	1.30×10^{-4}	KPIAPI
QPL.ta.NAM.ipbb-2D	*AX-94444526*	2D	30,405,035	9.41×10^{-5}	KPIAPI
QPL.ta.NAM.ipbb-4A	*AX-94945797*	4A	541,340,650	1.27×10^{-4}	both
QPL.ta.NAM.ipbb-4B.1	*AX-95129444*	4B	480,923,965	2.04×10^{-4}	SPCGF
QPL.ta.NAM.ipbb-4B.2	*AX-95630385*	4B	609,515,886	5.45×10^{-4}	both
QPL.ta.NAM.ipbb-6B	*AX-94793082*	6B	117,516,187	1.81×10^{-6}	KPIAPI
QPL.ta.NAM.ipbb-7A.1	*AX-94634646*	7A	23,238,304	5.31×10^{-4}	KPIAPI
QPL.ta.NAM.ipbb-7A.2	*AX-95179073*	7A	647,297,932	2.23×10^{-6}	both
QPL.ta.NAM.ipbb-7B.1	*AX-94503821*	7B	−1	4.60×10^{-4}	KPIAPI
QPL.ta.NAM.ipbb-7B.2	*AX-94587603*	7B	61,077,481	2.04×10^{-5}	SPCGF
QPL.ta.NAM.ipbb-7B.3	*AX-94545252*	7B	133,792,764	4.94×10^{-5}	both
QPL.ta.NAM.ipbb-7B.4	*AX-94505633*	7B	401,550,322	5.32×10^{-4}	both

Note: Chr—chromosome; Pos., bp—physical position of markers; UNK—unknown chromosome; −1—unknown positions; KRIAPI—Kazakh Research Institute of Agriculture and Plant Industry; SPCGF—Alexandr Barayev Scientific-Production Center for Grain Farming.

Among the identified QTLs for HD, a total of 26 were stable (QTLs found in two or more environments), with 8 and 5 of these detected at the KRIAPI and SPCGF, respectively. Thirteen of these QTLs were common to both regions. Notably, the most significant *p*-value of 3.56×10^{-20} was observed for chromosome 5A, detected in both regions (Tables 7 and S3). Furthermore, *AX-94654737* exhibited detection in both regions with a PVE of 19.33%. Table 7 and visual representations are provided in Manhattan plots and Q–Q plots in Figure 4A,B for further details.

The effect of each QTL varied significantly, with the highest value observed for *AX-94681430* (−4.3 days), explained by the phenotypic variation (PVE) of 14.25% detected at the SPCGF. Another notable QTL, *AX-95122517*, was identified at the KRIAPI and had a *p*-value from 3.04×10^{-6}, with a phenotypic variation of 30.83% (Tables 7 and S3).

For SMD, 22 stable QTLs were identified, with 8 and 2 detected at the KRIAPI and SPCGF, respectively. The most significant *p*-value (1.49×10^{-11}) was observed for chromosome 7B in both regions (Tables 7 and S3). Additionally, *AX-94510416* was detected in both regions with a PVE of 41.36%.

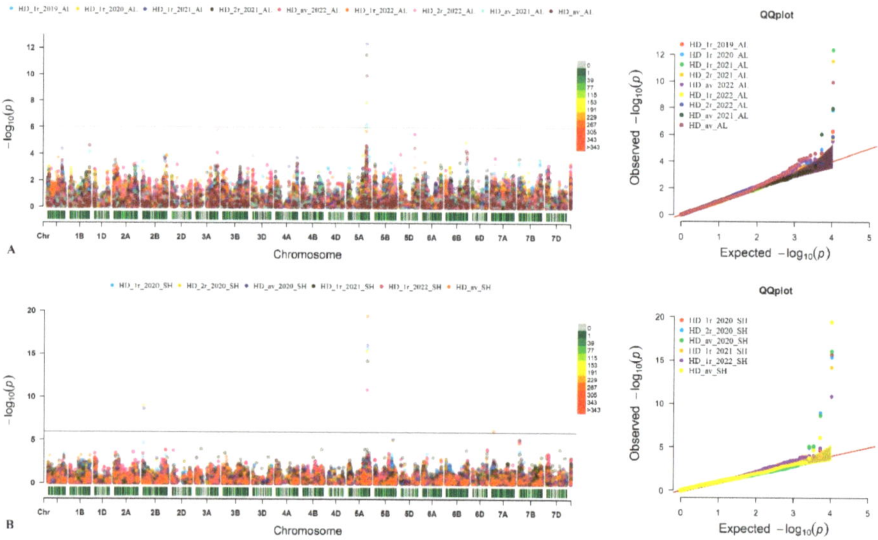

Figure 4. Manhattan and quantile–quantile plots (Q-Q) for the heading date (HD) in the genome-wide association study panel phenotyped in the Almaty region, KRIAPI (**A**), and Akmola region, SPCGF (**B**).

Regarding PL and PH, 14 and 12 QTLs were identified, respectively. The most significant QTL for PL was located on chromosome 6B and was significant in the 2022 season at the KRIAPI, with a PVE of 22.7% (Tables 7 and S3). Furthermore, *AX-95179073* was detected in both regions and mapped to chromosome 7A with a PVE of 20.11%. All results obtained were noted to hold significance and warrant consideration in breeding projects associated with plant adaptation and yield-related traits.

4. Discussion

Breeding programs focused on increased yield often include optimizing additional key agronomic traits like plant height (PH) and the date of heading (HD) [69,70]. The ideal combination for HD and PH can vary significantly depending on local environmental conditions. Therefore, managing these traits poses a challenge because of their interconnected nature, where changes in one trait can affect others [71–74].

In the present study, 290 RILs of the NAM population were tested in two different contrasting parts of Kazakhstan, at the KRIAPI (southeast) and SPCGF (north). The comparative analysis of climate conditions (rainfall and temperature) showed that higher precipitation significantly contributes to increased productivity (Table 2). The assessment of the studied traits noted a large grain yield difference between the two regions (Figure 1, Table 3). The Pearson correlation analysis for average data over three years (2020–2022) showed that early HD, late SMD, taller PH, and longer PL were significantly favorable for higher YM2 at the KRIAPI but insignificant at the SPCGF. The controversial correlations among the studied traits at the SPCGF were most probably affected by the amount of rainfall at the early plant developmental stages. In favorable plant growth conditions in 2020, the YM2 negatively correlated with HD. Also, the annual assessment of correlation results at the SPCGF suggested that YM2 positively correlates with PH in two out of three years (Table 5). Therefore, early flowering time and taller plants are more favorable for higher seed productivity in both contrasting regions.

The analysis of the average data of YM2 showed that 11 and 60 RILs showed higher values than local check cultivars at the KRIAPI and SPCGF, respectively. Two RILs, NAM-326 (Paragon × Wylakatchem-092) and NAM-138 (Paragon × Watkins349-027), showed higher yields in both regions than the local comparison check cultivars under all studied conditions (Table S1); these can be used for further wheat breeding projects in Kazakhstan. The phenotypic data presented in the current study showed that the NAM population is a valuable resource for improving agronomic traits.

The GWAS analyses of the NAM population in the two regions led to the identification of 74 QTLs in the four traits related to plant adaptation. Notably, the largest number of QTL was identified for HD, SMD, and PH, which shows a wide range of phenotypic variations in these traits in the two regions. The largest number of QTLs were identified for HD (26 QTLs) located on chromosomes 1A (2 QTLs), 1B, 2A (4 QTLs), 2B (2 QTLs), 3A, 3B, 3D, 4A, 5A (3 QTLs), 5B, 5D, 6A, 6B, 6D, 7A, and 7B (2 QTLs). The most significant QTL (*AX-94654737*) with a *p*-value of 3.56×10^{-20} was observed for chromosome 5A (588,761,524 bp) detected in both regions (Tables 7 and S3). The analysis aimed at identifying putative candidate genes using the reference genome in the Wheat Ensembl database revealed that the QTL was situated at 588,761,524 bp on chromosome 5A. Within this position, *TraesCS5A02G392700* was identified, encoding a protein annotated as an *ABC* transporter. Interestingly, in the literature, this protein has been associated with the wheat resistance gene *Lr34* [75]. This protein in the wheat ABC transporter *Lr34*, a member of the G subfamily, is known to confer partial, durable, and broad-spectrum resistance against several biotrophic fungi such as powdery mildew, leaf rust, or stem rust. Initially, it was assumed that the *Vrn-A1* gene was located at this position. Although, the physical position of *Vrn-A1* spans from 587,411,454 bp to 587,423,416 bp, with a difference of approximately 1,350,070 bp. LD analysis for chromosome 5A, which spans 6,057,956 bp, revealed that the Vrn-A1 gene is relatively close to other loci on the chromosome. Another *Vrn* gene, *Vrn-B3*, was identified within the QTL (*AX-94810990*) located at 9,702,461 bp on chromosome 7A. This position closely matches the physical position of *Vrn-B3*, which spans from 9,702,354 bp to 9,704,354 bp. *Vrn-B3* belongs to the *Vrn1* gene family, which comprises vernalization genes that regulate wheat flowering. Specifically, *Vrn-B3* promotes the transition from vegetative to reproductive growth in response to vernalization. It acts as a repressor of flowering, and its expression is downregulated by exposure to cold temperatures [76,77].

The literature survey suggested that 16 of the 74 QTLs identified in our study had also been detected in previous QTL mapping studies and GWASs for all studied traits (Table S4) [78–91]. The majority of these matches were found for PH (eight QTLs), followed by HD (five QTLs), SMD (two QTLs), and PL (one QTLs) (Table S4). Three associations associated with PH were identical to the genetic positions of QTLs identified in analyses of eight traits using 94 RILs of the mapping population of Pamyati Azieva × Paragon, tested in Kazakhstan's northern and southern regions [78]. Three QTNs (*AX-95255993*, *AX-94504542*, *AX-94654737*) associated with HD had similar physical positions to QTLs identified in the GWAS of agronomic and quality traits in a NAM population to exploit the genetic diversity of the USDA-ARS NSGC [44].

The significant SNPs in the detected QTLs were analyzed to identify putative candidate genes using the annotated Chinese Spring reference genome [92] in the Wheat Ensembl database [67]. The results showed that out of the 74 identified QTLs, 51 were located in genic positions (Table S3). An analysis of these 51 genes suggested that most were associated with controlling plant growth, development, and abiotic/biotic stress tolerance [93–100]. For example, two QTLs were associated with HD (*AX-94943644*) and SMD (*AX-94483125*), where significant SNPs were aligned with F-box-domain-containing proteins. The F-box proteins regulate plant development and control flowering time [93,94]. It was determined that *AX-94634646*, associated with PL, encodes *Domain of unknown function* (DUF)-domain-containing proteins, which play a role in plant development and fitness in rice [95]. The list of genes and proteins related to stress resistance/tolerance includes protein kinase superfamily protein (*TraesCS1B02G469400*, *TraesCS2B02G001600*, *TraesCS2D02G474300*, *TraesCS3B02G424200*) [96], a zinc finger protein (*TraesCS7A02G173200*) [97], and a CSC1-like protein RXW8 (*TraesCS5A02G012500*) associated with chilling tolerance [98]. In addition, a list of genes related to stress drought tolerance included *TraesCS2B02G608300* (Potassium efflux antiporter), *TraesCS3A02G493000* (EF-hand-domain-containing protein) [99], and *TraesCS1D02G072900* (*WRKY26* transcription factor) [100].

The alignment of associations identified in this study with previously published reports confirms the results' reliability. While the identified QTLs should undergo further validation in subsequent experiments, there is a promising indication that most presumably novel associations hold significance for plant adaptation-related traits. Consequently, the SNPs identified within the detected QTLs will likely have significant value for successful application in marker-assisted wheat breeding.

5. Conclusions

The analysis of 290 RILs of the NAM population in two contrasting regions of Kazakhstan (north and southeast) indicated that early heading time and taller plants are more favorable for grain productivity. The assessment of the average YM2 values suggested that 11 and 60 RILs showed higher values than local check cultivars in the southeast and north regions, respectively. Hence, the phenotypic data showed that the NAM population is valuable for improving agronomic traits. The GWAS of the NAM population in the two regions allowed the identification of 74 QTLs in the four traits related to plant adaptation (HD, SMD, PH, and PL). The largest number of QTLs were identified for HD (26 QTLs), including two QTLs in the vicinity of the physical positions of *Vrn-A1* (chromosome 5A) and *Vrn-B3* (chromosome 7A). The study provided a valuable data source for the search for new genes associated with wheat plant adaption.

Supplementary Materials: The following supporting information can be downloaded at: https://www.mdpi.com/article/10.3390/plants13182623/s1, Table S1: The raw field data at the Kazakh Research Institute of Agriculture and Plant Industry (KRIAPI, Almaty region, Southeast Kazakhstan) and Alexandr Barayev Scientific-Production Center for Grain Farming (SPCGF, Shortandy, Akmola region, Northern Kazakhstan); Table S2: Results of t-test for five studied traits between two regions in southeast and north of Kazakhstan; Table S3: The list of QTLs and genes for five studied traits identified using 290 RILs of the NAM population in the conditions of the Kazakh Research Institute of Agriculture and Plant Industry (KRIAPI, Almaty region, Southeast Kazakhstan, 2019–2022) and Alexandr Barayev Scientific-Production Center for Grain Farming (SPCGF, Shortandy, Akmola region, Northern Kazakhstan, 2020–2022); Table S4: List of identified QTLs based on the GWAS analysis of wheat collection compared to the associations revealed in previously published reports; Figure S1: Population structure of the NAM population based on 10,448 SNP markers: (A) STRUCTURE Harvester output for delta K; (B) separation of samples into clusters based on the STRUCTURE package at K = 4. The colors in the boxes represent the clusters identified in Figure S1B; Figure S2: Chromosome-wide linkage disequilibrium (LD) decay estimated for 10,448 SNPs of 290RILs of the NAM population.

Author Contributions: Conceptualization, Y.T.; Formal analysis, A.A.; Funding acquisition, Y.T.; Investigation, A.A., S.G., A.B., S.A. and Y.T.; Methodology, S.G., Y.T. and S.A.; Project administration, S.A.; Resources, S.G. and C.P.; Supervision, Y.T.; Visualization, A.A.; Writing—original draft, A.A., S.G., S.A. and Y.T.; Writing—review and editing, A.A., S.G., S.A. and Y.T. All authors have read and agreed to the published version of the manuscript.

Funding: This research was funded by the Science Committee of the Ministry of Science and Higher Education (former Ministry of Education and Science) of the Republic of Kazakhstan (Grant No. BR24992903, AP14871383).

Data Availability Statement: The original contributions presented in the study are included in the article/supplementary material, further inquiries can be directed to the corresponding author/s.

Conflicts of Interest: The authors declare no conflict of interest.

References

1. Braun, H.J.; Atlin, G.; Payne, T.; Reynolds, M.P. *Climate Change and Crop Production*; CABI Publishers: Wallingford, UK, 2010; pp. 115–138.
2. Food and Agriculture Organization of the United Nations. Available online: https://www.fao.org/faostat/en/#home (accessed on 25 February 2024).
3. Shiferaw, B.; Smale, M.; Braun, H.J.; Duveiller, E.; Reynolds, M.; Muricho, G. Crops that feed the world 10. Past successes and future challenges to the role played by wheat in global food security. *Food Secur.* **2013**, *5*, 291–317. [CrossRef]
4. Li, H.; Zhou, Y.; Xin, W.; Wei, Y.; Zhang, J.; Guo, L. Wheat breeding in northern China: Achievements and technical advances. *Crop J.* **2019**, *7*, 718–729. [CrossRef]
5. Hyles, J.; Bloomfield, M.T.; Hunt, J.R.; Trethowan, R.M.; Trevaskis, B. Phenology and related traits for wheat adaptation. *Heredity* **2020**, *125*, 417–430. [CrossRef] [PubMed]
6. Dowla, M.N.U.; Edwards, I.; O'Hara, G.; Islam, S.; Ma, W. Developing wheat for improved yield and adaptation under a changing climate: Optimization of a few key genes. *Engineering* **2018**, *4*, 514–522. [CrossRef]
7. Curtis, B.C.; Rajaram, S.; Gómez Macpherson, H. (Eds.) *Bread Wheat: Improvement and Production*; Food and Agriculture Organization of the United Nations: Rome, Italy, 2002; 554p.
8. Yan, L.; Fu, D.; Li, C.; Blechl, A.; Tranquilli, G.; Bonafede, M.; Sanchez, A.; Valarik, M.; Yasuda, S.; Dubcovsky, J. The wheat and barley vernalization gene VRN3 is an orthologue of FT. *Proc. Natl. Acad. Sci. USA* **2006**, *103*, 19581–19586. [CrossRef]
9. Bonnin, I.; Rousset, M.; Madur, D.; Sourdille, P.; Dupuits, C.; Brunel, D.; Goldringer, I. FT genome A and D polymorphisms are associated with the variation of earliness components in hexaploid wheat. *Theor. Appl. Genet.* **2008**, *116*, 383–394. [CrossRef]
10. Trevaskis, B. The central role of the *VERNALIZATION1* gene in the vernalization response of cereals. *Funct. Plant Biol.* **2010**, *37*, 479–487. [CrossRef]
11. Foulkes, M.J.; Sylvester-Bradley, R.; Worland, A.J.; Snape, J.W. Effects of a photoperiod-response gene *Ppd-D1* on yield potential and drought resistance in UK winter wheat. *Euphytica* **2004**, *135*, 63–73. [CrossRef]
12. Beales, J.; Turner, A.; Griffiths, S.; Snape, J.W.; Laurie, D.A. A pseudo-response regulator is misexpressed in the photoperiod insensitive *Ppd-D1a* mutant of wheat (*Triticum aestivum* L.). *Theor. Appl. Genet.* **2007**, *115*, 721–733. [CrossRef]
13. Kumar, S.; Sharma, V.; Chaudhary, S.; Tyagi, A.; Mishra, P.; Priyadarshini, A.; Singh, A. Genetics of flowering time in bread wheat Triticum aestivum: Complementary interaction between vernalization-insensitive and photoperiod-insensitive mutations imparts very early flowering habit to spring wheat. *J. Genet.* **2012**, *91*, 33–47. [CrossRef]
14. Griffiths, S.; Simmonds, J.; Leverington, M.; Wang, Y.; Fish, L.; Sayers, L.; Alibert, L.; Orford, S.; Wingen, L.; Herry, L.; et al. Meta-QTL analysis of the genetic control of ear emergence in elite European winter wheat germplasm. *Theor. Appl. Genet.* **2012**, *119*, 383–395. [CrossRef] [PubMed]
15. Zikhali, M.; Griffiths, S. The effect of Earliness per se (*Eps*) genes on flowering time in bread wheat. In *Advances in Wheat Genetics: From Genome to Field*; Springer: Tokyo, Japan, 2015; pp. 339–345.
16. Zikhali, M.; Wingen, L.U.; Griffiths, S. Delimitation of the Earliness per se D1 (*Eps-D1*) flowering gene to a subtelomeric chromosomal deletion in bread wheat (*Triticum aestivum*). *J. Exp. Bot.* **2016**, *67*, 287–299. [CrossRef] [PubMed]
17. Peng, J.; Richards, D.E.; Hartley, N.M.; Murphy, G.P.; Devos, K.M.; Flintham, J.E.; Beales, J.; Fish, L.J.; Worland, A.J.; Pelica, F.; et al. 'Green revolution' genes encode mutant gibberellin response modulators. *Nature* **1999**, *400*, 256–261. [CrossRef] [PubMed]
18. Hedden, P. The genes of the Green Revolution. *Trends Genet.* **2003**, *19*, 5–9. [CrossRef] [PubMed]
19. Ellis, M.H.; Rebetzke, G.J.; Azanza, F.; Richards, R.A.; Spielmeyer, W. Molecular mapping of gibberellin-responsive dwarfing genes in bread wheat. *Theor. Appl. Genet.* **2005**, *111*, 423–430. [CrossRef]
20. van Eeuwijk, F.A.; Bink, M.C.; Chenu, K.; Chapman, S.C. Detection and use of QTL for complex traits in multiple environments. *Curr. Opin. Plant Biol.* **2010**, *13*, 193–205. [CrossRef]
21. Collard, B.C.; Jahufer, M.Z.Z.; Brouwer, J.B.; Pang, E.C.K. An introduction to markers, quantitative trait loci (QTL) mapping and marker-assisted selection for crop improvement: The basic concepts. *Euphytica.* **2005**, *142*, 169–196. [CrossRef]
22. Xu, Y.; Li, P.; Yang, Z.; Xu, C. Genetic mapping of quantitative trait loci in crops. *Crop J.* **2017**, *5*, 175–184. [CrossRef]
23. Sukumaran, S.; Dreisigacker, S.; Lopes, M.; Chavez, P.; Reynolds, M.P. Genome-wide association study for grain yield and related traits in an elite spring wheat population grown in temperate irrigated environments. *Theor. Appl. Genet.* **2015**, *128*, 353–363. [CrossRef]
24. Gao, F.; Wen, W.; Liu, J.; Rasheed, A.; Yin, G.; Xia, X.; Wu, X.; He, Z. Genome-wide linkage mapping of QTL for yield components, plant height and yield-related physiological traits in the Chinese wheat cross zhou 8425B/Chinese spring. *Front. Plant Sci.* **2015**, *6*, 1099. [CrossRef]
25. Zhang, H.; Chen, J.; Li, R.; Deng, Z.; Zhang, K.; Liu, B.; Tian, J. Conditional QTL mapping of three yield components in common wheat (*Triticum aestivum* L.). *Crop J.* **2016**, *4*, 220–228. [CrossRef]
26. Jin, J.; Liu, D.; Qi, Y.; Ma, J.; Zhen, W. Major QTL for seven yield-related traits in common wheat (*Triticum aestivum* L.). *Front. Genet.* **2020**, *11*, 1012. [CrossRef]
27. Pang, Y.; Liu, C.; Wang, D.; St Amand, P.; Bernardo, A.; Li, W.; He, F.; Li, L.; Wang, L.; Yuan, X.; et al. High-resolution genome-wide association study identifies genomic regions and candidate genes for important agronomic traits in wheat. *Mol. Plant* **2020**, *13*, 1311–1327. [CrossRef] [PubMed]

28. Isham, K.; Wang, R.; Zhao, W.; Wheeler, J.; Klassen, N.; Akhunov, E.; Chen, J. QTL mapping for grain yield and three yield components in a population derived from two high-yielding spring wheat cultivars. *Theor. Appl. Genet.* **2021**, *134*, 2079–2095. [CrossRef] [PubMed]
29. Kang, C.S.; Mo, Y.J.; Kim, K.M.; Kim, K.H.; Chun, J.B.; Park, C.S.; Cho, S.W. Mapping quantitative trait loci for yield potential traits in wheat recombinant inbred lines. *Agronomy* **2021**, *11*, 22. [CrossRef]
30. Li, T.; Li, Q.; Wang, J.; Yang, Z.; Tang, Y.; Su, Y.; Zhang, J.; Qiu, X.; Pu, X.; Pan, Z.; et al. High-resolution detection of quantitative trait loci for seven important yield-related traits in wheat (*Triticum aestivum* L.) using a high-density SLAF-seq genetic map. *BMC Genom. Data* **2022**, *23*, 37. [CrossRef]
31. Korte, A.; Farlow, A. The advantages and limitations of trait analysis with GWAS: A review. *Plant Methods* **2013**, *9*, 29. [CrossRef]
32. Myles, S.; Peiffer, J.; Brown, P.J.; Ersoz, E.S.; Zhang, Z.; Costich, D.E.; Buckler, E.S. Association mapping: Critical considerations shift from genotyping to experimental design. *Plant Cell* **2009**, *21*, 2194–2202. [CrossRef]
33. Poland, J.A.; Bradbury, P.J.; Buckler, E.S.; Nelson, R.J. Genome-wide nested association mapping of quantitative resistance to northern leaf blight in maize. *Proc. Natl. Acad. Sci. USA* **2011**, *108*, 6893–6898. [CrossRef]
34. Yu, J.; Holland, J.B.; McMullen, M.D.; Buckler, E.S. Genetic Design and Statistical Power of Nested Association Mapping in Maize. *Genetics* **2008**, *178*, 539–551. [CrossRef]
35. McMullen, M.D.; Kresovich, S.; Villeda, H.S.; Bradbury, P.; Li, H.; Sun, Q.; Flint-Garcia, S.; Thornsberry, J.; Acharya, C.; Bottoms, C.; et al. Genetic properties of the maize nested association mapping population. *Science* **2009**, *325*, 737–740. [CrossRef] [PubMed]
36. Scott, M.F.; Ladejobi, O.; Amer, S.; Bentley, A.R.; Biernaskie, J.; Boden, S.A.; Clark, M.; Acqua, M.D.; Dixon, L.E.; Filippi, C.V.; et al. Multiparent populations in crops: A toolbox integrating genomics and genetic mapping with breeding. *Heredity* **2020**, *125*, 396–416. [CrossRef] [PubMed]
37. Mackay, I.J.; Bansept-Basler, P.; Barber, T.; Bentley, A.R.; Cockram, J.; Gosman, N.; Greenland, A.J.; Horsnell, R.; Howells, R.; O'Sullivan, D.M.; et al. An eight-parent multiparent advanced generation inter-cross population for winter-sown wheat: Creation, properties, and validation. *G3* **2014**, *4*, 1603–1610. [CrossRef]
38. Maurer, A.; Draba, V.; Jiang, Y.; Schnaithmann, F.; Sharma, R.; Schumann, E.; Killian, B.; Reif, J.; Pillen, K. Modelling the genetic architecture of flowering time control in barley through nested association mapping. *BMC Genom.* **2015**, *16*, 290. [CrossRef]
39. Nice, L.M.; Steffenson, B.J.; Brown-Guedira, G.L.; Akhunov, E.D.; Liu, C.; Kono, T.; Morrell, P.L.; Blake, T.; Horsley, R.; Smith, K.; et al. Development and genetic characterization of an advanced backcross-nested association mapping (AB-NAM) population of wild × cultivated barley. *Genetics* **2016**, *203*, 1453–1467. [CrossRef]
40. Kidane, Y.G.; Gesesse, C.A.; Hailemariam, B.N.; Desta, E.A.; Mengistu, D.K.; Fadda, C.; Enrico Pè, M.; Dell'Acqua, M. A large nested association mapping population for breeding and quantitative trait locus mapping in Ethiopian durum wheat. *Plant Biotechnol. J.* **2019**, *17*, 1380–1393. [CrossRef] [PubMed]
41. Fragoso, C.A.; Moreno, M.; Wang, Z.; Heffelfinger, C.; Arbelaez, L.J.; Aguirre, J.A.; Franco, N.; Romero, L.; Labadie, K.; Zhao, H.; et al. Genetic architecture of a rice nested association mapping population. *G3* **2017**, *7*, 1913–1926. [CrossRef] [PubMed]
42. Bouchet, S.; Olatoye, M.O.; Marla, S.R.; Perumal, R.; Tesso, T. Increased power to dissect adaptive traits in global sorghum diversity using a nested association mapping population. *Genetics* **2017**, *206*, 573–585. [CrossRef]
43. Xavier, A.; Xu, S.; Muir, W.M.; Rainey, K.M. NAM: Association studies in multiple populations. *Bioinformatics* **2015**, *31*, 3862–3864. [CrossRef]
44. Sallam, A.H.; Manan, F.; Bajgain, P.; Martin, M.; Szinyei, T.; Conley, E.; Brown-Guedira, G.; Muehlbauer, G.J.; Anderson, J.A.; Steffenson, B.J. Genetic architecture of agronomic and quality traits in a nested association mapping population of spring wheat. *Plant Genome* **2020**, *13*, e20051. [CrossRef]
45. Hu, J.; Xiao, G.; Jiang, P.; Zhao, Y.; Zhang, G.; Ma, X.; Yao, J.; Xue, L.; Su, P.; Bao, Y. QTL detection for bread wheat processing quality in a nested association mapping population of semi-wild and domesticated wheat varieties. *BMC Plant Biol.* **2022**, *22*, 129. [CrossRef] [PubMed]
46. Sandhu, K.S.; Mihalyov, P.D.; Lewien, M.J.; Pumphrey, M.O.; Carter, A.H. Genomic selection and genome-wide association studies for grain protein content stability in a nested association mapping population of wheat. *Agronomy* **2021**, *11*, 2528. [CrossRef]
47. Chidzanga, C.; Mullan, D.; Roy, S.; Baumann, U.; Garcia, M. Nested association mapping-based GWAS for grain yield and related traits in wheat grown under diverse Australian environments. *Theor. Appl. Genet.* **2022**, *135*, 4437–4456. [CrossRef] [PubMed]
48. Christopher, M.; Paccapelo, V.; Kelly, A.; Macdonald, B.; Hickey, L.; Richard, C.; Verbyla, A.; Chenu, K.; Borrell, A.; Amin, A.; et al. QTL identified for stay-green in a multi-reference nested association mapping population of wheat exhibit context dependent expression and parent-specific alleles. *Field Crops Res.* **2021**, *270*, 108181. [CrossRef]
49. Ren, D.; Fang, X.; Jiang, P.; Zhang, G.; Hu, J.; Wang, X.; Meng, Q.; Cui, W.; Lan, S.; Ma, X.; et al. Genetic architecture of nitrogen-deficiency tolerance in wheat seedlings based on a nested association mapping (NAM) population. *Front. Plant Sci.* **2018**, *9*, 845. [CrossRef]
50. Chidzanga, C.; Fleury, D.; Baumann, U.; Mullan, D.; Watanabe, S.; Kalambettu, P.; Pontre, R.; Edwards, J.; Forrest, K.; Wong, D.; et al. Development of an australian bread wheat nested association mapping population, a new genetic diversity resource for breeding under dry and hot climates. *Int. J. Mol. Sci.* **2021**, *22*, 4348. [CrossRef]
51. Schnaithmann, F.; Kopahnke, D.; Pillen, K. A first step toward the development of a barley NAM population and its utilization to detect QTLs conferring leaf rust seedling resistance. *Theor. Appl. Genet.* **2014**, *127*, 1513–1525. [CrossRef]

52. Bajgain, P.; Rouse, M.N.; Tsilo, T.J.; Macharia, G.K.; Bhavani, S.; Jin, Y.; Anderson, J.A. Nested association mapping of stem rust resistance in wheat using genotyping by sequencing. *PLoS ONE* **2016**, *11*, e0155760. [CrossRef]
53. Mackay, T.F. The genetic architecture of quantitative traits. *Annu. Rev. Genet.* **2001**, *35*, 303–339. [CrossRef]
54. Nicotra, A.B.; Atkin, O.K.; Bonser, S.P.; Davidson, A.M.; Finnegan, E.J.; Mathesius, U.; Poot, P.; Purugganan, M.D.; Richards, C.L.; Valladares, F.; et al. Plant phenotypic plasticity in a changing climate. *Trends Plant Sci.* **2010**, *15*, 684–692. [CrossRef]
55. Andrés, F.; Coupland, G. The genetic basis of flowering responses to seasonal cues. *Nat. Rev. Genet.* **2012**, *13*, 627–639. [CrossRef] [PubMed]
56. Lobell, D.B.; Gourdji, S.M. The influence of climate change on global crop productivity. *Plant Physiol.* **2012**, *160*, 1686–1697. [CrossRef] [PubMed]
57. Wingen, L.U.; West, C.; Leverington-Waite, M.; Collier, S.; Orford, S.; Goram, R.; Yang, C.; King, J.; Allen, A.; Burridge, A.; et al. Wheat landrace genome diversity. *Genetics* **2017**, *205*, 1657–1676. [CrossRef]
58. Wingen, L.U.; Orford, S.; Goram, R.; Leverington-Waite, M.; Bilham, L.; Patsiou, T.S.; Ambrose, M.; Dicks, J.; Griffiths, S. Establishing the AE Watkins landrace cultivar collection as a resource for systematic gene discovery in bread wheat. *Theor. Appl. Genet.* **2014**, *127*, 1831–1842. [CrossRef]
59. Dospekhov, B. *Methods of Field Experience*; Kolos: Moscow, Russia, 1985.
60. Allen, A.M.; Barker, G.L.; Berry, S.T.; Coghill, J.A.; Gwilliam, R.; Kirby, S.; Robinson, P.; Brenchley, R.C.; D'Amore, R.; McKenzie, N.; et al. Transcript-specific. single-nucleotide polymorphism discovery and linkage analysis in hexaploid bread wheat (*Triticum aestivum* L.). *Plant Biotechnol. J.* **2011**, *9*, 1086–1099. [CrossRef]
61. Bradbury, P.J.; Zhang, Z.; Kroon, D.E.; Casstevens, T.M.; Ramdoss, Y. Buckler ES. TASSEL: Software for association mapping of complex traits in diverse samples. *Bioinformatics* **2007**, *23*, 2633–2635. [CrossRef]
62. Lipka, A.E.; Tian, F.; Wang, Q.; Peiffer, J.; Li, M.; Bradbury, P.J.; Core, M.; Buckler, E.; Zhang, Z. GAPIT: Genome association and prediction integrated tool. *Bioinformatics* **2012**, *28*, 2397–2399. [CrossRef]
63. Pritchard, J.K.; Stephens, M.; Donnelly, P. Inference of population structure using multilocus genotype data. *Genetics* **2000**, *155*, 945–959. [CrossRef] [PubMed]
64. Evanno, G.; Regnaut, S.; Goudet, J. Detecting the number of clusters of individuals using the software STRUCTURE: A simulation study. *Mol. Ecol.* **2005**, *14*, 2611–2620. [CrossRef]
65. Yin, L.; Zhang, H.; Tang, Z.; Xu, J.; Yin, D.; Zhang, Z.; Yuan, X.; Zhu, M.; Zhao, S.; Li, X.; et al. rMVP: A memory-efficient. visualization-enhanced. and parallel-accelerated tool for genome-wide association study. *Genom. Proteom. Bioinform.* **2021**, *19*, 619–628. [CrossRef]
66. Ensembl Plants. Available online: https://plants.ensembl.org/Triticum_aestivum/Info/Index (accessed on 15 February 2024).
67. RStudio Team. *RStudio: Integrated Development for R*; RStudio Inc.: Boston, MA, USA, 2015. Available online: http://www.maizegenetics.net/GAPIT (accessed on 20 January 2024).
68. Genievskaya, Y.; Turuspekov, Y.; Rsaliyev, A.; Abugalieva, S. Genome-wide association mapping for resistance to leaf, stem, and yellow rusts of common wheat under field conditions of South Kazakhstan. *PeerJ* **2020**, *8*, e9820. [CrossRef] [PubMed]
69. Reynolds, M.P.; Langridge, P. Physiological breeding. *Curr. Opin. Plant Biol.* **2016**, *31*, 162–171. [CrossRef] [PubMed]
70. Zwer, P.K.; Ram, P.C. Marker assisted breeding: A novel strategy for enhancing productivity in wheat. *J. Plant Biochem. Physiol.* **2017**, *5*, 1000185.
71. Aisawi, K.A.B.; Reynolds, M.P.; Singh, R.P.; Foulkes, M.J. The physiological basis of the genetic progress in yield potential of CIMMYT spring wheat cultivars from 1966 to 2009. *Crop Sci.* **2015**, *55*, 1749–1764. [CrossRef]
72. Martinez, A.F.; Lister, C.; Freeman, S.; Ma, J.; Berry, S.; Wingen, L.; Griffiths, S. Resolving a QTL complex for height, heading, and grain yield on chromosome 3A in bread wheat. *J. Exp. Bot.* **2021**, *72*, 2965–2978. [CrossRef]
73. Lopes, M.S.; El-Basyoni, I.; Baenziger, P.S.; Singh, S.; Royo, C.; Ozbek, K.; Aktas, H.; Ozer, E.; Ozdermir, F.; Manickavelu, A.; et al. Exploiting genetic diversity from landraces in wheat breeding for adaptation to climate change. *J. Exp. Bot.* **2015**, *66*, 3477–3486. [CrossRef]
74. Shewry, P.R.; Hey, S.J. The contribution of wheat to human diet and health. *Food Energy Secur.* **2015**, *4*, 178–202. [CrossRef]
75. Deppe, J.P.; Rabbat, R.; Hörtensteiner, S.; Keller, B.; Martinoia, E.; Lopéz-Marqués, R.L. The wheat ABC transporter Lr34 modifies the lipid environment at the plasma membrane. *J. Biol. Chem.* **2018**, *293*, 18667–18679. [CrossRef]
76. Yan, L.; Helguera, M.; Kato, K.; Fukuyama, S.; Sherman, J.; Dubcovsky, J. Allelic variation at the VRN-1 promoter region in polyploid wheat. *Theor. Appl. Genet.* **2004**, *109*, 1677–1686. [CrossRef]
77. Shaw, L.M.; Turner, A.S.; Laurie, D.A.; Snape, J.W. Association mapping of flowering time genes in elite European winter wheat varieties. *Theor. Appl. Genet.* **2012**, *125*, 721–735.
78. Amalova, A.; Yermekbayev, K.; Griffiths, S.; Abugalieva, S.; Babkenov, A.; Fedorenko, E.; Abugalieva, A.; Turuspekov, Y. Identification of quantitative trait loci of agronomic traits in bread wheat using a Pamyati Azieva × Paragon mapping population harvested in three regions of Kazakhstan. *PeerJ* **2022**, *10*, e14324. [CrossRef] [PubMed]
79. Amalova, A.; Abugalieva, S.; Chudinov, V.; Sereda, G.; Tokhetova, L.; Abdikhalyk, A.; Turuspekov, Y. QTL mapping of agronomic traits in wheat using the UK Avalon × Cadenza reference mapping population grown in Kazakhstan. *PeerJ* **2021**, *9*, e10733. [CrossRef] [PubMed]
80. Tahmasebi, S.; Heidari, B.; Pakniyat, H.; McIntyre, C.L. Mapping QTLs associated with agronomic and physiological traits under terminal drought and heat stress conditions in wheat (*Triticum aestivum* L.). *Genome* **2017**, *60*, 26–45. [CrossRef] [PubMed]

81. Hu, J.; Wang, X.; Zhang, G.; Jiang, P.; Chen, W.; Hao, Y.; Ma, X.; Xu, S.; Lia, J.; Kong, L.; et al. QTL mapping for yield-related traits in wheat based on four RIL populations. *Theor. Appl. Genet.* **2020**, *133*, 917–933. [CrossRef] [PubMed]
82. El-Feki, W.M.; Byrne, P.F.; Reid, S.D.; Haley, S.D. Mapping quantitative trait loci for agronomic traits in winter wheat under different soil moisture levels. *Agronomy* **2018**, *8*, 133. [CrossRef]
83. Mason, R.E.; Hays, D.B.; Mondal, S.; Ibrahim, A.M.; Basnet, B.R. QTL for yield, yield components and canopy temperature depression in wheat under late sown field conditions. *Euphytica* **2013**, *194*, 243–259. [CrossRef]
84. Sherman, J.D.; Martin, J.M.; Blake, N.K.; Lanning, S.P.; Talbert, L.E. Genetic basis of agronomic differences between a modern and a historical spring wheat cultivar. *Crop Sci.* **2014**, *54*, 1–13. [CrossRef]
85. Zhai, H.; Feng, Z.; Li, J.; Liu, X.; Xiao, S.; Ni, Z.; Sun, Q. QTL analysis of spike morphological traits and plant height in winter wheat (*Triticum aestivum* L.) using a high-density SNP and SSR-based linkage map. *Front. Plant Sci.* **2016**, *7*, 1617. [CrossRef]
86. Zhao, C.; Zhang, N.; Wu, Y.; Sun, H.; Liu, C.; Fan, X.; Yan, X.; Li, J.; Cui, F. QTL for spike-layer uniformity and their influence on yield-related traits in wheat. *BMC Genet.* **2019**, *20*, 23. [CrossRef]
87. Carter, A.H.; Garland-Campbell, K.; Kidwell, K.K. Genetic mapping of quantitative trait loci associated with important agronomic traits in the spring wheat (*Triticum aestivum* L.) cross 'Louise' × 'Penawawa'. *Crop Sci.* **2011**, *51*, 84–95. [CrossRef]
88. Assanga, S.O.; Fuentealba, M.; Zhang, G.; Tan, C.; Dhakal, S.; Rudd, J.C.; Ibrahim, A.M.H.; Xue, Q.; Haley, S.; Chen, J.; et al. Mapping of quantitative trait loci for grain yield and its components in a US popular winter wheat TAM 111 using 90K SNPs. *PLoS ONE* **2017**, *12*, e0189669. [CrossRef] [PubMed]
89. Guan, P.; Lu, L.; Jia, L.; Kabir, M.R.; Zhang, J.; Lan, T.; Zhao, Y.; Xi, M.; Hu, Z.; Yao, Y.; et al. Global QTL analysis identifies genomic regions on chromosomes 4A and 4B harboring stable loci for yield-related traits across different environments in wheat (*Triticum aestivum* L.). *Front. Plant Sci.* **2018**, *9*, 529. [CrossRef] [PubMed]
90. Gahlaut, V.; Jaiswal, V.; Tyagi, B.S.; Singh, G.; Sareen, S.; Balyan, H.S.; Cupta, P.K. QTL mapping for nine drought-responsive agronomic traits in bread wheat under irrigated and rain-fed environments. *PLoS ONE* **2017**, *12*, e0182857. [CrossRef] [PubMed]
91. Heidari, B.; Saeidi, G.; Sayed, T.B.; Suenaga, K. QTLs involved in plant height, peduncle length and heading date of wheat (*Triticum aestivum* L.). *J. Agric. Sci. Technol.* **2012**, *14*, 1093–1104.
92. International Wheat Genome Sequencing Consortium (IWGSC); Appels, R.; Eversole, K.; Stein, N.; Feuillet, C.; Keller, B.; Rogers, J.; Pozniak, C.J.; Choulet, F.; Distelfeld, A.; et al. Shifting the limits in wheat research and breeding using a fully annotated reference genome. *Science* **2018**, *361*, eaar7191.
93. Jain, M.; Nijhawan, A.; Arora, R.; Agarwal, P.; Ray, S.; Sharma, P.; Kapoor, S.; Tyagi, A.K.; Khurana, J.P. F-box proteins in rice. Genome-wide analysis, classification, temporal and spatial gene expression during panicle and seed development, and regulation by light and abiotic stress. *Plant Physiol.* **2007**, *143*, 1467–1483. [CrossRef]
94. Hong, M.J.; Kim, D.Y.; Kang, S.Y.; Kim, D.S.; Kim, J.B.; Seo, Y.W. Wheat F-box protein recruits proteins and regulates their abundance during wheat spike development. *Mol. Biol. Rep.* **2012**, *39*, 9681–9696. [CrossRef]
95. Wang, L.; Shen, R.; Chen, L.T.; Liu, Y.G. Characterization of a novel DUF1618 gene family in rice. *J. Integr. Plant Biol.* **2014**, *56*, 151–158. [CrossRef]
96. Wei, K.; Li, Y. Functional genomics of the protein kinase superfamily from wheat. *Mol. Breed.* **2019**, *39*, 141. [CrossRef]
97. Han, G.; Qiao, Z.; Li, Y.; Yang, Z.; Wang, C.; Zhang, Y.; Liu, L.; Wang, B. RING zinc finger proteins in plant abiotic stress tolerance. *Front. Plant Sci.* **2022**, *13*, 1055. [CrossRef]
98. Vaitkevičiūtė, G.; Chawade, A.; Lillemo, M.; Liatukas, Ž.; Aleliūnas, A.; Armonienė, R. Genome-Wide Association Analysis of Freezing Tolerance and Winter Hardiness in Winter Wheat of Nordic Origin. *Plants* **2023**, *12*, 4014. [CrossRef] [PubMed]
99. Kaur, A.; Sharma, A.; Verma, P.C.; Upadhyay, S.K. EF-hand domain-containing proteins in Triticum aestivum: Insight into their roles in stress response and signalling. *S. Afr. J. Bot.* **2022**, *149*, 663–681. [CrossRef]
100. He, G.H.; Xu, J.Y.; Wang, Y.X.; Liu, J.M.; Li, P.S.; Chen, M.; Ma, Y.Z.; Xu, Z.S. Drought-responsive WRKY transcription factor genes TaWRKY1 and TaWRKY33 from wheat confer drought and/or heat resistance in Arabidopsis. *BMC Plant Biol.* **2016**, *16*, 116. [CrossRef] [PubMed]

plants

MDPI

Article

The F-Box Protein TaFBA1 Positively Regulates Drought Resistance and Yield Traits in Wheat

Qinxue Li [1,2], Xiaoyu Zhao [2], Jiajie Wu [2], Huixia Shou [1,*] and Wei Wang [2,*]

1 The Provincial International Science and Technology Cooperation Base on Engineering Biology, International Campus of Zhejiang University, Haining 314400, China; chen979402575@126.com
2 National Key Laboratory of Wheat Improvement, Shandong Agricultural University, Tai'an 271018, China; zxyaworld95@163.com (X.Z.); jiajiewu@sdau.edu.cn (J.W.)
* Correspondence: huixia@zju.edu.cn (H.S.); wangw@sdau.edu.cn (W.W.)

Abstract: Environmental stresses, including drought stress, seriously threaten food security. Previous studies reported that wheat F-box protein, TaFBA1, responds to abiotic stresses in tobacco. Here, we generated transgenic wheat with enhanced (overexpression, OE) or suppressed (RNA interference, RNAi) expression of *TaFBA1*. The *TaFBA1*-OE seedlings showed enhanced drought tolerance, as measured by survival rate and fresh weight under severe drought stress, whereas the RNAi plants showed the opposite phenotype. Furthermore, the OE plants had stronger antioxidant capacity compared to WT and RNAi plants and maintained stomatal opening, which resulted in higher water loss under drought stress. However, stronger water absorption capacity in OE roots contributed to higher relative water contents in leaves under drought stress. Moreover, the postponed stomatal closure in OE lines helped to maintain photosynthesis machinery to produce more photoassimilate and ultimately larger seed size. Transcriptomic analyses conducted on WT and OE plants showed that genes involved in antioxidant, fatty acid and lipid metabolism and cellulose synthesis were significantly induced by drought stress in the leaves of OE lines. Together, our studies determined that the F-box protein TaFBA1 modulated drought tolerance and affected yield in wheat and the *TaFBA1* gene could provide a desirable target for further breeding of wheat.

Keywords: wheat (*Triticum aestivum* L.); F-box protein; drought tolerance; antioxidant capacity; water absorption

Citation: Li, Q.; Zhao, X.; Wu, J.; Shou, H.; Wang, W. The F-Box Protein TaFBA1 Positively Regulates Drought Resistance and Yield Traits in Wheat. *Plants* **2024**, *13*, 2588. https://doi.org/10.3390/plants13182588

Academic Editor: Vladimir V. Kuznetsov

Received: 19 July 2024
Revised: 6 September 2024
Accepted: 14 September 2024
Published: 16 September 2024

1. Introduction

As a food crop that is a major source of starch and calories throughout the world, wheat (*Triticum aestivum* L.) plays an important role in food security in many countries, including China [1]. Although wheat yields have significantly increased due to the breeding of new varieties in recent years, extreme environmental conditions, like drought, are still major factors limiting production [2–4]. Therefore, identifying the key genes and molecular mechanisms that participate in the response to drought is of great significance to maintaining high yields of this essential crop under adverse conditions. There have been more and more studies reporting that some genes are involved in wheat drought tolerance [5,6]. Qiu et al. [7] showed that overexpression of TaASR1-D in wheat could improve osmotic and drought tolerance by affecting reactive oxygen species (ROS) accumulation and ABA signaling. Overexpressing ABA receptor TaPYL1-1B increased the water-use efficiency of transgenic wheat by regulating ABA and drought response genes under drought conditions [8]. Moreover, the wheat DREB transcription factor TaDTG6-B also functioned as a positive regulation factor of wheat drought tolerance [9]. All of these studies on wheat drought tolerance provide important theoretical bases for breeding wheat to handle various abiotic stresses, especially drought.

Extreme and unpredictable environments usually result in severe damage to plants, which cannot move to escape outside stresses [10]. To adapt to stress conditions, plants can only react at the molecular and biochemical levels, such as rapidly decreasing undesirable proteins and increasing protective functions [11]. The ubiquitin-26S proteasome system (UPS) utilizes three types of enzymes, Ubiquitin (Ub)-activating enzymes (E1s), Ub-conjugating enzymes (E2s), and Ub ligases (E3s), to rapidly and effectively select intracellular proteins for degradation. E3s recognize the target proteins and label them for later degradation, and many E3s are reported to be involved in abiotic stress response in plants [12,13]. In wheat, overexpression of the U-box E3 ligase TaPUB1 enhances salt stress tolerance through interacting with α-mannosidase protein TaMP [12]. The SKP1/CUL1/F-box (SCF) complex is the best-characterized class of E3 ligases, among which the F-box-containing protein is responsible for recognizing the substrates and initiating responses to abiotic stresses [13]. Researchers have reported various roles of F-box proteins during abiotic stresses by different regulation mechanisms. In *Arabidopsis*, the auxin-mediated stress response factor AtFBA1, an F-box protein, conferred tolerance to salt and osmotic stress by triggering an ABA-mediated plant response [14]. And F-box protein AtPP2-B11 influences the expression of Na^+ homeostasis genes under salt stress, and AtPP2-B11-OE lines exhibited lower Na^+ accumulation in *Arabidopsis* [15]. Moreover, Sharma et al. [16] showed that *OsFBX257*, a rice F-box protein-coding gene, influenced leaf and grain length, and number of panicles, while significantly increasing the grain yield under drought stress. However, some F-box proteins function in regulating plant abiotic stress negatively. For example, luciferase and Yeast-2-Hybrid (Y2H) assay revealed that GhTULP34, a protein containing the F-box domain, interacted with GhSKP1A, suggesting its negative role in osmotic stress regulation [17]. Zhang et al. reported that the overexpression of DOR1 led to increased drought sensitivity, indicating that DOR1 acted as a negative regulator of drought stress tolerance [18].

Wheat is a crop mainly grown in arid and semi-arid areas, where it can easily be subject to drought, which leads to severe yield losses [2]. Despite an increasing number of genes recently reported to regulate drought tolerance in wheat, only a small number of E3 ligases are known to be involved in wheat drought response, with almost all either U-box or Ring finger E3 ligases [3,7]. As an important member of the SCF complexes, F-box proteins play important roles in response to abiotic and biotic stresses in plants [19,20]. Genes including the F-box domain have been identified in many species of plants, including *Arabidopsis*, tobacco, rice, and wheat. However, there is limited research available on F-box proteins in wheat exposed to abiotic stresses, especially drought stress [21,22]. TaFBA1 is an F-box protein and a previous study reported that it improved the drought and salt stress tolerance of tobacco by increasing the antioxidant ability and maximizing intracellular Na^+ compartmentalization, respectively [23,24]. An et al. [6] first indicated that TaFBA1-overexpressing *Arabidopsis* was insensitive to ABA and TaFBA1 regulation on drought tolerance may be independent of ABA synthesis. Given that *TaFBA1* is a gene from wheat. In this study, the drought tolerance of wheat was investigated by overexpressing and repressing the F-box-encoding gene *TaFBA1*. The results demonstrated that the overexpression of *TaFBA1* significantly enhanced drought tolerance and affected grain yield in wheat. These findings provide new insights into the roles of TaFBA1 in drought tolerance and its potential application in the improvement of abiotic stress resistance in wheat.

2. Results

2.1. Generation and Identification of TaFBA1-Overexpressing and TaFBA1-RNAi Wheat Lines

The previous works showed that the wheat F-box gene *TaFBA1* could improve plant tolerance to several abiotic stresses, including salt, drought, and heat in transgenic tobacco [13,23,24]. To explore the biological functions of *TaFBA1* in wheat, the cultivar CB037 (wild type) was used for transformation and transgenic lines either overexpressing or knocking down *TaFBA1* were generated (Figure 1A,B). We obtained more than 50 *TaFBA1* RNA interference (*TaFBA1*-RNAi) lines and 80 *TaFBA1*-overexpression (*TaFBA1*-OE) lines. Quantitative real-time PCR (qRT-PCR) analysis confirmed that the transcription of *TaFBA1* was reduced in the RNAi lines (FR) and increased in the overexpression lines (FO; Figure 1C,D).

Two overexpression lines (FO3, FO5) and knockdown lines (FR2, FR8) were chosen for the next experiments. E3 ligase activities were examined in CB037 and transgenic lines (FO and FR) under normal and drought conditions. The results showed that the OE lines had higher E3 ligase activity than that of WT with or without drought stress, while the RNAi lines exhibited the inverse trend. All the lines showed higher enzyme activities after drought treatment (Figure S1). These results indicated that *TaFBA1* was successfully overexpressed or silenced, both transcriptionally and translationally, in the transgenic wheat plants.

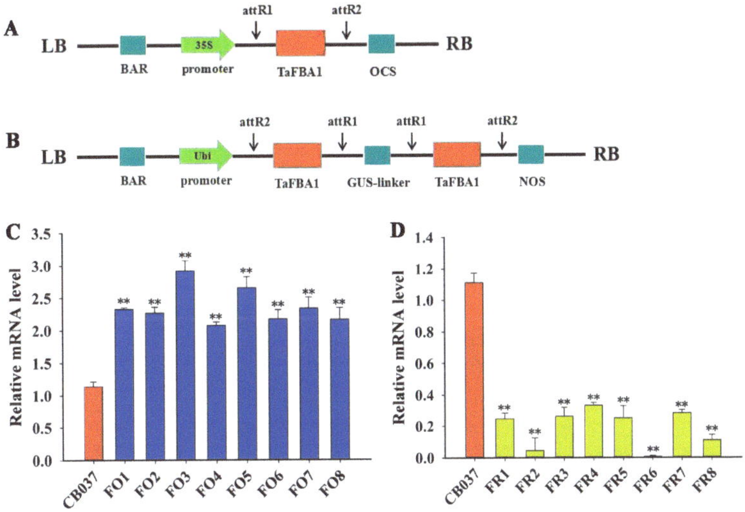

Figure 1. Molecular identification of TaFBA1 transgenic lines. Schematic diagrams of constructs used for (**A**) TaFBA1 overexpression (FO) and (**B**) RNAi-mediated knockdown (FR). The expression levels of TaFBA1 in (**C**) FO and (**D**) FR lines were assessed using qRT-PCR. The data represent the mean ± SD of three biological replicates. ** $p < 0.01$.

2.2. TaFBA1 Overexpression Confers Drought Tolerance to Wheat at Seedling and Heading Stages

To dissect the function of *TaFBA1* in wheat drought tolerance, the phenotype of WT, FO3 and FR2 seedlings were photographed, and the OE plants showed significantly better growth status than WT and RNAi plants. Since there are many differences between growth in hydroponics and soil, the responses of all the lines were examined to drought stress in a potting mix and obtained similar phenotypes (Figure S2B). At this stage, no morphological or developmental abnormalities were apparent in any of the lines under normal conditions. However, the OE lines showed greater tolerance to the drought stress (Figure S2B). After dehydration for 25 days, the OE lines were slightly wilted, while over 50% of leaves in the RNAi lines were severely wilted and ~41% were wilted in WT (Figure S2C). With continuous growth for 28 days, the fresh weight of all seedlings did increase, even under drought stress, but the fresh weight of OE lines was significantly higher than those of WT, while that of RNAi lines and WT were quite similar (Figure S2D). These results indicated that the OE lines were less sensitive to drought compared with WT and RNAi plants and that *TaFBA1* overexpression could significantly improve drought tolerance in wheat at the seedling stage.

For mature plants, all genotypes showed similar growth under well-watered conditions, and there was no difference in plant height. However, when exposed to 20% PEG6000, signs of stress were less prominent in the OE lines after being exposed to drought stress for 2 or 3 weeks, with fewer leaf-wilting symptoms and taller plants. However, the RNAi lines showed a higher rate of leaf wilting and were shorter (Figure 2). The above results suggested the overexpression of *TaFBA1* could confer drought tolerance at both the wheat seedling and heading stages.

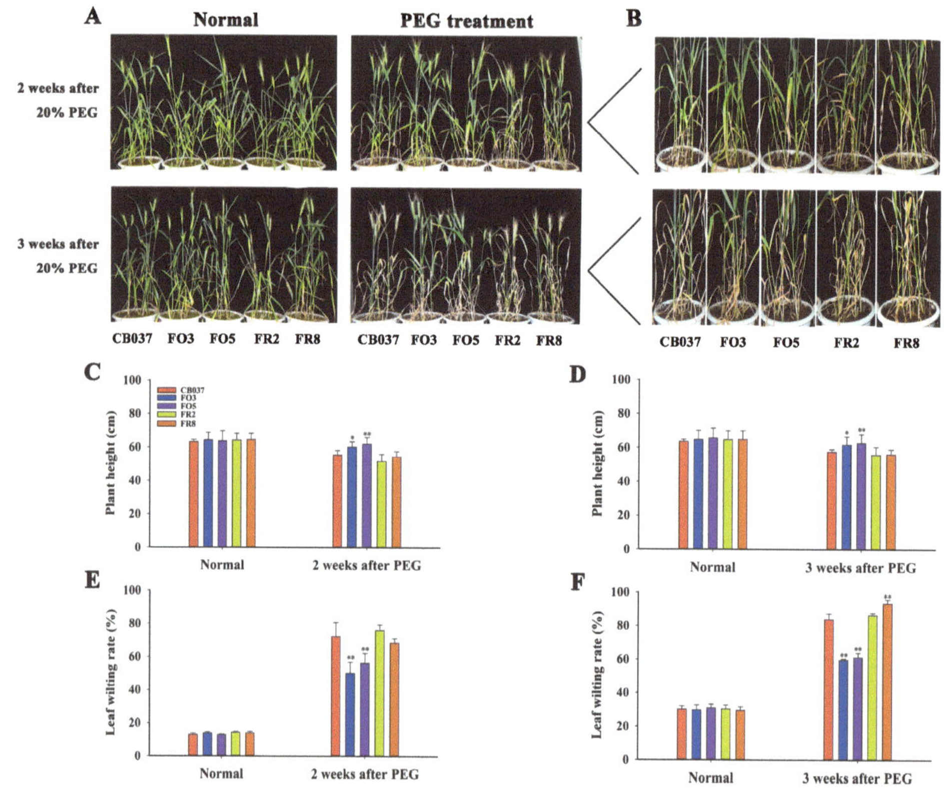

Figure 2. Analysis of drought tolerance in WT and *TaFBA1* transgenic wheat lines at the heading stage. (**A**) Phenotype and (**B**) magnifying local picture of 32-day-old wheat exposed to 20% PEG6000 for 2 and 3 weeks. Plant height after drought stress for (**C**) 2 weeks and (**D**) 3 weeks, respectively. Leaf wilting rate (%) after drought stress for (**E**) 2 weeks and (**F**) 3 weeks. The data represent the mean ± SD of three biological replicates. * $p < 0.05$; ** $p < 0.01$.

2.3. *TaFBA1* Overexpression Impacted Spike Weight and Grain Size

Next, it was observed that yield-related traits of the WT, *TaFBA1*-OE and *TaFBA1*-RNAi lines under normal and dehydration conditions. As shown in Figure 3, under normal growth condition, the grains of the OE lines were a little longer and wider than WT (Figure 3A,C,D), causing a slight increase in the 100-grain volume and weight (Figure 3B(right),G) compared with WT. The grains of the RNAi lines were slightly thinner than WT, while the grain length was comparable to WT (Figure 3A–D,G). The differences in grain size among the lines were even more striking after drought treatment (Figure 3A–D). Examination of grain weight per spike revealed a significant loss in the WT and RNAi lines but not the OE lines (Figure 3F), while the number of grains per spike was similar between all lines under either normal or dehydration conditions (Figure 3E). The above results suggested that *TaFBA1* played positive roles in the yield traits of wheat, especially under drought conditions.

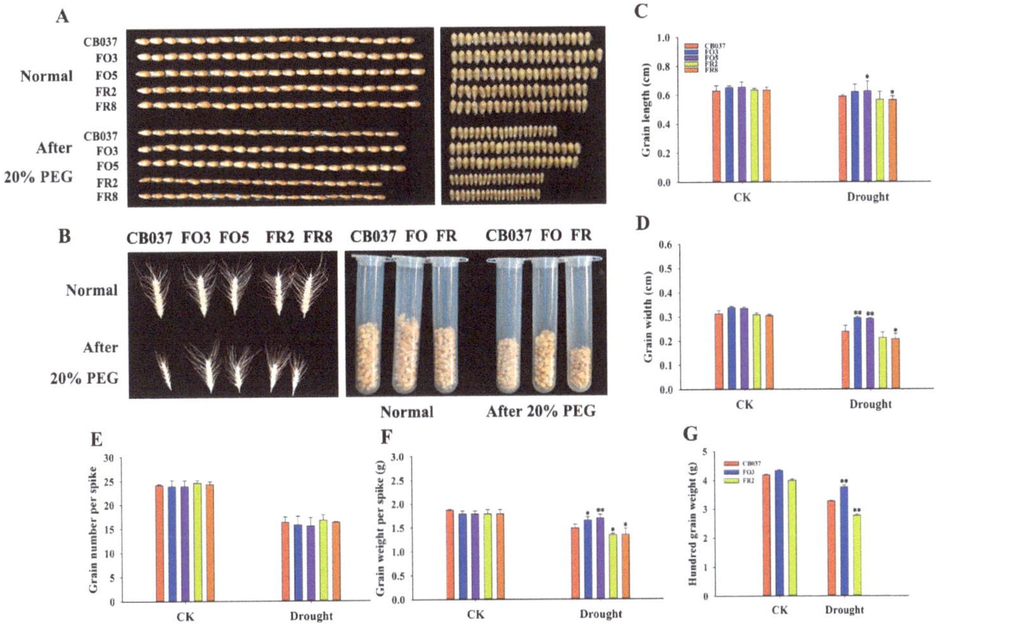

Figure 3. Effects of *TaFBA1* overexpression on wheat grain traits under normal and drought stress conditions. (**A**) Grain length and width phenotype of 20 mature wheat grains harvested under normal and 20% PEG6000 treatment conditions. (**B**) Spike size and hundred-grain volume of mature wheat grains. (**C**) Average grain length. (**D**) Average grain width. (**E**) Grain number per spike. (**F**) Grain weight per spike. (**G**) Hundred-grain weight. The data represent the mean ± SD of three biological replicates. * $p < 0.05$; ** $p < 0.01$.

2.4. *TaFBA1 Overexpression Improved the Photosynthetic Capacity of Transgenic Wheat under Drought Stress*

To evaluate the effects of drought on plant photosynthetic activity, the flag leaves of all the lines were used to assess eight photosynthetic parameters. Under control conditions, there were no obvious differences in these photosynthetic parameters between the transgenic and WT plants. After PEG treatment, the net photosynthesis (Pn) rate decreased in all lines, although the Pn rate of the OE lines was still higher than WT and RNAi lines (Figure 4A). The variations among the transpiration rate (E, Figure 4B) and stomatal conductance (Gs, Figure 4C) were consistent with those of the Pn values. Similar results were also observed for photosystem II photochemical potential (Fv/Fm) and quantum yield of electron transfer through PSII (ΦPSII), but the lines exhibited minor differences (Figure 4D,E). The intercellular CO_2 concentration (Ci) was similar in all lines, although the value was slightly lower in the OE lines, opposite the effect on the Pn rate (Figure 4F). The chlorophyll content often reflects the degree of leaf chlorosis [25,26]. We found that the contents of chlorophyll a and chlorophyll b were reduced by an average of 45% and 43% in OE lines, respectively, but by 56% and 54% in WT and 67% and 63% in RNAi plants after dehydration stress (Figure 4G,H). These results further demonstrated that *TaFBA1* played positive roles in improving drought tolerance by supporting photosynthetic capacity in wheat.

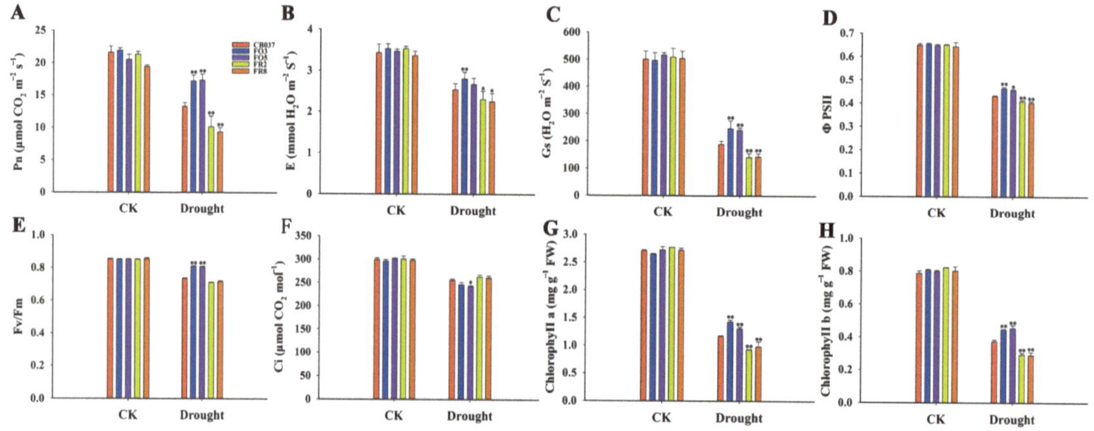

Figure 4. Photosynthetic capacity of WT and transgenic wheat lines under drought stress. WT, *TaFBA1*-OE and *TaFBA1*-RNAi wheat lines were grown under normal condition and under dehydration with 20% PEG6000 for 2 weeks before determination of the (**A**) Net photosynthetic rate (Pn), the (**B**) transpiration rate (E), the (**C**) stomatal conductance (Gs), the (**D**) actual PSII efficiency (ΦPSII), the (**E**) maximum photochemical efficiency of PSII (Fv/Fm), the (F) intercellular CO_2 concentration (Ci), the contents of (**G**) chlorophyll a and (**H**) chlorophyll b. The data represent the mean ± SD of three biological replicates. * $p < 0.05$; ** $p < 0.01$.

2.5. *TaFBA1 Overexpression Alleviated Oxidative Damage in Wheat under Drought Stress*

NBT and DAB staining are often used as indicators of ROS (mainly O^{2-} and H_2O_2) accumulation. Under well-watered conditions, there was no significant difference in nitroblue tetrazolium (NBT) and 3,3′-diaminobenzidine (DAB) staining between WT and transgenic plants. After drought treatment, the staining was deepest in the leaves of the RNAi lines, while that of the OE lines was the lightest (Figure 5A). Quantitation of the O^{2-} and H_2O_2 levels were consistent with the staining results (Figure 5B,C). This indicated that *TaFBA1*-OE plants accumulated less ROS under drought stress. Moreover, following PEG treatment for 5 days, the protein carbonylation level of the OE lines was less than the WT and RNAi lines (Figure S3). This indicated that the proteins suffered less damage from oxidation under drought stress.

The level of malondialdehyde (MDA) is an indicator of oxidation of lipids, and oxidation of lipids can lead to disruption of the cell membrane, which can be measured as electrolyte leakage. From Figure 5D,E, the drought-treated *TaFBA1*-OE lines had significantly lower MDA contents and less electrolyte leakage relative to the WT, while the RNAi lines showed an opposite trend, which suggested that the cell membranes in the *TaFBA1*-RNAi lines are more severely damaged by ROS.

2.6. *TaFBA1 Overexpression Enhanced the ROS Scavenging Capacity of Wheat under Drought Stress*

To determine whether ROS were detoxified more rapidly in the OE lines or whether plants had a stronger metabolic capacity to cope with elevated ROS, the activities of several key antioxidase, including super oxide dismutase (SOD), catalase (CAT), ascorbate peroxidase (APX), peroxidase (POD), and glutathione peroxidase (GPx) were examined in leaves exposed to drought or PEG for 5 days (Figure 6). After drought stress, the SOD, POD and GPx activities increased in all lines, with higher values in the OE lines than WT and RNAi lines (Figure 6A,D,E). Although the levels of both CAT and APX activity decreased after drought stress, the values were still higher in the OE lines (Figure 6B,C). These results suggested that the *TaFBA1*-OE lines were more effective in terms of ROS detoxification than the WT and RNAi lines.

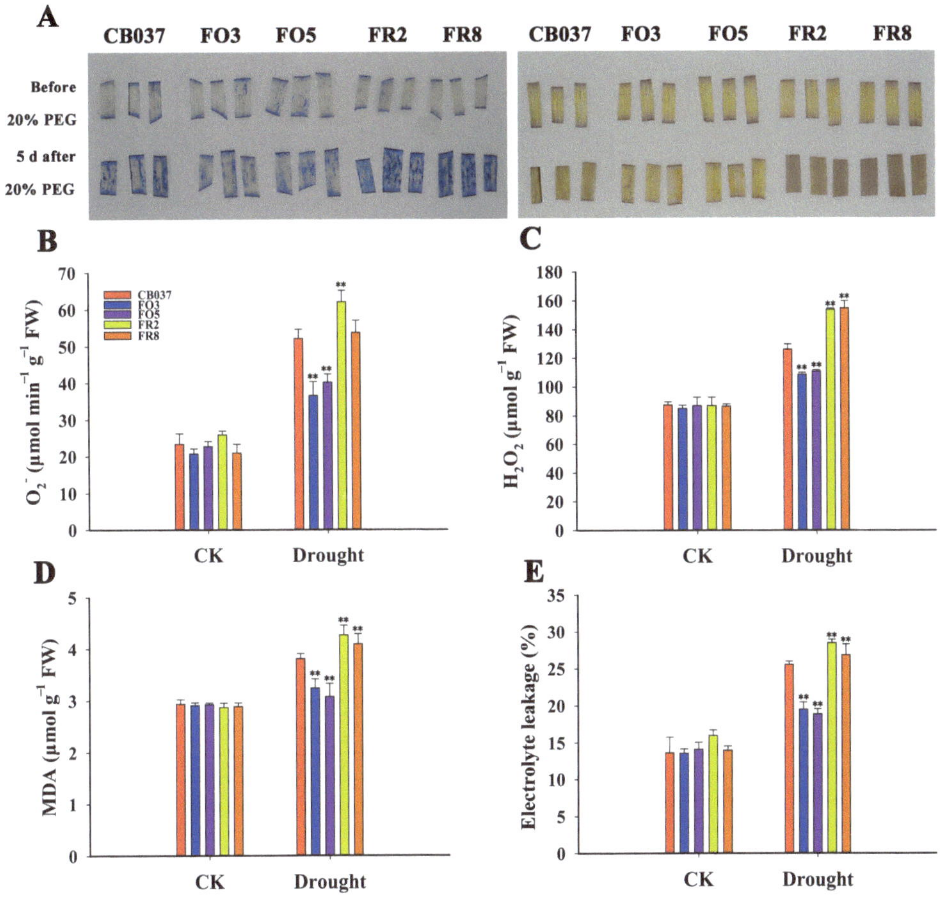

Figure 5. Reactive Oxygen Species (ROS) accumulation and cell membrane oxidative damage in WT and transgenic lines under drought stress. Two-week-old seedlings of WT, *TaFBA1*-OE and *TaFBA1*-RNAi wheat exposed to PEG6000 as dehydration stress for about 5 days. Leaves were stained with (**A**) NBT and (**B**) DAB staining for detecting O^{2-} and H_2O_2 levels. Quantification of (**B**) O^{2-} and (**C**) H_2O_2 levels in leaves as above. The (**D**) MDA content and (**E**) relative electrolyte leakage were determined in the lines grown in normal and dehydration conditions. Protein carbonylation, as another measure of oxidative stress is presented in Figure S3. The data represent the mean ± SD of three biological replicates. ** $p < 0.01$.

The same approach was taken regarding the activities of enzymes in the ascorbate (AsA)-GSH cycle, including monodehydroascorbate reductase (MDAR), dehydroascorbate reductase (DHAR), and glutathione reductase (GR). The activity of GR was slightly decreased after drought treatment, but there was no significant difference between the lines (Figure 6F). The MDAR and DHAR activities were up-regulated significantly after dehydration, with a greater increase in the OE lines than WT and RNAi lines (Figure 6G,H). Together, these results suggested that *TaFBA1* promoted the activity of the enzymes in the AsA-GSH cycle, which are involved in cellular redox homeostasis.

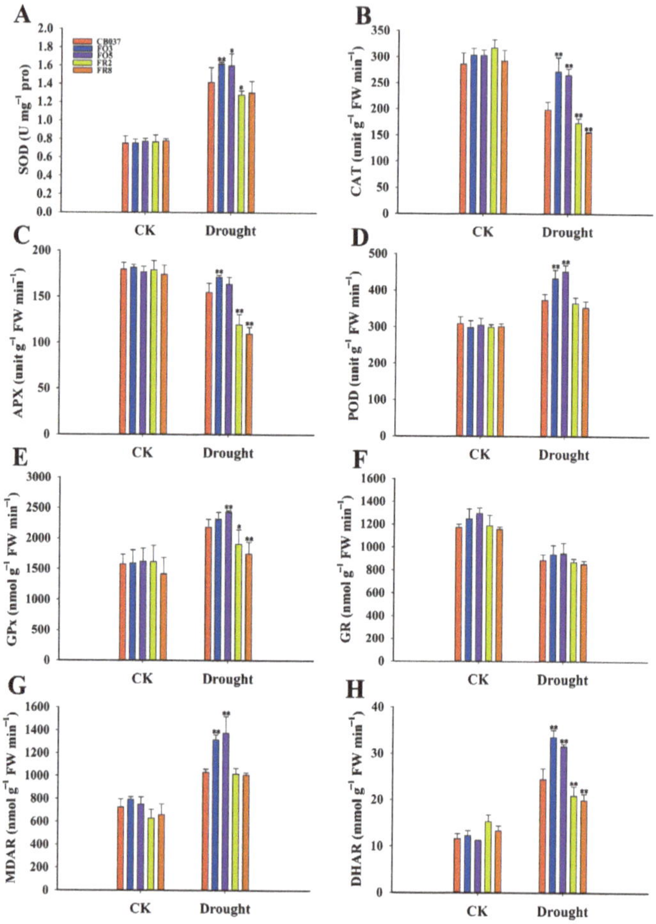

Figure 6. Antioxidative abilities of WT and transgenic wheat lines under normal and drought stress. Two-week-old seedlings of WT, *TaFBA1*-OE and *TaFBA1*-RNAi wheat exposed to PEG6000 as dehydration stress for about 5 days. Leaves were sampled for the determination of antioxidase enzyme activities. SOD (**A**), CAT, (**B**) APX (**C**), POD (**D**), GPx (**E**), GR (**F**), MDAR (**G**), and DHAR (**H**) activities in wheat grown under normal and drought stress conditions. The data represent the mean ± SD of three biological replicates. * p <0.05; ** p< 0.01.

The transcripts of some antioxidation-related genes, such as *TaCu/Zn-SOD*, *TaMn-SOD*, *TaFe-SOD*, *TaCAT*, *TaAPX*, *TaPOD*, *TaGPx*, *TaDHAR* and *TaMDAR* were monitored in the flag leaves of plants exposed to PEG (Figure 7). The results showed that the expression trends of the most genes discussed above were consistent with their enzyme activities, with the exception of *TaAPX*, whose expression level was increased in all lines after dehydration. However, different from the GR activity (Figure 6F), the expression of *TaGR* increased in all lines after PEG treatment and had significantly higher levels in the OE lines (Figure 7H). Collectively, the gene transcripts showed a greater increase in the OE lines than WT and RNAi lines under drought stress. Additionally, we determined the transcript levels of several stress-related genes, including, *TaLEA7* (late embryo genesis abundant protein), *TaRD29B* (responsive to desiccation 29B), *TaDREB6* (dehydration responsive element binding protein6), *TaFER-5B* (Ferritin), *TaSAPK2* (sucrose non-fermenting1-type Ser/Thr protein kinase), and *TaP5CS* (Delta 1-pyrroline-5-carboxylate synthetase) (Figure 7K–P). There were no notable differences in transcript levels among all the lines under normal conditions. Five of these stress-responsive genes, *TaLEA7*, *TaRD29B*, *TaDREB6*, *TaFER-5B*, and *TaSAPK2*, were up-regulated in all lines when plants were exposed to drought stress, but the transcript levels increased to a larger extent in the OE lines compared to the WT. These results suggested that the higher drought tolerance observed in the OE lines might result from increased transcript levels of stress-related and antioxidant-related genes, which would reduce the ROS content due to the enhancement of antioxidant capability by *TaFBA1* overexpression. However, the expression level of *TaP5CS*, which encodes a proline synthesis or catabolism enzyme, was the lowest in the OE lines (Figure 7P). This result is consistent with the trend of proline content in Figure 10G.

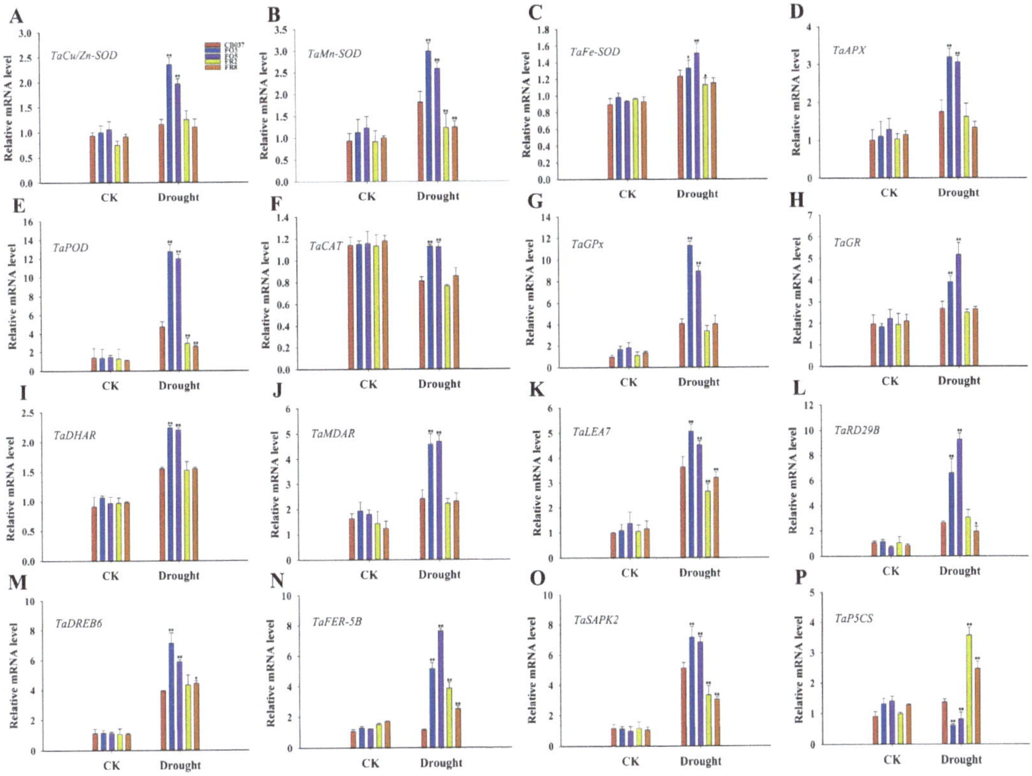

Figure 7. Relative expression of antioxidant-related and stress-responsive genes in WT and transgenic wheat lines. Relative expression of antioxidant-related genes, namely *TaCu/Zn-SOD* (**A**), *TaMn-SOD* (**B**), *TaFe-SOD* (**C**), *TaAPX* (**D**), *TaPOD* (**E**), *TaCAT* (**F**), and *TaGPX* (**G**), *TaGR* (**H**), *TaDHAR* (**I**), and *TaMDAR* (**J**) and stress-responsive genes, namely *TaLEA7* (**K**), *TaRD29B* (**L**), *TaDREB6* (**M**), *TaFER-5B* (**N**), *TaSAPK2* (**O**), and *TaP5CS* (**P**), in the flag leaves of WT and transgenic wheat under drought stress. The data represent the mean ± SD of three biological replicates. * $p < 0.05$; ** $p < 0.01$.

Further, methyl viologen (MV) was sprayed on the wheat leaves as an external oxidative stress. Fourteen days after treatment with MV, the OE lines exhibited less leaf wilting and higher total Chl contents than that of the WT and RNAi lines, which suggested that the overexpression of *TaFBA1* significantly increased the antioxidant capacity of transgenic wheat under drought stress (Figure S4).

2.7. *TaFBA1* Overexpression Supported Stomatal Opening

To verify whether the drought-tolerant phenotype of OE lines was derived from better water maintenance capacity, the stomatal aperture was first categorized as three levels (completely open, partially open, and completely closed) in the WT, OE and RNAi lines under normal and drought conditions via microscopy. Following drought stress, 45% of the stoma were completely closed in the OE lines but 56% were closed in WT, while 15% were completely open in the OE lines while only 12% were completely open in WT. The percentage of partially open stoma was 40% and 32% in OE lines and WT, respectively (Figure 8A,B). Moreover, the water loss rates of detached leaves showed that the TaFBA1-OE lines lost water much more rapidly than WT (Figure 8C).

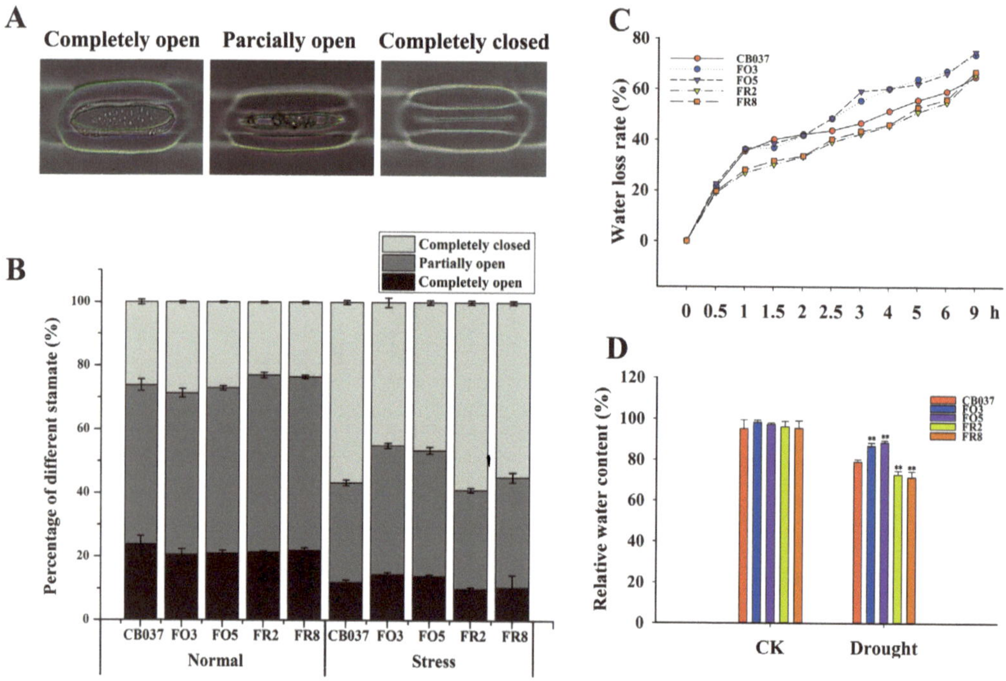

Figure 8. Stomatal aperture on the surface of WT and transgenic wheat lines leaves. (**A**) Images of stoma with different aperture on the leaves of WT and transgenic wheat after PEG treatment obtained using a fluorescence microscope. (**B**) Percentage of stoma of different aperture. (**C**) Water loss rate of detached leaves of WT and transgenic wheat lines. (**D**) Relative water content (RWC) in leaves of all lines grown under normal and drought stress conditions. The data represent the mean ± SD of three biological replicates. ** $p < 0.01$.

The sensitivity of the wheat lines to ABA was determined using germinating seeds. As shown in Figure 9, the shoot length and root length of the OE lines were distinctly greater than the WT and RNAi lines under 1 μM ABA. With increasing ABA concentration, seedling growth of all lines was further limited, although root growth of the OE lines remained significantly greater than the WT and RNAi lines. This lack of ABA inhibition during germination indicated that *TaFBA1*-OE plants were less sensitive to ABA.

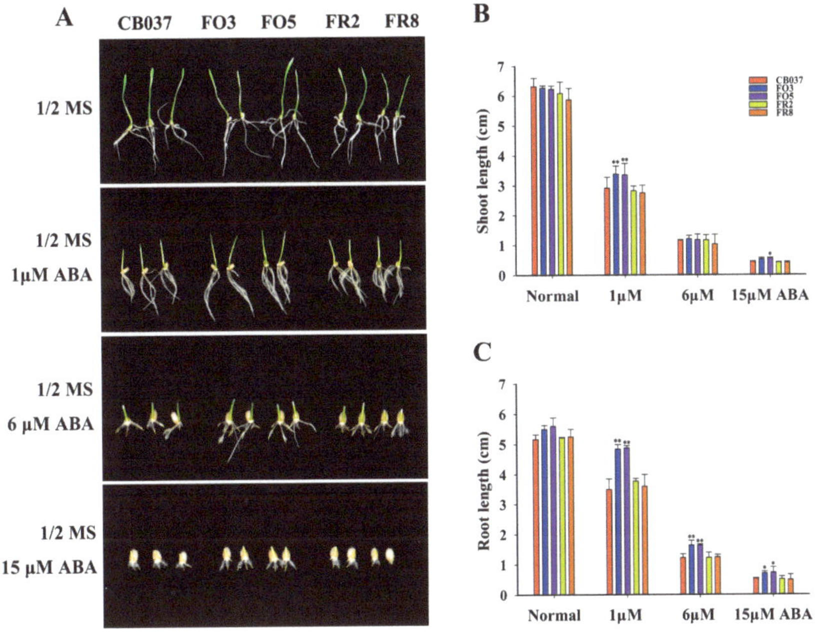

Figure 9. Response of shoot and root growth in germinating WT and transgenic wheat lines to ABA treatment. The (**A**) growth phenotype, (**B**) shoot length, and (**C**) primary root length of wheat seedlings germinated in the presence of different concentrations of ABA for 5 days. The data represent the mean ± SD of three biological replicates. * $p < 0.05$; ** $p < 0.01$.

2.8. *TaFBA1 Overexpression Enhanced Root Water Absorption Capacity of Wheat under Drought Stress*

The root system is the main organ of the plant that absorbs water from soil [27,28]. To explore whether the increased drought tolerance of the OE lines was related to their root development, we observed the root architecture and counted the root length of all lines. The results showed that there was no significant difference in root growth status of all the lines before or after PEG treatment (Figure 10A–C). Surprisingly, the root vitality and aquaporin (AQP) activity of the OE lines were higher than that of WT after dehydration treatment (Figure 10D,E). It was speculated that the higher root vitality and AQP activity under drought stress enhanced the water absorption capacity of the OE lines, which could compensate for the water loss caused by the large stomatal opening while maintaining a high relative water content (RWC) under drought condition (Figure 8D).

Further, we determined the proline and soluble sugar contents in wheat, as these compounds are considered important metabolite contributors to osmotic adjustment [29]. The results showed that under normal conditions, both of soluble sugars and proline were at similar levels in all genotypes. After drought treatment, the contents of proline and soluble sugar increased in all lines, but the OE lines accumulated more soluble sugar than the WT and RNAi lines. However, the proline content showed an opposite trend after drought stress, with lower levels in the OE lines and higher levels in the RNAi lines (Figure 10F,G). These results suggested that it was soluble sugar but not proline that contributed to the increased drought tolerance of the OE lines. Overall, these observations supported the idea that *TaFBA1* could reduce wheat sensitivity to drought by increasing water absorption capacity and the accumulation of osmoprotectants.

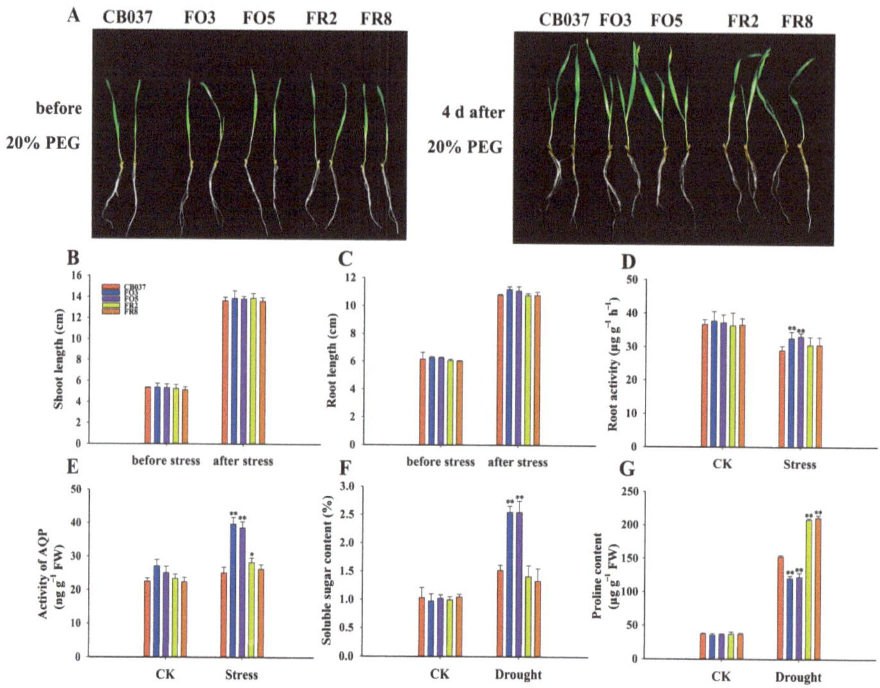

Figure 10. Effects of drought stress on root growth, water absorption ability, and osmotic adjustment substances contents in WT and transgenic wheat lines. The (**A**) root growth phenotype, (**B**) shoot length, and (**C**) root length of wheat seedlings before and after 20% PEG6000 treatment. (**D**) Root vitality and (**E**) AQP activity of wheat seedlings with and without 20% PEG6000 treatment for 4 days. The contents of (**F**) soluble sugar (**G**) and proline in all lines under drought stress for 5 days in soil mixture. The data represent the mean ± SD of three biological replicates. * $p < 0.05$; ** $p < 0.01$.

2.9. Transcriptomic Analyses of CB037 and TaFBA1-OE Revealed Functions of TaFBA1 in Drought Tolerance

To gain insight into the molecular mechanisms underlying the *TaFBA1*-mediated response to drought stress, RNA sequencing was conducted on the FO3 and CB037 plants under normal and drought conditions. Before RNA sequencing, we tested the expression pattern of *TaFBA1* in leaves of CB037 grown in soil in response to 20% PEG6000 treatment. The results showed that *TaFBA1* was induced by drought stress, reaching peak expression at 6 h, and then decreasing over the next 18 h (Figure S5). This suggested that *TaFBA1* fairly functions early in the response to drought stress.

Based on the expression pattern of *TaFBA1*, we stressed wheat plants with a 20% PEG6000 soil drench for 6 h (SCB037; SFO3) in parallel to growing plants without stress treatment (CB037; FO3) before sampling leaves for RNA extraction and sequencing. After the 6-h PEG6000 treatment, we identified 1751 differentially expressed genes (DEGs) between CB037 and SCB037 (Figure 11A) and constructed a heatmap visualizing the expression profiles of these DEGs (Figure 11B). However, PEG6000 treatment caused a less dramatic transcriptomic change in FO3 plants relative to CB037 (Figure 11A,B). These data suggested that *TaFBA1*-OE lines were less sensitive in terms of transcriptomic changes to drought stress than WT.

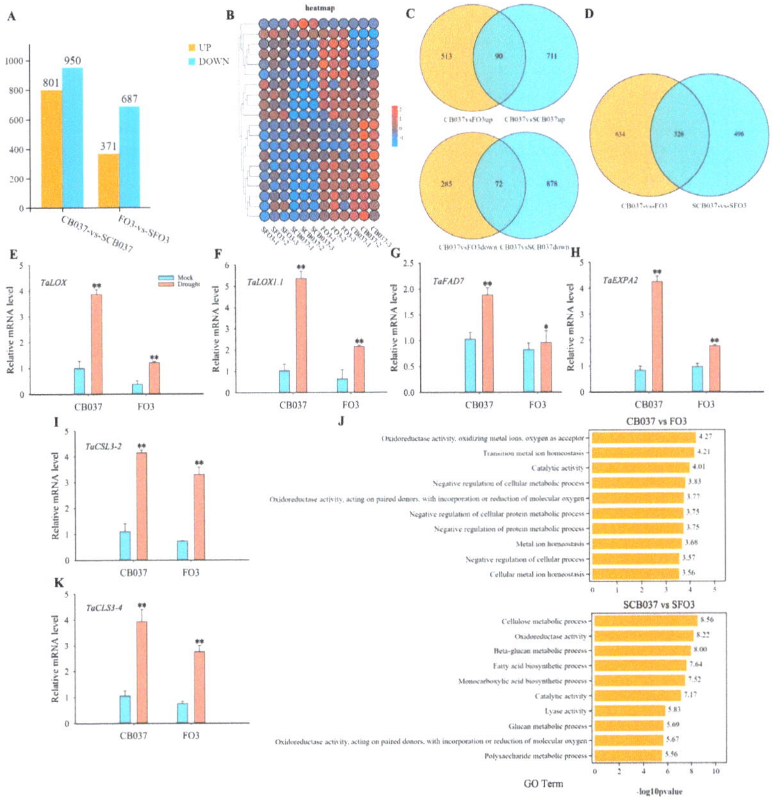

Figure 11. Transcriptome analysis of *TaFBA1*-OE wheat with or without drought treatment. (**A**) The number of up-regulated and down-regulated genes and (**B**) heatmap of expression profiles of DEGs in CB037 and SCB037 and in FO3 and SFO3. These comparisons are between plants of the same genotype grown under normal and dehydration conditions. (**C**) Venn diagrams showing the DEGs between different comparisons; CB037 vs. FO3up and CB037 vs. FO3down mean genes up-regulated and down-regulated in the well-watered OE line (FO3) compared to the well-watered WT (CB037); CB037 vs. SCB037up and CB037 vs. SCB037down mean genes up-regulated and down-regulated in dehydrated CB037 compared to well-watered CB037. (**D**) The DEGs in CB037 and FO3, SCB037 and SFO3. Relative expression levels of (**E**) *TaLOX*, (**F**) *TaLOX1.1*, (**G**) *TaFAD7*, (**H**) *TaEXPA2*, (**I**) *TaCSL3-2*, and (**K**) *TaCSL3-4*. (**J**) Top 10 enriched terms among the DEGs identified from CB037/FO3 and SCB037/SFO3, respectively. The data represent the mean ± SD of three biological replicates. * $p < 0.05$; ** $p < 0.01$.

Venn diagram analyses indicated that 90 up-regulated genes were common between the comparisons of *TaFBA1*-OE to WT under the well-watered condition and of dehydrated versus well-watered WT plants, while 72 genes were shared between the down-regulated genes in these two comparisons (Figure 11C). These genes were differentially expressed not only in response to drought stress in CB037 but also in FO3 compared with the CB037 under normal conditions. These overlapping DEGs might have primed the *TaFBA1*-OE lines to better respond to drought stress, correlating to the stronger drought tolerance phenotype (Figure 2 and Figure S2). In addition, 326 DEGs were shared between the FO3/CB037 and SFO3/SCB037 comparisons (Figure 11D), suggesting that these genes were stably regulated by *TaFBA1* under both normal and drought conditions.

The enrichment analysis was performed to classify the DEGs from the unstressed (FO3/CB037) and stressed (SFO3/SCB037) comparisons into various functional pathways based on gene ontology (GO). The top 10 enriched terms from each pairwise comparison were selected for further analyses (Figure 11J). Both the FO3/CB037 and SFO3/SCB037 comparisons included the terms "oxidoreductase activity" and "catalytic activity". However, the DEGs from the SFO3/SCB037 comparison contained genes involved in the "fatty acid biosynthetic process" and "cellulose metabolic process", while the FO3/CB037 comparison did not. This suggested that under normal conditions, *TaFBA1* mainly influences metal ion homeostasis and oxidoreductase activity, whereas under drought stress *TaFBA1* may affect drought tolerance through the regulation of oxidoreductase activity, fatty acid synthesis and lipid metabolism, as well as cellulose and beta-glucan metabolism.

To further validate the transcriptomic data, six DEGs were selected among the genes included in the GO terms "oxidoreductase activity", "fatty acid biosynthetic process" and "cellulose metabolic process", namely *TaLOX* (lipoxygenase), *TaLOX1.1* (lipoxygenase 1.1), *TaFAD7* (omega-3 fatty acid desaturase 7), *TaEXPA2* (expansin-A2-like), *TaCSL3-2*, (mixed-linked glucan synthase 3-2), and *TaCSL3-4* (mixed-linked glucan synthase 3-4), for analysis by qRT-PCR. The results showed that these abiotic stress response genes were significantly up-regulated after drought treatment. Interestingly, the fold change of up-regulation was less in CB037 than in FO3 (Figure 11E–I,K). Meanwhile, the consistent results between this microarray data and qRT-PCR data above also confirmed the reliability of the transcriptome data in this paper.

3. Discussion

3.1. *TaFBA1 Positively Regulated the Drought Tolerance of Wheat*

As global warming accelerates, droughts are likely to be more frequent and longer lasting. Wheat is a crop with a high risk of exposure to drought due to the geographical regions in which it is cultivated [2]. In order to survive under adverse environments, plants have evolved sophisticated mechanisms, including the degradation of proteins by UPS [13,30]. We transformed *TaFBA1* into wheat to generate overexpression and RNAi lines, which we used to explore *TaFBA1* response to drought (Figure 1). As shown in Figure 2 and Figure S2, all the *TaFBA1*-OE lines showed enhanced drought tolerance at both the seedling and heading stages. The photosynthetic system is usually susceptible to damage induced by drought. The effects are either direct, such as diffusion limitations through the stoma and the mesophyll and alterations in photosynthetic metabolism, or secondary, such as oxidative stress arising from the superimposition of multiple stresses [25]. In this study, the stronger photosynthetic capacity of the OE lines under drought stress implied greater production of photoassimilates, which may be an important factor in the bigger grains (Figures 3 and 4). The results above suggested that TaFBA1 positively regulated drought tolerance in wheat.

3.2. *TaFBA1 Improved Drought Tolerance of Wheat by Enhanced Antioxidant Ability and Stress-Related Genes Regulation*

Under drought stress, plants easily produce ROS, and excessive accumulation of ROS will cause oxidative stress, inhibit plant growth, and even cause cell death, so the ROS scavenging system plays an important role in reducing the harmful effects of drought stress [31,32]. Here, we found that the activities of some antioxidant enzymes were significantly higher in the OE lines than in WT and RNAi lines after drought stress (Figure 6), implying that these antioxidant enzymes played a major role in ROS clearance in OE lines. Plants respond to drought by regulating the transcription of corresponding genes [33]. In our research, many antioxidant-related and stress-related genes affected by *TaFBA1* also functioned in regulating the tolerance of *TaFBA1*-OE wheat (Figure 7). However, the expression trend of *TaP5CS* (Figure 7P) and the proline content (Figure 10G) in the OE lines indicated that proline is not the main permeable substance in TaFBA1-OE lines under drought stress.

From the RNA-Seq analysis, we also found that genes related to cellulose metabolism, fatty acid synthesis and lipid metabolism showed significantly differential expression when the wheat seedlings were exposed to dehydration stress (Figure 11J). The plant cell wall plays an important role in the flexibility and stability of cell structures during plant development, while some cellulose synthesis-related genes regulate plant tolerance to abiotic stress [34–36]. Lipids are one of the major components of biological membranes, including the plasma membrane, the interface between the cell and the environment. Abiotic stresses like water deficit and temperature stress trigger lipid-dependent signaling cascades that activate plant adaptation processes to deal with stressful environments [37–41]. It was reported that *LOX* genes are involved in various growth and development processes and play important roles in plant resistance to abiotic stress, including root growth and plant development [42–45]. Based on the functions of lipid metabolism in plant abiotic stress tolerance above and the results in this research, we speculated that maybe some transcription factors or function proteins involved in cellulose metabolism, lipid synthesis and metabolism could interact with TaFBA1 to mediate the drought tolerance in wheat. Therefore, that will be the next aim to explore the underlying mechanisms in relation to the *TaFBA1*-mediated response to drought.

3.3. *Enhanced Root Water Absorption Capacity of TaFBA1-OE Wheat Made Contribution to Increased Drought Tolerance*

Maintaining sufficient available water is crucial for plants to survive during drought stress. Water uptake and transport directly influence plant growth and development, especially under dehydration stress, and directly affect the normal metabolism of plants [46]. Therefore, plants will continuously optimize their roots to ensure their water and nutrient supplies, especially under drought stress [27,28]. For example, root volume and vitality affect nutrient and water absorption, and thus, yield [47,48]. The transmembrane transport of water is mainly mediated by aquaporins (AQPs), which regulate plant osmotic balance by controlling water transport and can interact with stress response proteins to cope with drought stress [49]. In *Arabidopsis*, Jang et al. [50] found that several aquaporin genes were significantly up-regulated or down-regulated under drought stress. In our study, there was no significant difference in the root growth status of all lines after drought treatment, but the root vitality and AQP activity were significantly higher in the OE lines than in the WT and RNAi lines (Figure 10D,E), indicating that the roots of the OE lines had strong water absorption ability and could transfer more water into plant cells to improve their drought tolerance.

A stoma consists of a pair of guard cells that control the flow of water and gas in and out of the leaves. Stomatal aperture is extremely sensitive to environmental conditions such as light, gas, and temperature [51]. Under drought conditions, larger stomatal aperture increases transpiration and the consumption of available water in plants, which is unfavorable to water accumulation. But opened stoma contributes to the absorption of CO_2 for photosynthesis and the accumulation of biomass [52]. Therefore, it is worthwhile to explore the balance between photosynthesis and transpiration. We found that the *TaFBA1*-OE lines had a greater proportion of fully and half-opened stomas (Figure 8A,B) and a higher water loss rate (Figure 8C) than WT. Surprisingly, the water content of the *TaFBA1*-OE leaves after drought stress was higher. Based on the analyses above, we believed that the stronger root vitality (Figure 10D) and higher AQP activity (Figure 10E) in the *TaFBA1*-OE lines supported the ability of the root system to absorb and transport water. At the same time, the higher soluble sugar content in the OE lines (Figure 10F) could maintain the osmotic potential in the leaf cells, helping them to hold on to intracellular water (Figure 8D).

The phytohormone ABA induces stomatal closure and is considered to be an important response factor to drought stress. When suffering from drought stress, plants usually accelerate stomatal closure to hold water content for normal plant growth [4]. However, An et al. [6] reported that *TaFBA1*-overexpressing *Arabidopsis* was insensitive to ABA and that the TaFBA1 regulation on drought tolerance may be independent of ABA synthesis.

Here, the wheat OE lines closed their stoma slower than the WT and RNAi lines, and the response to ABA was also similar to the study from An et al. [6]. This suggested the overexpression of *TaFBA1* decreased the sensitivity of wheat to ABA. In this respect, the physiological mechanisms through which *TaFBA1* responds to drought stress differ from many other drought response genes.

3.4. *TaFBA1 Increased Wheat Grain Size, Especially under Drought Stress*

Maintaining yield is often the long-term aim of crop studies related to abiotic stress [8]. Wheat is a globally important food crop, and years of research and breeding have led to increased stable yields [53,54]. Seed size is closely related to wheat yield and is regulated by many signaling pathways, including the UPS pathway [55]. The studies in model plants *Arabidopsis* and other food or oil crops indicated that E3 ligases play important roles in yield traits but that different genes function differently in different species [19,56]. Here, we found that under normal conditions, the seeds of OE lines were fuller than WT and the RNAi lines and the difference became more significant after drought treatment (Figure 3). The corresponding statistical results of yield traits in Figure 3 indicated that it was the filling status rather than the number of grains per spike that directly affected the yield of the OE lines. The grain weight of wheat is determined by the sink capacity and accumulation of dry matter and photosynthetic capacity and filling intensity affect substance accumulation [57,58]. Our examination of the photosynthetic capacity of all lines found that the Pn of the OE lines was significantly higher than that of the WT and RNAi lines after drought stress (Figure 4), which supported higher seed filling in the *TaFBA1*-OE lines. The molecular mechanisms underlying larger seed size in the *TaFBA1*-OE lines need to be further explored then.

4. Materials and Methods

4.1. *Plant Materials and Growth Conditions*

Wheat (*Triticum aestivum* L.) plants were used for physiological and molecular analyses. For germination, wheat grains were surface sterilized in 2% (*w*/*v*) sodium hypochlorite (NaClO) and then laid on moistened filter paper for 2 d at 25 °C. Seedlings of uniform size were transplanted into steam-sterilized soil mix (matrix, vermiculite, and nutrient soil, 1:1:1, *v*/*v*/*v*) and then placed into a growth chamber (light/dark temperature 25 °C/23 °C) under 300 μmol m^{-2} s^{-1} illumination and a relative humidity of 75% [59]. The wheat cultivar Shannong 16 was used to amplify cDNA sequences of *TaFBA1* [20]. The wheat cultivar CB037 was used for the generation of *TaFBA1*-OE and RNAi transgenic plants.

4.2. *Vector Construction, Generation of Transgenic Wheat and Verification of Transgenic Lines*

To generate the overexpression vector for wheat transformation, the cDNA of *TaFBA1* was amplified and cloned into the pEarleyGate101 vector after the 35S promoter [6]. To construct the RNAi vector for knocking down TaFBA1, two truncated fragments of *TaFBA1* CDS (△TaFBA1) were amplified and inserted into the pC336 vector, driven under the Ubiquitin promoter, to form an intron-containing hairpin RNA constructs (△TaFBA1-intron-△TaFBA1). The OE and RNAi constructs are shown in Figure 1A,B. The constructs were introduced into immature embryos of wheat cv. CB037 by particle bombardment as described [12]. Primers used for these amplifications are listed in Table S1.

Verification of transgenic wheat by herbicide, PCR, and qRT-PCR analyses. For selection of the bar gene, 80 mg/L glyphosate was applied to 7-day-old wheat seedlings to initially screen transgenic T0 wheat plants. The herbicide-tolerant seedlings were then sampled for DNA extraction and PCR analysis. For qRT-PCR analysis, total RNA was extracted from the leaves of wheat seedlings using TRIZOL reagent (Vazyme, Nanjing, China). The cDNA was synthesized, and qRT-PCR was performed using a ChamQ Universal SYBR qPCR Master Mix Kit (Vazyme, Nanjing, China). The relevant primers are listed in Table S1. All reactions were run in triplicate. The confirmed transgenic lines were grown, and their progeny were screened for positive segregants. Homozygous transgenic plants were obtained in the T3 generation and used for stress-tolerant test experiments.

4.3. Drought Stress Treatment

Transgenic wheat seeds (T3) from two *TaFBA1*-overexpressing lines (FO3, FO5), two *TaFBA1*-RNAi lines (FR2, FR8) and their wild-type (WT; CB037) were germinated for about 3 days on moist filter paper in a culture dish (25 °C).

For the analysis of wheat seedling phenotypes following drought stress, seeds were then put in a bottomless 96-well plate that was placed in a shallow container and incubated in a growth chamber under a 16-h light (25 °C) /8-h dark (23 °C) photoperiod in Hoagland's nutrient solution. After growth for 3 weeks, the seedlings were at the two-leaf stage and the nutrient solution was adjusted to contain 20% (*w*/*v*) PEG6000. After 21 days of growth under osmotic stress, all lines were grown again in Hoagland's solution for a week.

For natural drought experiments in soil, the wheat plants were grown in a soil-containing potting mix for 2 weeks under normal conditions before drought stress was created by withholding water for 25 days. All seedlings were then re-watered. The leaf wilting rate and fresh weight of all lines were recorded. For PEG6000 treatment, 3-day-old seedlings were transferred to containers with soil mix in containers, and when grown up for 11 days, after which they were watered with 20% PEG6000 for about 5 days. The control plants were irrigated with 200 mL water every 3 days. Plants were visualized and sampled to measure some physiological parameters. Three independent experiments were performed to obtain data for statistical analyses [4,53].

For drought stress treatment of wheat at the heading stage, six germinated seeds were grown in the same pot (matrix, vermiculite, and nutrient soil (1:1:3, *v*/*v*/*v*), 4 L, 6 plants per pot) under imitated natural conditions (25 °C daytime and 23 °C nighttime temperatures). Water was replaced by 20% PEG6000 from 32-day-old plants for 14 days and 21 days, while the control plants were irrigated with 400 mL water every 3 days [29,53]. Plants were photographed and relative physiological and photosynthetic parameters were monitored as reported [60]. Agronomic traits including hundred grain weight, grain number per spike, grain length and grain width were measured in the harvested plants.

For analysis of root morphology following drought stress, one-leaf seedlings were cultured in hydroponics for 4 days in 20% PEG6000 solution. The roots phenotype was photographed before and after exposure to the PEG, and root length and shoot length of all lines were recorded. Root physiological parameters, such as root vitality and AQP activity, were measured for lines grown under normal and drought conditions. The AQP activity was assessed by enzyme-linked immunosorbent assay (ELISA). Fresh root tips (0.5 g, 2 cm) were used to measure root activity by the 2,3,5-triphenyltetrazolium chloride (TTC) method [24].

4.4. Chlorophyll Content and Photosynthetic Parameter Assays

The chlorophyll content was measured using a UV spectrophotometric method as described previously [60]. The photosynthetic gas exchange parameters of the flag leaves and ΦPSII, Fv/Fm were measured using the procedures described by Wang et al. [60].

4.5. Malondialdehyde Content, Electrolyte Leakage, Proline and Soluble Sugar Contents

Two-week-old plants were treated with 20% PEG6000 for about 5 days. The MDA content was measured in accordance with the described method by Zhou et al. [20]. Electrolyte leakage was measured according to the literature [61]. Proline content was determined according to previously described methods [13]. The soluble sugar content was determined using the anthrone method [62].

4.6. Antioxidant Analysis Experiments

Two-week-old seedlings grown in soil mix were watered with 20% PEG6000 for about 5 days. The leaves from normal growth and dehydration conditions were collected to conduct

ROS analysis and antioxidative enzyme activities assays. DAB and NBT staining, as well as H_2O_2 concentration and O^{2-} production were conducted as described previously [61]. The activities of antioxidant enzymes, including CAT, POD, SOD, APX, MDAR, DHAR, GR, and GPx, were measured as described by the literature [59].

Wheat seedlings grown in soil mix under-watered conditions for 2 weeks were sprayed with 10 μM MV. Phenotypes of wheat seedlings before and 14 days after spraying with 10 μM MV were photographed and the leaf wilting rate was recorded.

Two-week-old transgenic wheat and WT plants were subjected to water and 20% PEG6000 solution for 5 days. The leaves of these plants were then used for the immunological analysis of protein carbonylation. Briefly, total protein was extracted from the wheat leaves followed by the determination of protein concentration according to Bradford [63]. The separation of protein by SDS-PAGE and detection of carbonylation level were carried out according to the description [13].

4.7. *Stomatal Aperture*

To analyze the stomatal aperture of WT and transgenic wheat leaves, seedlings were exposed to 20% PEG6000 or water for 5 days. The middle part of the nether leaf epidermis was covered with transparent nail polish and allowed to dry for about 3 min. The dried nail polish was peeled off, removing a layer of epidermis, and placed on a glass slide for microscopic observation. The aperture status stoma was observed and imaged under an AXIO microscope (Zeiss, Jena, Germany) at 10 × 40 amplification [64]. Stomatal types were defined and analyzed according to the method [65].

4.8. *Seed Germination and Seedling Growth Assays*

To assay the response of seed germination to ABA, wheat seeds from two *TaFBA1*-overexpressing lines (FO3, FO5), two TaFBA1-RNAi lines (FR2, FR8), and WT (CB037) were sterilized using 70% ethanol for 2 min and then disinfected with 4% sodium hypochlorite (NaClO) for 10 min. All sterilized seeds were rinsed in sterilized deionized water at least five times and germinated on 1/2 Murashige-Skoog (MS) medium containing different concentrations of ABA (0, 1, 6, 15 μM) under a 16-h light/8-h dark photoperiod at 25 °C. The root length and shoot length were measured after they grew on the different 1/2 MS media for 5 days.

4.9. *Water Loss Rate, Relative Water Content*

To examine water loss in the WT and transgenic lines, the leaves of 6 plants from each line grown under either control or drought stress conditions were harvested. The leaves were retained at room temperature and weighed at 0, 0.5, 1, 1.5, 2, 2.5, 3, 4, 5, 6 and 9 h. Water loss was represented as the proportion of fresh weight lost (calculated using the initial weight of the plant samples) over time [29,66]. Relative water content was determined using the following formula: RWC = (fresh weight − dry weight)/(rehydrated weight − dry weight) [20].

4.10. *RNA-Seq Assay*

Two-week-old CB037 and TaFBA1-overexpressing wheat plants were maintained in a well-watered state or subjected to drought for 6 h before leaves were collected for RNA extraction. RNA-seq was performed using an Illumina HiseqTM 2500/4000 by Gene Denovo Biotechnology Co., Ltd. (Guangzhou, China) with three biological replicates per sample. Bioinformatic analysis was performed using Omicsmart, a real-time interactive online platform for data analysis (http://www.omicsmart.com). DEGs were identified using the method described in Kan et al. [67]. DEGs were identified as genes with a fold change ≥2 and a false discovery rate (FDR) < 0.05 by comparison. GO term enrichment analysis of the gene sets of interest was performed as described [68]. The same samples were used for the validation of transcription data via qRT-PCR.

4.11. qRT-PCR Assay

Samples were collected from wheat seedlings and leaves treated or not treated with PEG6000 for RNA extraction and reverse transcription using HiScript Q RT SuperMix for qPCR (Vazyme, Nanjing, China) according to the manufacturer's instructions. qRT-PCR was performed using the AceQ qPCR SYBR Green Master Mix (Vazyme, Nanjing, China) and the CFX96 Real-Time System (Bio-Rad, Hercules, CA, USA). All primers used for qRT-PCR experiments are listed in Table S2.

4.12. Statistical Analysis

All experiments were performed in at least three independent biological replicates. Data points represent mean ± SD (standard deviation) of three replicates. The differences between non-transformed control and transformed plants were analyzed using Student's *t*-test, and asterisks above columns in the figures indicated statistical differences (* $p < 0.05$; ** $p < 0.01$).

5. Conclusions

According to the results in this paper, we speculate that *TaFBA1* regulates multiple systems that support wheat tolerance to drought stress. On the one hand, TaFBA1 enhanced the antioxidant capacity of transgenic wheat by inducing the expression of genes related to the antioxidant system to reduce cellular ROS accumulation which would reduce the oxidative damage to the chloroplast, and, as a result, improve the drought tolerance of OE lines. On the other hand, *TaFBA1* negatively regulated wheat sensitivity to ABA, resulting in slower stomatal closure and increased water loss under drought stress, but this was negated by the stronger water absorption capacity of the roots which allowed the cells to maintain a higher water content despite the increased transpiration. Meanwhile, the larger stomatal aperture helped the plant leaves to absorb more CO_2, reduced the production of excess reducing power and ensured higher carbon assimilation.

Together, the results showed that TaFBA1 played a positive role in drought tolerance in wheat. RNA-seq analysis showed that lipid metabolism and fatty acid synthesis genes, as well as genes from cellulose synthesis, might also play important roles in the *TaFBA1*-regulated wheat tolerance to drought. Our next exploration of the mechanisms underlying drought resistance in this important crop species is promising and we hope to impact variety breeding with our ongoing discoveries.

Supplementary Materials: The following supporting information can be downloaded at: https://www.mdpi.com/article/10.3390/plants13182588/s1, Figure S1. E3 ligase activity of transgenic wheat lines. Figure S2. Analysis of drought tolerance in WT 'CB037'and transgenic wheat lines at the seedling stage. Figure S3. Effects of drought stress on protein carbonylation levels in WT and transgenic wheat lines. Figure S4. Effects of methyl viologen (MV) treatment on WT and transgenic wheat lines. Figure S5. The response of *TaFBA1* to 20% PEG6000 in the leaves of CB037. Table S1. The primers used for PCR amplification. Table S2. The primers used for qRT-PCR.

Author Contributions: Conceptualization, Q.L. and W.W.; methodology, Q.L.; software, Q.L.; validation, H.S. and W.W.; formal analysis, Q.L. and H.S.; investigation, H.S. and W.W.; resources, Q.L., X.Z. and J.W.; data curation, Q.L. and X.Z.; writing—original draft preparation, Q.L.; writing—review and editing, Q.L., H.S. and W.W.; supervision, H.S. and W.W.; funding acquisition, W.W. All authors have read and agreed to the published version of the manuscript.

Funding: This research was supported by the National Natural Science Foundation of China (31370304) and by Funds of the Shandong "Double Tops" Program and the National Key Research and Development Program of China (2021YFF1001204).

Data Availability Statement: The data is contained within the manuscript and Supplementary Materials.

Acknowledgments: We thank Jiajie Wu (Shandong Agricultural University) for assistance with the transgenic wheat. We thank Yong Wang (Shandong Agricultural University) for the advice on the research methods.

Conflicts of Interest: The authors declare no conflicts of interest.

References

1. Langridge, P.; Reynolds, M. Breeding for drought and heat tolerance in wheat. *Theor. Appl. Genet.* **2021**, *134*, 1753–1769. [CrossRef]
2. Ma, J.H.; Geng, Y.K.; Liu, H.; Zhang, M.Q.; Liu, S.J.; Hao, C.Y.; Hou, J.; Zhang, Y.F.; Zhang, D.J.; Zhang, W.J.; et al. TaTIP41 and TaTAP46 positively regulate drought tolerance in wheat by inhibiting PP2A activity. *J. Integr. Plant Biol.* **2023**, *65*, 2056–2070. [CrossRef] [PubMed]
3. Zhang, L.L.; Zheng, Y.; Xiong, X.X.; Li, H.; Zhang, X.; Song, Y.L.; Zhang, X.H.; Min, D.H. The wheat VQ motif-containing protein TaVQ4-D positively regulates drought tolerance in transgenic plants. *J. Exp. Bot.* **2023**, *74*, 5591–5605. [CrossRef]
4. Cui, L.G.; Shan, J.X.; Shi, M.; Gao, J.P.; Lin, H.X. DCA1 Acts as a Transcriptional Co-activator of DST and Contributes to Drought and Salt Tolerance in Rice. *PLoS Genet.* **2015**, *11*, e1005617. [CrossRef]
5. Vishwakarma, K.; Upadhyay, N.; Kumar, N.; Yadav, G.; Singh, J.; Mishra, R.K.; Kumar, V.; Verma, R.; Upadhyay, R.G.; Pandey, M.; et al. Abscisic acid signaling and abiotic stress tolerance in plants: A review on current knowledge and future prospects. *Front. Plant Sci.* **2017**, *8*, 161. [CrossRef]
6. An, J.; Li, Q.X.; Yang, J.J.; Zhang, G.Q.; Zhao, Z.X.; Wu, Y.Z.; Wang, Y.; Wang, W. Wheat F-box Protein TaFBA1 Positively Regulates Plant Drought Tolerance but Negatively Regulates Stomatal Closure. *Front. Plant Sci.* **2019**, *10*, 1242–1261. [CrossRef] [PubMed]
7. Qiu, D.; Hu, W.; Zhou, Y.; Xiao, J.; Hu, R.; Wei, Q.H.; Zhang, Y.; Feng, J.L.; Sun, F.S.; Sun, J.T.; et al. TaASR1-D confers abiotic stress resistance by affecting ROS accumulation and ABA signalling in transgenic wheat. *Plant Biotech. J.* **2021**, *19*, 1588–1601. [CrossRef] [PubMed]
8. Mao, H.D.; Jian, C.; Cheng, X.X.; Chen, B.; Mei, F.M.; Li, F.F.; Zhang, Y.F.; Li, S.M.; Du, L.Y.; Li, T.; et al. The wheat ABA receptor gene *TaPYL1-1B* contributes to drought tolerance and grain yield by increasing water-use efficiency. *Plant Biotechnol. J.* **2022**, *20*, 846–861. [CrossRef]
9. Mei, F.M.; Chen, B.; Du, L.Y.; Li, S.M.; Zhu, D.H.; Chen, N.; Zhang, Y.F.; Li, F.F.; Wang, Z.X.; Cheng, X.X.; et al. A gain-of-function allele of a DREB transcription factor gene ameliorates drought tolerance in wheat. *Plant Cell* **2022**, *34*, 4472–4494. [CrossRef]
10. Mao, H.D.; Li, S.M.; Chen, B.; Jian, C.; Mei, F.M.; Zhang, Y.F.; Li, F.F.; Chen, N.; Li, T.; Du, L.Y.; et al. Variation in cis-regulation of a NAC transcription factor contributes to drought tolerance in wheat. *Mol. Plant* **2022**, *15*, 276–292. [CrossRef]
11. Finley, D.; Ulrich, H.D.; Sommer, T.; Kaiser, P. The ubiquitin–proteasome system of saccharomyces cerevisiae. *Genetics* **2012**, *192*, 319–360. [CrossRef] [PubMed]
12. Wang, W.L.; Wang, W.Q.; Wu, Y.Z.; Li, Q.X.; Zhang, G.Q.; Shi, R.R.; Yang, J.J.; Wang, Y.; Wang, W. The involvement of wheat U-box E3 ubiquitin ligase TaPUB1 in salt stress tolerance. *J. Integr. Plant Biol.* **2020**, *62*, 631–651. [CrossRef]
13. Li, Q.X.; Wang, W.Q.; Wang, W.L.; Zhang, G.Q.; Liu, Y.; Wang, Y.; Wang, W. Wheat F-Box protein gene *TaFBA1* is involved in plant tolerance to heat stress. *Front. Plant Sci.* **2018**, *9*, 521. [CrossRef] [PubMed]
14. Kim, Y.Y.; Cui, M.H.; Noh, M.S.; Jung, K.W.; Shin, J.S. The FBA motif-containing protein AFBA1 acts as a novel positive regulator of ABA response in *Arabidopsis*. *Plant Cell Physiol.* **2017**, *58*, 574–586. [CrossRef]
15. Jia, F.J.; Wang, C.Y.; Huang, J.G.; Yang, G.D.; Wu, C.A.; Zheng, C.C. SCF E3 ligase PP2-B11 plays a positive role in response to salt stress in *Arabidopsis*. *J. Exp. Bot.* **2015**, *66*, 4683–4697. [CrossRef]
16. Sharma, E.; Bhatnagar, A.; Bhaskar, A.; Majee, S.M.; Kieffer, M.; Kepinski, S.; Khurana, P.; Khurana, J.P. Stress-induced F-Box protein-coding gene *OsFBX257* modulates drought stress adaptations and ABA responses in rice. *Plant Cell Environ.* **2023**, *46*, 1207–1231. [CrossRef]
17. Li, Z.S.; Wang, X.Y.; Cao, X.C.; Chen, B.Z.; Ma, C.K.; Lv, J.Y.; Sun, Z.M.; Qiao, K.K.; Zhu, L.F.; Zhang, C.J.; et al. GhTULP34, a member of tubby-like proteins, interacts with GhSKP1A to negatively regulate plant osmotic stress. *Genomics* **2021**, *113*, 462–474. [CrossRef]
18. Zhang, Y.E.; Xu, W.; Li, Z.; Deng, X.W.; Wu, W.; Xue, Y. F-box protein DOR functions as a novel inhibitory factor for abscisic acid-induced stomatal closure under drought stress in *Arabidopsis*. *Plant Physiol.* **2008**, *148*, 2121–2133. [CrossRef] [PubMed]
19. Chen, Y.; Xu, Y.Y.; Luo, W.; Li, W.X.; Chen, N.; Zhang, D.J.; Chong, K. The F-Box Protein OsFBK12 Targets OsSAMS1 for Degradation and Affects Pleiotropic Phenotypes, Including Leaf Senescence, in Rice. *Plant Physiol.* **2013**, *163*, 1673–1685. [CrossRef]
20. Zhou, S.; Sun, X.; Yin, S. The role of the F-box gene *TaFBA1* from wheat (*Triticum aestivum* L.) in drought tolerance. *Plant Physiol. Biochem.* **2014**, *84*, 213–223. [CrossRef]
21. Al-Saharin, R.; Hellmann, H.; Mooney, S. Plant E3 Ligases and Their Role in Abiotic Stress Response. *Cells* **2022**, *11*, 890. [CrossRef] [PubMed]
22. Li, B.W.; Gao, S.; Yang, Z.M.; Song, J.B. The F-box E3 ubiquitin ligase AtSDR is involved in salt and drought stress responses in *Arabidopsis*. *Gene* **2022**, *809*, 146011. [CrossRef] [PubMed]

23. Kong, X.Z.; Zhou, S.M.; Yin, S.H.; Zhao, Z.X.; Han, Y.Y.; Wang, W. Stress-Inducible Expression of an F-box Gene *TaFBA1* from Wheat Enhanced the Drought Tolerance in Transgenic Tobacco Plants without Impacting Growth and Development. *Front. Plant Sci.* **2016**, *7*, 1295. [CrossRef] [PubMed]
24. Zhao, Z.X.; Zhang, G.Q.; Zhou, S.M.; Ren, Y.Q.; Wang, W. The improvement of salt tolerance in transgenic tobacco by overexpression of wheat F-box gene *TaFBA1*. *Plant Sci.* **2017**, *259*, 71–85. [CrossRef] [PubMed]
25. Chaves, M.M.; Flexas, J.; Pinheiro, C. Photosynthesis under drought and salt stress: Regulation mechanisms from whole plant to cell. *Ann. Bot.* **2009**, *103*, 551–560. [CrossRef]
26. Xu, N.N.; Chu, Y.L.; Chen, H.L.; Li, X.X.; Wu, Q.; Jin, L.; Wang, G.X.; Huang, J.L. Rice transcription factor OsMADS25 modulates root growth and confers salinity tolerance via the ABA-mediated regulatory pathway and ROS scavenging. *PLoS Genet.* **2018**, *14*, e1007662. [CrossRef]
27. Yu, H.; Chen, X.; Hong, Y.Y.; Wang, Y.; Xu, P.; Ke, S.D.; Liu, H.Y.; Zhu, J.K.; Oliver, D.J.; Xiang, C.B. Activated expression of an *Arabidopsis* HD-START protein confers drought tolerance with improved root system and reduced stomatal density. *Plant Cell* **2008**, *20*, 1134–1151. [CrossRef]
28. Ma, H.Z.; Liu, C.; Li, Z.X.; Ran, Q.J.; Xie, G.N.; Wang, B.M.; Fang, S.; Chu, J.F.; Zhang, J.R. ZmbZIP4 contributes to stress resistance in maize by regulating ABA synthesis and root development. *Plant Physiol.* **2018**, *178*, 753–770. [CrossRef]
29. Yang, J.J.; Zhang, G.Q.; An, J.; Li, Q.X.; Chen, Y.H.; Zhao, X.Y.; Wu, J.J.; Wang, Y.; Hao, Q.Q.; Wang, W.Q.; et al. Expansin gene *TaEXPA2* positively regulates drought tolerance in transgenic wheat (*Triticum aestivum* L.). *Plant Sci.* **2020**, *298*, 110596. [CrossRef]
30. Ding, S.; Zhang, B.; Qin, F. *Arabidopsis* RZFP34/CHYR1, a ubiquitin E3 ligase, regulates stomatal movement and drought tolerance via SnRK2.6-mediated phosphorylation. *Plant Cell* **2015**, *27*, 3228–3244. [CrossRef]
31. Guo, C.K.; Yao, L.Y.; You, C.J.; Wang, S.S.; Cui, J.; Ge, X.C.; Ma, H. MID1 plays an important role in response to drought stress during reproductive development. *Plant J.* **2016**, *88*, 280–293. [CrossRef] [PubMed]
32. Kaouthar, F.; Ameny, F.K.; Yosra, K.; Walid, S.; Ali, G.; Faiçal, B. Responses of transgenic *Arabidopsis* plants and recombinant yeast cells expressing a novel durum wheat manganese superoxide dismutase TdMnSOD to various abiotic stresses. *J. Plant Physiol.* **2016**, *198*, 56–68. [CrossRef]
33. Shinozaki, K.; Yamaguchi-Shinozaki, K. Gene networks involved in drought stress response and tolerance. *J. Exp. Bot.* **2007**, *58*, 221–227. [CrossRef]
34. Endler, A.; Kesten, C.; Schneider, R.; Zhang, Y.; Ivakov, A.; Froehlich, A.; Funke, N.; Persson, S. A mechanism for sustained cellulose synthesis during salt stress. *Cell* **2015**, *162*, 1353–1364. [CrossRef]
35. Wang, T.; McFarlane, H.E.; Persson, S. The impact of abiotic factors on cellulose synthesis. *J. Exp. Bot.* **2016**, *67*, 543–552. [CrossRef]
36. Lou, H.Y.; Tucker, M.R.; Shirley, N.J.; Lahnstein, J.; Yang, X.J.; Ma, C.; Schwerdt, J.; Fusi, R.; Burton, R.A.; Band, L.R.; et al. The *cellulose synthase-like F3* (*CslF3*) gene mediates cell wall polysaccharide synthesis and affects root growth and differentiation in barley. *Plant J.* **2022**, *110*, 1681–1699. [CrossRef] [PubMed]
37. Hou, Q.C.; Ufer, G.; Bartels, D. Lipid signalling in plant responses to abiotic stress. *Plant Cell Environ.* **2016**, *39*, 1029–1048. [CrossRef]
38. He, M.; He, C.Q.; Ding, N.Z. Abiotic stresses: General defenses of land plants and chances for engineering multistress tolerance. *Front. Plant Sci.* **2018**, 9. [CrossRef] [PubMed]
39. He, M.; Ding, N.Z. Plant Unsaturated Fatty Acids: Multiple Roles in Stress Response. *Front. Plant Sci.* **2020**, *11*, 562785. [CrossRef]
40. Zhang, M.; Barg, R.; Yin, M.G.; Gueta-Dahan, Y.; Leikin-Frenkel, A.; Salts, Y.; Shabtai, S.; Ben-Hayyim, G. Modulated fatty acid desaturation via overexpression of two distinct ω-3 desaturases differentially alters tolerance to various abiotic stresses in transgenic tobacco cells and plants. *Plant J.* **2005**, *44*, 361–371. [CrossRef]
41. Fan, J.L.; Yu, L.H.; Xu, C.C. A central role for triacylglycerol in membrane lipid breakdown, fatty acid β-oxidation, and plant survival under extended darkness. *Plant Physiol.* **2017**, *174*, 1517–1530. [CrossRef] [PubMed]
42. Wozniak, A.; Kesy, J.; Glazinska, P.; Glinkowski, W.; Narozna, D.; Bocianowski, J.; Rucinska-Sobkowiak, R.; Mai, V.C.; Krzesinski, W.; Samardakiewicz, S.; et al. The Influence of Lead and Acyrthosiphon pisum (Harris) on Generation of *Pisum sativum* Defense Signaling Molecules and Expression of Genes Involved in Their Biosynthesis. *Int. J. Mol. Sci.* **2023**, *24*, 10671. [CrossRef] [PubMed]
43. López, M.A.; Vicente, J.; Kulasekaran, S.; Vellosillo, T.; Martínez, M.; Irigoyen, M.L.; Cascón, T.; Bannenberg, G.; Hamberg, M.; Castresana, C. Antagonistic role of 9-lipoxygenase-derived oxylipins and ethylene in the control of oxidative stress, lipid peroxidation and plant defence. *Plant J.* **2011**, *67*, 447–458. [CrossRef] [PubMed]
44. Upadhyay, R.K.; Handa, A.K.; Mattoo, A.K. Transcript abundance patterns of 9-and 13-lipoxygenase subfamily gene members in response to abiotic stresses (heat, cold, drought or salt) in tomato (*Solanum lycopersicum* L.) highlights member-specific dynamics relevant to each stress. *Genes* **2019**, *10*, 683–697. [CrossRef] [PubMed]
45. Sun, Q.; Zhang, B.; Yang, C.L.; Wang, W.L.; Xiang, L.; Wang, Y.P.; Chan, Z.L. Jasmonic acid biosynthetic genes *TgLOX4* and *TgLOX5* are involved in daughter bulb development in tulip (*Tulipa gesneriana*). *Hortic. Res.* **2022**, *11*, 6–18. [CrossRef]
46. Mega, R.; Abe, F.; Kim, J.S.; Tsuboi, Y.; Tanaka, K.; Kobayashi, H.; Sakata, Y.; Hanada, K.; Tsujimoto, H.; Kikuchi, J.; et al. Tuning water-use efficiency and drought tolerance in wheat using abscisic acid receptors. *Nat. Plants* **2019**, *5*, 153–159. [CrossRef]
47. Khandal, H.; Gupta, S.K.; Dwivedi, V.; Mandal, D.; Sharma, N.K.; Vishwakarma, N.K.; Pal, L.; Choudhary, M.; Francis, A.; Malakar, P.; et al. Root-specific expression of chickpea cytokinin oxidase/ dehydrogenase 6 leads to enhanced root growth, drought tolerance and yield without compromising nodulation. *Plant Biotechnol. J.* **2020**, *18*, 2225–2240. [CrossRef]

48. Zhao, S.; Gao, H.B.; Jia, X.M.; Li, X.W.; Mao, K.; Ma, F.W. The γ-clade HD-Zip I transcription factor MdHB-7 regulates salt tolerance in transgenic apple (*Malus domestica*). *Plant Soil* **2021**, *463*, 509–522. [CrossRef]
49. Aroca, R.; Porcel, R.; Ruiz-Lozano, J.M. Regulation of root water uptake under abiotic stress conditions. *J. Exp. Bot.* **2012**, *63*, 43–57. [CrossRef]
50. Jang, J.Y.; Kim, D.J.; Kim, Y.O.; Kim, J.S.; Kang, H. An expression analysis of a gene family encoding plasma membrane aquaporins in response to abiotic stresses in *Arabidopsis thaliana*. *Plant Mol. Biol.* **2004**, *54*, 713–725. [CrossRef]
51. Sugiyama, Y.; Uraji, M.; Watanabe-Sugimoto, M.; Okuma, E.; Munemasa, S.; Shimoishi, Y. FIA functions as an early signal component of abscisic acid signal cascade in Vicia faba guard cells. *J. Exp. Bot.* **2012**, *63*, 1357–1365. [CrossRef] [PubMed]
52. Saradadevi, R.; Bramley, H.; Palta, J.A.; Siddique, K.H.M. Stomatal behaviour under terminal drought affects post-anthesis water use in wheat. *Funct. Plant Biol.* **2017**, *44*, 279–289. [CrossRef] [PubMed]
53. Zhang, N.; Yin, Y.J.; Liu, X.Y.; Tong, S.M.; Xing, J.W.; Zhang, Y.; Pudake, R.N.; Izquierdo, E.M.; Peng, H.R.; Xin, M.M.; et al. The E3 ligase TaSAP5 alters drought stress responses by promoting the degradation of DRIP proteins. *Plant Physiol.* **2017**, *175*, 1878–1892. [CrossRef] [PubMed]
54. Li, S.N.; Lin, D.X.; Zhang, Y.W.; Deng, M.; Chen, Y.X.; Lv, B.; Li, B.S.; Lei, Y.; Wang, Y.P.; Zhao, L.; et al. Genome-edited powdery mildew resistance in wheat without growth penalties. *Nature* **2022**, *602*, 455–460. [CrossRef] [PubMed]
55. Li, N.; Li, Y. Signaling pathways of seed size control in plants. *Curr. Opin. Plant Biol.* **2016**, *33*, 23–32. [CrossRef]
56. Miller, C.; Wells, R.; McKenzie, N.; Trick, M.; Ball, J.; Fatihi, A.; Dubreucq, B.; Chardot, T.; Lepiniec, L.; Bevan, M.W. Variation in expression of the HECT E3 ligase UPL3 modulates LEC2 levels, seed size, and crop yields in *Brassica napus*. *Plant Cell* **2019**, *31*, 2370–2385. [CrossRef]
57. Liu, L.C.; Tong, H.G.; Xiao, Y.H.; Che, R.H.; Xu, F.; Hu, B.; Liang, C.; Chu, J.; Li, J.; Chu, C. Activation of Big Grain1 significantly improves grain size by regulating auxin transport in rice. *Proc. Natl. Acad. Sci. USA* **2015**, *112*, 11102–11107. [CrossRef]
58. Brinton, J.; Simmonds, J.; Minter, F.; Leverington-Waite, M.; Snape, J.; Uauy, C. Increased pericarp cell length underlies a major quantitative trait locus for grain weight in hexaploid wheat. *New Phytol.* **2017**, *215*, 1026–1038. [CrossRef]
59. Liu, S.T.; Liu, S.W.; Wang, M.; Wei, T.D.; Meng, C.; Wang, M.; Xia, G.M. A Wheat *SIMILAR To RCD-ONE* Gene Enhances Seedling Growth and Abiotic Stress Resistance by Modulating Redox Homeostasis and Maintaining Genomic Integrity. *Plant Cell* **2014**, *26*, 164–180. [CrossRef]
60. Wang, W.Q.; Li, Q.X.; Tian, F.X.; Deng, Y.M.; Wang, W.L.; Wu, Y.Z.; Yang, J.J.; Wang, Y.; Hao, Q.Q.; Wang, W. Wheat NILs contrasting in grain size show different expansin expression, carbohydrate and nitrogen metabolism that are correlated with grain yield. *Field Crop. Res.* **2019**, *241*, 0378–4290. [CrossRef]
61. Ma, Y.M.; Yang, C.; He, Y.; Tian, Z.H.; Li, J.X. Rice OVATE family protein 6 regulates plant development and confers resistance to drought and cold stresses. *J. Exp. Bot.* **2017**, *68*, 4885–4898. [CrossRef] [PubMed]
62. Wang, N.; Zhang, W.X.; Qin, M.Y.; Li, S.; Qiao, M.; Liu, Z.H.; Xiang, F.N. Drought tolerance conferred in Soybean (*Glycine max*. L) by GmMYB84, a novel R2R3-MYB transcription factor. *Plant Cell Physiol.* **2017**, *58*, 1764–1776. [CrossRef] [PubMed]
63. Bradford, M.M. A rapid and sensitive method for the quantitation of microgram quantities of protein utilizing the principle of protein-dye binding. *Anal. Biochem.* **1976**, *72*, 248–254. [CrossRef] [PubMed]
64. Huang, L.C.; Chen, L.; Wang, L.; Yang, Y.L.; Rao, Y.C.; Ren, D.Y.; Dai, L.P.; Gao, Y.H.; Zou, W.W.; Lu, X.L.; et al. A Nck-associated protein 1-like protein affects drought sensitivity by its involvement in leaf epidermal development and stomatal closure in rice. *Plant J.* **2019**, *98*, 884–897. [CrossRef]
65. Matsuda, S.; Takano, S.; Sato, M.; Furukawa, K.; Nagasawa, H.; Yoshikawa, S.; Kasuga, J.; Tokuji, Y.; Yazaki, K.; Nakazono, M.; et al. Rice stomatal closure requires guard cell plasma membrane ATP-binding cassette transporter RCN1/OsABCG5. *Mol. Plant* **2016**, *9*, 417–427. [CrossRef]
66. Wang, Z.Y.; Tian, X.J.; Zhao, Q.Z.; Liu, Z.Q.; Li, X.F.; Ren, Y.K.; Tang, J.Q.; Fang, J.; Xu, Q.J.; Bu, Q.Y. The E3 ligase DROUGHT HYPERSENSITIVE negatively regulates cuticular wax biosynthesis by promoting the degradation of transcription factor ROC4 in rice. *Plant Cell* **2018**, *30*, 228–244. [CrossRef]
67. Kan, L.; Liao, Q.C.; Chen, Z.P.; Wang, S.Y.; Ma, Y.F.; Su, Z.Y.; Zhang, L. Dynamic Transcriptomic and Metabolomic Analyses of *Madhuca pasquieri* (Dubard) H. J. Lam During the Post-germination Stages. *Front. Plant Sci.* **2021**, *12*, 731203. [CrossRef]
68. Young, M.D.; Wakefield, M.J.; Smyth, G.K.; Oshlack, A. 2010. Gene ontology analysis for RNA-seq: Accounting for selection bias. *Genome Biol.* **2010**, *11*, R14. [CrossRef]

plants

MDPI

Article

Regulation of Anthocyanin Accumulation in Tomato *Solanum lycopersicum* L. by Exogenous Synthetic dsRNA Targeting Different Regions of *SlTRY* Gene

Andrey R. Suprun *, Artem Yu. Manyakhin, Evgeniya V. Trubetskaya, Konstantin V. Kiselev and Alexandra S. Dubrovina

Federal Scientific Center of the East Asia Terrestrial Biodiversity, Far Eastern Branch of the Russian Academy of Sciences, 690022 Vladivostok, Russia; dubrovina@biosoil.ru (A.S.D.)
* Correspondence: suprun@biosoil.ru

Abstract: RNA interference (RNAi) is a regulatory and protective mechanism that plays a crucial role in the growth, development, and control of plant responses to pathogens and abiotic stresses. In spray-induced gene silencing (SIGS), exogenous double-stranded RNAs (dsRNA) are used to efficiently regulate target genes via plant surface treatment. In this study, we aimed to evaluate the effect of specific exogenous dsRNAs on silencing different regions (promoter, protein-coding and intron) of the target *SlTRY* tomato gene, encoding an R3-type MYB repressor of anthocyanin biosynthesis. We also assessed the impact of targeting different *SlTRY* regions on the expression of genes involved in anthocyanin and flavonoid biosynthesis. This study demonstrated the critical importance of selecting the appropriate gene target region for dsRNA action. The highest inhibition of the *SlTRY* gene expression and activation of anthocyanin biosynthesis was achieved by dsRNA complementary to the protein-coding region of *SlTRY* gene, compared with dsRNAs targeting the *SlTRY* promoter or intron regions. Silencing the *SlTRY* gene increased the content of anthocyanins and boosted levels of other substances in the phenylpropanoid pathway, such as caffeoyl putrescine, chlorogenic acid, ferulic acid glucoside, feruloyl quinic acid, and rutin. This study is the first to examine the effects of four different dsRNAs targeting various regions of the *SlTRY* gene, an important negative regulator of anthocyanin biosynthesis.

Keywords: *Solanum lycopersicum*; exogenous dsRNA; gene silencing; RNA interference; plant foliar treatment; plant gene regulation; tomato; transcription factors

Citation: Suprun, A.R.; Manyakhin, A.Y.; Trubetskaya, E.V.; Kiselev, K.V.; Dubrovina, A.S. Regulation of Anthocyanin Accumulation in Tomato *Solanum lycopersicum* L. by Exogenous Synthetic dsRNA Targeting Different Regions of *SlTRY* Gene. *Plants* **2024**, *13*, 2489. https://doi.org/10.3390/plants13172489

Academic Editors: Zhiyong Li and Jian Zhang

Received: 14 August 2024
Revised: 31 August 2024
Accepted: 3 September 2024
Published: 5 September 2024

1. Introduction

Tomato (*Solanum lycopersicum* L.) is one of the most widely grown and consumed vegetable crops in the world [1]. Tomato is also widely used in molecular biology as a model object, including for studying the mechanism of anthocyanin biosynthesis [2–4]. Anthocyanins are a class of flavonoids formed by the phenylpropanoid biochemical pathway in plants. They have a common basal structure consisting of two aromatic benzene rings separated by an oxygen-containing heterocycle consisting of three carbon atoms [5]. Glycosylated forms of anthocyanins are stored in the vacuoles of plant cells, contributing to the diverse colors observed in flowers and fruits, such as red, blue, and purple. These vibrant hues play a crucial role in attracting pollinators and animals, which aids in both pollination and seed dispersal [2,5]. In plant tissues, anthocyanins help protect plants from a variety of biotic and abiotic stressors, including those caused by insects, pathogenic fungi and bacteria, as well as drought, UV radiation and low temperatures [6–8]. Notably, anthocyanins are extremely beneficial for human health due to their antioxidant activity and ability to influence signaling pathways in animal cells [2]. Studies have shown that eating foods that reach these metabolites reduces the risk of cancer, coronary heart disease, atherosclerosis, obesity and diabetes [9–12].

Anthocyanins, the pigments responsible for the vibrant reds, purples, and blues in fruits, vegetables, and flowers, are produced through a complex biochemical pathway involving a series of enzymatic reactions. This process, known as anthocyanin biosynthesis, begins with the amino acid phenylalanine, which undergoes a series of transformations catalyzed by specific enzymes, such as phenylalanine ammonia lyase (PAL), 4-coumaryl–CoA ligase (4CL), chalcone synthase (CHS), chalcone isomerase (CHI), flavanone 3-hydroxylase (F3H), flavonoid 3′-hydroxylase (F3′H), flavonoid 3′5′-hydroxylase (F3′5′H), dihydroflavonol-4-reductase (DFR), anthocyanidin synthase (ANS), flavonol 3-glucosyltransferase (3GT) and rhamnosyltransferase (RT) [13] (Figure 1). The expression of early biosynthetic genes (EBG) (CHS, CHI, F3H) is modulated by transcription factors of the R2R3-MYB subgroup. Late biosynthetic genes (LBGs) (DFR, ANS, UFGT) are regulated by three transcription factors known as the ternary MBW complex (R2R3-MYB, bHLH and WD40) [13,14]. MYB transcription factors play a crucial role in regulating the production of anthocyanins and act as molecular switches, either activating or repressing the expression of genes involved in anthocyanin biosynthesis. To exert their influence, MYB factors team up with other protein partners, namely bHLH (basic helix–loop–helix) and WD40 factors forming the MBW complex [15]. In recent years, an increasing number of regulatory proteins have been characterized. These also include negative regulators, either R2R3-MYB or R3-MYB proteins, which have one or two MYB domain repeats, respectively, that can disrupt the activity of the MBW complex [14,16,17]. Tomato plants carrying a mutation in the *SlMYBATV* gene showed an increased expression of biosynthetic genes and accumulated increased amounts of anthocyanins [16]. MYBATV has also been shown to interact with tomato bHLH factors involved in MBW complexes, resulting in the disruption of its activity and suppression of anthocyanin production. Colanero et al. [17], using a genome-wide screen, identified two genes of transcription factors (TFs) (*SlTRY* and *SlMYBATV*) as R3-type MYB repressors, and four genes of TFs (*SlMYB3, SlMYB7, SlMYB32* and *SlMYB76*) were identified as R2R3-type MYB repressors. The overexpression of *SlTRY* in Arabidopsis was also shown to result in decreased anthocyanin accumulation [18]. The involvement of *SlTRY* as a negative regulator is also supported by our previous study, in which the inhibition of this gene via the RNA interference mechanism resulted in increased accumulation of anthocyanins in tomato [4].

RNA interference is a regulatory and protective mechanism involved in growth and development processes, as well as in the control of plant responses to pathogens or abiotic stresses [19]. RNA interference is known to be involved in plant responses to unwanted nucleic acids and transposons, and in regulating the expression of endogenous protein-coding genes. The process of RNA interference in a plant begins with exogenous dsRNA entering the cell and binding to DCL ribonuclease, which cleaves it into small fragments 20–25 bp long, with two unpaired bases at the 5′ and 3′ ends. The fragments interact with the RISC complex, which cleaves one of the RNA strands. Preference is given to the fragment whose 5′ end is less tightly conjugated. The second RNA strand is cleaved by endonucleases. The resulting complex moves around the cell in search of homologous messenger RNA. Having found it, the AGO protein from the RISC complex cuts the messenger RNA [19,20].

Increasingly, the RNA interference mechanism is used as a tool for molecular biology research to regulate gene expression in order to study their functions and improve plant quality [21]. SIGS allows the use of exogenous RNAs for the simple and efficient regulation of target genes via the RNAi mechanism [21,22]. This method allows us to combat fungal or bacterial pathogens and viral diseases, and also to regulate the expression of the plants' own genes [4,22–24]. External treatment of plants with dsRNA complementary to target plant genes resulted in a decreased expression of target genes, namely the *EPSPS* gene (3-phosphate synthase gene) in the leaves of tobacco and amaranth [25]; Myb1 gene in orchid flower buds [26]; *Mob1A, WRKY23* and actin genes in *Arabidopsis thaliana* [27]; and *LBDIf7* and *GST40* genes in grapevine [23,28]. The research findings indicate that the external treatment of plant surfaces with gene-specific dsRNA resulted in alterations to the phenotype and biochemical processes, thereby enhancing the plants' resistance to fungal infection and improving their resilience to abiotic stress [23,28]. In our previous study, we showed that the efficiency of silencing the *A. thaliana NPTII* transgene using exogenous dsRNA is affected by factors such as plant treatment time, soil moisture, plant age, the method of applying dsRNA to the plant surface and various abiotic factors. However, there is currently no specific information on the effect of the region of the target gene to which the dsRNA is complementary on the efficiency of silencing [29].

The objective of the present study was to evaluate the effect of specific exogenous dsRNA on different regions of the tomato anthocyanin biosynthesis negative regulator gene *SlTRY* at the mRNA level. Additionally, we assessed the impact on the expression of several genes involved in anthocyanin and other flavonoid biosynthesis, including 4-coumarate-CoA ligase (4CL), chalcone synthase (SlCHS1, SlCHS2), chalcone isomerase (CHI), flavanone-3-hydroxylase (F3H), flavonoid-3′-hydroxylase (F3′H), flavonol synthase (FLS), anthocyanidin synthase (ANS), flavonoid 3-O-glucoside-rhamnosyltransferase (RT) and 4-coumaroyl shikimate/quinate 3′-hydroxylase (C3′H). We also investigated the effect of dsRNA-TRY on the content and profile of anthocyanins and other secondary metabolites. Through this comprehensive analysis, we aimed to better understand the role of *SlTRY*, the regulatory mechanisms influencing anthocyanin biosynthesis and the broader flavonoid pathway in tomato plants.

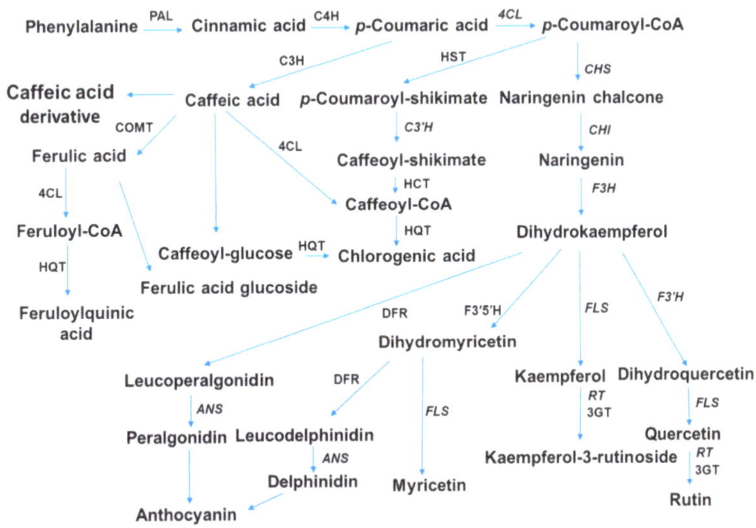

Figure 1. Schematic phenylpropanoid pathway in tomato. Phenylalanine-ammonia-lyase (PAL), cinnamate 4-hydroxylase (C4H), 4-coumarate-CoA ligase (4CL), chalcone synthase (SlCHS1, SlCHS2), chalcone isomerase (CHI), flavanone-3-hydroxylase (F3H), flavonoid-3′-hydroxylase (F3′H), flavonol synthase (FLS), flavonoid-3′-5′-hydroxylase (F3′5′H), dihydroflavonol reductase (DFR), anthocyanidin synthase (ANS), flavonoid 3-O-glucoside-rhamnosyltransferase (RT), flavonoid-3-O-glucosyltransferase (3GT), 4-coumaroyl shikimate/quinate 3′-hydroxylase (C3′H), cinnamoyl-CoA shikimate/quinate transferase (HCT), p-coumaroyl ester 3-hydroxylase (C3H), hydroxycinnamoyl-CoA quinate transferase (HQT), hydroxycinnamoyl-CoA shikimate/quinate hydroxycinnamoyl-transferase (HCT), caffeate O-methyltransferase (COMT). Lilac color represents the biosynthesis of anthocyanins, and yellow color represents the biosynthesis of flavonols and flavonol glycosides. Genes analyzed in this work are shown in italics. Based on the article of Rigano et al. [30].

2. Results

2.1. Exogenous dsRNAs Downregulate mRNA Levels SlTRY Transcription Factors

We obtained in vitro synthesized dsRNAs complementary to different regions of the *SlTRY* gene, which negatively regulates anthocyanin biosynthesis in tomato leaves. First, dsRNA was generated to target solely the promoter region of the SlTRY gene, designated as dsRNA-Prom1 (Figure 2a,b). Next, we synthesized dsRNA that targeted both the promoter and the beginning of the protein coding region, designated as dsRNA-Prom2 (Figure 2a,b). dsRNA-TRY targeted the entire coding sequence (Figure 2b). It is known that the *SlTRY* gene contains one intron, which is removed during splicing. We were interested in whether dsRNA targeting an intron is able to initiate the RNA interference of immature mRNA, namely pre-mRNA. For this purpose, we obtained dsRNA targeting the intron of the *SlTRY* gene, designated as dsRNA-Intron (Figure 2a).

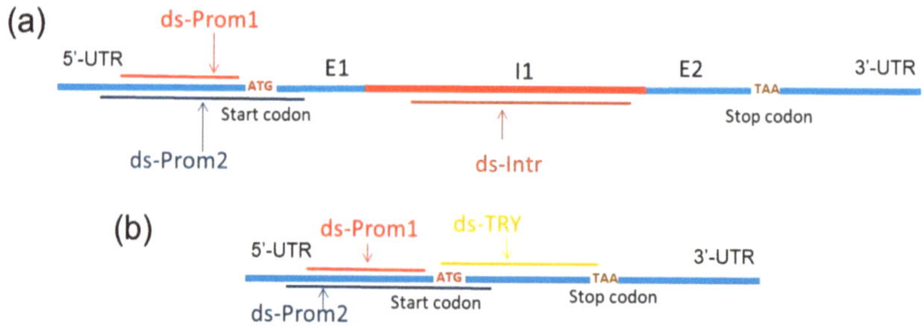

Figure 2. Schematic representation of the dsRNAs positions on the *SlTRY* gene. (**a**) Positions of dsRNA-Prom1 (308 bp), dsRNA-Prom2 (386 bp) and dsRNA-Intron (483 bp) on pre-mRNA; (**b**) positions of dsRNA-Prom1 (308 bp), dsRNA-Prom2 (386 bp) and dsRNA-TRY (285 bp) on mRNA. E1—Exon 1, I1—Intron, E2—Exon 2.

Then, we analyzed whether the exogenous application of our prepared dsRNA-Prom1, dsRNA-Prom2, dsRNA-TRY and dsRNA-Intron double-stranded RNAs to the leaf surface of five-week-old *S. lycopersicum* could result in any changes in the mRNA transcript levels of the *SlTRY* gene compared to water-treated controls seven days after dsRNA treatment (Figure 3). Since anthocyanin production and the expression of anthocyanin biosynthesis genes in *S. lycopersicum* were low under standard cultivation conditions, we divided the treated plants into two groups for post-treatment incubation. The first group consisted of plants grown under control conditions (22 °C, 16 h light), while the second group was grown under anthocyanin-inducing conditions (12 °C, 23 h light) for seven days. This was carried out to induce the expression of anthocyanin biosynthetic genes and the accumulation of anthocyanins in tomato leaves. Additionally, we aimed to analyze the effects of dsRNA on these processes.

qRT–PCR analysis showed that the mRNA level of the *SlTRY* gene was significantly lower after leaf treatment with dsRNA-Prom2 and dsRNA-TRY under both standard and anthocyanin-inducing conditions (Figure 3a). The treatment of plants with dsRNA-Prom1 and dsRNA-Intron did not significantly decrease the expression of the *SlTRY* gene. Moreover, dsRNA-Intron treatment increased the expression of the *SlTRY* gene compared to the control, but this effect was observed only under standard conditions (Figure 3a).

dsRNA-Intron treatment can cause an increase in *SlTRY* gene expression, potentially through an intron-mediated expression enhancement mechanism [31]. The fact that this effect was only observed under standard conditions should suggest that stress conditions may negate the regulatory effects of dsRNA-Intron treatment, but this requires further investigation. Furthermore, the changes in *SlTRY* gene expression following dsRNA treatments corresponded with changes in the color intensity of *S. lycopersicum* leaves. The darkest tomato leaves were observed after treatment with dsRNA-Prom2 and dsRNA-

TRY under both standard and anthocyanin-inducing conditions (Figure 3b). These results suggest that the efficient silencing of a plant target gene requires dsRNA to target a protein-coding region of at least 50 base pairs in length (like dsRNA-Prom2).

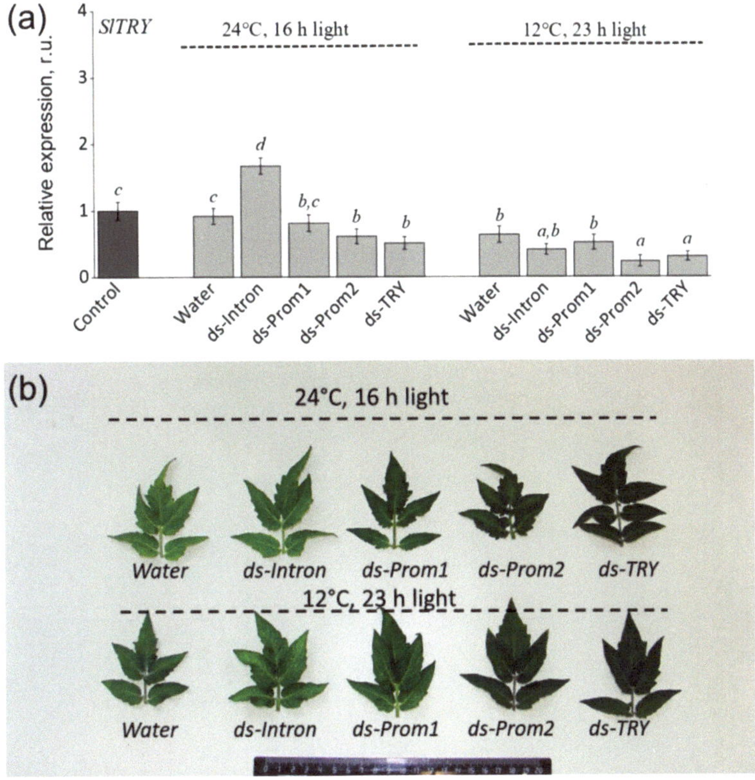

Figure 3. (**a**) Relative fold change in *Solanum lycopersicum* mRNA levels after treatment with dsRNA compared to untreated plants. (**b**) Photograph of leaves on day 7 after treatment of plants with water and dsRNAs under normal conditions and conditions inducing anthocyanin accumulation. Control—average gene expression before treatment; Water—*S. lycopersicum* treated with sterile water; ds-Prom1—S. lycopersicum treated with dsRNA-Prom1; ds-Prom2—*S. lycopersicum* treated with dsRNA-Prom2; ds-Intron—*S. lycopersicum* treated with dsRNA-Intron; dsTRY—*S. lycopersicum* treated with dsRNA-TRY. Data are presented as mean ± standard error. The means in each figure with the same letters were not different from each other according to one-way analysis of variance (ANOVA) followed by Tukey's multiple comparison test.

2.2. Exogenous dsRNAs Upregulate mRNA Levels of Phenylpropanoid Biosynthesis Pathway Genes

Then, we studied the effect of exogenous Prom1-, Prom2-, Intron- and SlTRY-dsRNAs on the expression of phenylpropanoid biosynthesis pathway genes, such as 4-coumaroyl–CoA ligase (*Sl4CL*), chalcone synthase (*SlCHS1* and *SlCHS2*), chalcone isomerase (*SlCHI*), flavanone 3-hydroxylase (*SlF3H*), flavonoid 3′-hydroxylase (*SlF3′H*), flavonol synthase (*SlFLS*), anthocyanidin synthase (*SlANS*), rhamnosyltransferase (*SlRT*) and 4-coumaroyl shikimate/quinate 3′-hydroxylase (*SlC3′H*) (Figure 4). These genes are key enzymes in the early and late biosynthesis of flavonoids, including anthocyanins and flavonols [13].

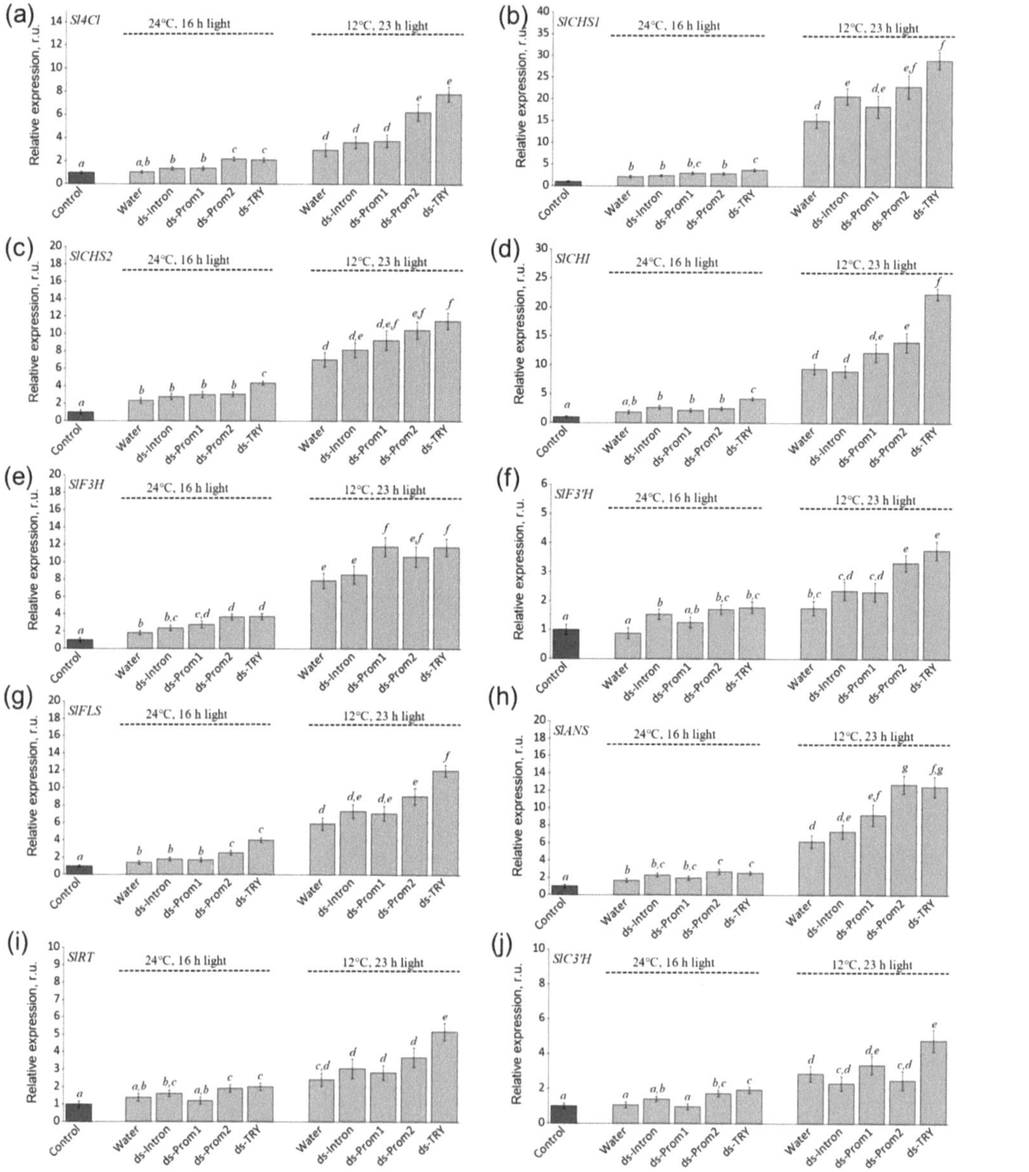

Figure 4. Relative fold changes in (**a**) *Sl4CL*, (**b**) *SlCHS1*, (**c**) *SlCHS2*, (**d**) *SlCHI*, (**e**) *SlF3H*, (**f**) *SlF3′H*, (**g**) *SlFLS*, (**h**) *SlANS*, (**i**) *SlRT* and (**j**) *SlC3′H* mRNA levels after dsRNA treatments of *Solanum lycopersicum* compared to untreated plants. Control—average gene expression before treatment; Water—*S. lycopersicum* treated with sterile water; ds-Prom1—*S. lycopersicum* treated with dsRNA-Prom1; ds-Prom2—*S. lycopersicum* treated with dsRNA-Prom2; ds-Intron—*S. lycopersicum* treated with dsRNA-Intron; dsTRY—*S. lycopersicum* treated with dsRNA-TRY. Data are presented as mean ± standard error. The means in each figure with the same letters were not different from each other according to one-way analysis of variance (ANOVA) followed by Tukey's multiple comparison test.

The analysis revealed a significant increase in the expression of all analyzed genes in *S. lycopersicum* grown under anthocyanin-inducing conditions compared to control conditions (Figure 4). This suggests that stress conditions activate the phenylpropanoid pathway. The expression levels of *Sl4CL, SlCHS1, SlCHS2, SlCHI, SlF3H, SlF3′H, SlFLS, SlANS, SlRT* and *SlC3′H* were notably higher (1.4–2.9 and 1.5–2.6 times under normal conditions and conditions inducing anthocyanin accumulation) in dsRNA-TRY-treated plants than in water-treated *S. lycopersicum* plants (Figure 4). Treatment with dsRNA-Prom2 also significantly upregulated the expression of most of the studied genes by 1.4 to 2.1 times compared to water-treated plants under anthocyanin-inducing conditions, with the exception of the *SlF3H, SlRT* and *SlC3′H* genes (Figure 4). Under control conditions, dsRNA-Prom2 also led to gene activation by 1.5 to 2 times compared to water-treated plants, though no significant activation was observed for *SlCHS1, SlCHS2* and *SlCHI* genes (Figure 4b–d).

Interestingly, using dsRNA-Prom1 targeting the promoter of the *SlTRY* gene resulted in a 1.5-fold increase in *SlF3H* gene expression compared to water-treated plants under both conditions, and an increase in *SlANS* gene expression under anthocyanin-inducing conditions (Figure 4e,h). However, the application of exogenous dsRNA-Intron did not lead to significant changes in the expression of phenylpropanoid pathway genes under either condition (Figure 4).

2.3. *Exogenous dsRNA Upregulates Secondary Metabolism*

We further studied the effects of exogenous Prom1-, Prom2-, Intron-, and SlTRY-dsRNAs on secondary metabolism using HPLC–mass spectrometry (HPLC-MS). According to HPLC-MS analysis, seven anthocyanin compounds were shown to be present in *S. lycopersicum* leaves, namely petunidin-3.5-O-diglucoside (1), petunidin-3-(caffeoyl)-rutinoside-5-glucoside (2), petunidin-3-(p-coumaroyl)-rutinoside-5-glucoside (3), delphinidin-3-O-(6″-O-p-coumaroyl)-glucoside (4), delphinidin-3-O-glucoside (5), malvidin-3-(p-coumaroyl)-rutinoside-5-glucoside (6) and cyanidin-3-O-(6″-O-p-coumaroyl)-glucoside (7) (Figure 5a, Table S2). It is possible that other anthocyanins were also present in the analyzed tissues of *S. lycopersicum*, but in trace amounts. Moreover, we have shown the presence of five substances in the leaves of *S. lycopersicum* that are related to caffeic acid derivatives and flavonols, namely caffeoyl putrescine (a), chlorogenic acid (b), ferulic acid glucoside (c), feruloyl quinic acid (d) and rutin (e) (Figure 5a–g). In addition, the content of most of the analyzed compounds was higher in plants grown under stress conditions (Figure 5c–h).

According to HPLC-MS analysis, this dsRNA-SlTRY-induced decrease in the *SlTRY* gene expression and increase in anthocyanin biosynthetic gene expression correlated with a significant increase in anthocyanin accumulation, reaching 2.2 and 6.2 mg/g FW under control and anthocyanin accumulation-inducing conditions, respectively (Figure 5h). A significant increase in tissue content was also observed after dsRNA-Prom2 treatment, but at lower levels than after dsRNA-SlTRY treatment, namely 1.3 and 3.6 mg/g FW under control and anthocyanin accumulation-inducing conditions, respectively. Importantly, the increase in anthocyanin content was achieved by petunidin-3-(caffeoyl)-rutinoside-5-glucoside and petunidin-3-(p-coumaroyl)-rutinoside-5-glucoside, which accounted for up to 85% of the total anthocyanins. The application of dsRNA-Intron and dsRNA-Prom1 did not lead to a significant increase in anthocyanin content compared to water-treated plants under both types of conditions, which is generally consistent with the expression of biosynthetic genes (Figure 5h).

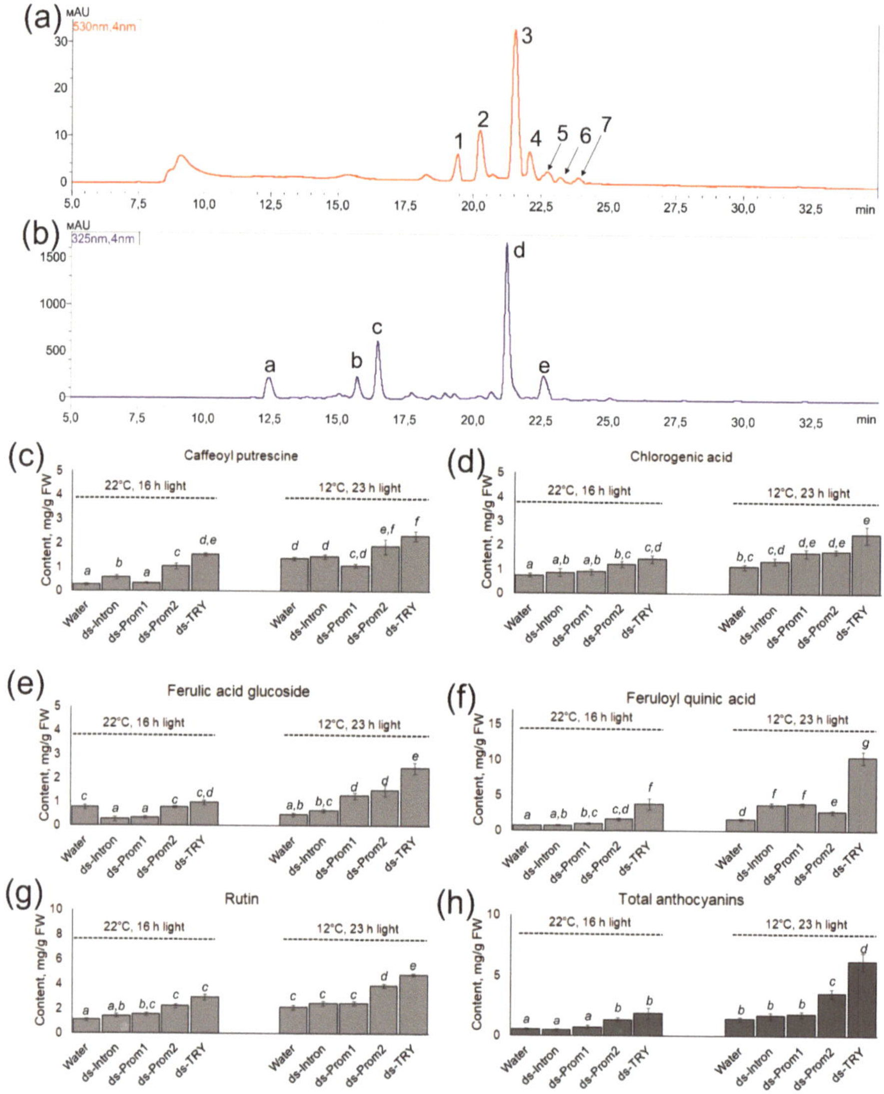

Figure 5. Content of secondary metabolites in the leaves of *Solanum lycopersicum* seven days after treatment with the dsRNAs under control conditions and anthocyanin modulatory conditions. (**a**) A representative HPLC-UV profile (530 nm) of anthocyanins; (**b**) a representative HPLC-UV profile (325 nm) of other secondary metabolites; (**c**) caffeoyl putrescine content; (**d**) chlorogenic acid content; (**e**) ferulic acid glucoside content; (**f**) feruloyl quinic acid content; (**g**) rutin content; (**h**) total anthocyanin content. (1) Petunidin-3.5-O-diglucoside; (2) petunidin-3-(caffeoyl)-rutinoside-5-glucoside; (3) petunidin-3-(p-coumaroyl)-rutinoside-5-glucoside; (4) delphinidin-3-O-(6″-O-p-coumaroyl)-glucoside; (5) delphinidin-3-O-glucoside; (6) malvidin-3-(p-coumaroyl)-rutinoside-5-glucoside; (7) cyanidin-3-O-(6″-O-p-coumaroyl)-glucoside; (a) caffeoyl putrescine; (b) chlorogenic acid; (c) ferulic acid glucoside; (d) feruloyl quinic acid; (e) rutin. *S. lycopersicum* was grown under control or anthocyanin-inducing conditions. Data are presented as mean ± standard error. The means in each figure with the same letters were not different from each other according to one-way analysis of variance (ANOVA) followed by Tukey's multiple comparison test.

Interestingly, the application of exogenous dsRNA-SlTRY and dsRNA-Prom2 affected not only the increase in the content of anthocyanins but also other secondary metabolites such as caffeoyl putrescine, chlorogenic acid, ferulic acid glucoside, feruloyl quinic acid and rutin, in contrast to dsRNA-Intron and dsRNA-Prom1, which did not have a significant effect (Figure 5a–g). The content of caffeoyl putrescine, which is involved in plant defense against insects [32], after the application of dsRNA-SlTRY, increased by 5.8 and 2 times in control and stress conditions, respectively, compared to water-treated plants (Figure 5c). The amount of chlorogenic acid increased by 2 times after the application of dsRNA-SlTRY and by 1.5 times after the application of dsRNA-Prom2 compared to the control in both types of conditions (Figure 5c). An increase in the content of ferulic acid glucoside was noted only under stress conditions up to 5.2 times after the application of dsRNA-SlTRY (Figure 5e). A pronounced positive effect of dsRNA-SlTRY application was exerted on the content of feruloyl quinic acid, namely by 5 and 6.2 times in control and stress conditions, respectively, compared to the control (Figure 5f). The amount of rutin also increased by 2.2 times in both types of conditions (Figure 5g). Our findings, which reveal a rise in the content of the analyzed substances as a result of dsRNA application, align well with the gene expression data. It can be assumed that the activation of early biosynthetic genes of the phenylpropanoid pathway, as a result of the inhibition of the negative regulator of biosynthesis TRY, has a positive effect on the entire secondary metabolism, and not only on the biosynthesis of anthocyanins.

3. Discussion

In today's world, it is essential to create innovative and effective methods to enhance and safeguard plant productivity through safe and eco-friendly technologies. The growing human population and negative impacts of environmental stresses necessitate the development of new molecular tools to improve and protect crops without modifying their genomes. Recently, a novel method known as SIGS has gained traction for directing plant traits in the desired way [19,33,34]. SIGS technology enables the protection of plants from insect pests, fungal infections, and viral pathogens by manipulating the expression of key genes in either the plants themselves or the pathogens [35–37]. For plant protection, foliar spray application is one of the most effective methods for delivering dsRNA in terms of cost, time, and labor. However, many questions remain about the effectiveness of SIGS application compared to traditional plant protection methods such as chemical treatments.

In particular, there is currently limited knowledge about the ability to regulate plant gene expression by applying dsRNA or siRNA to the plant surface. Research has demonstrated that treating plants with dsRNA targeting endogenous plant genes can suppress the mRNA levels of these genes. For instance, the transcript levels of the *EPSPS* gene were reduced in tobacco and amaranth leaves [25], and in orchid buds, the *Myb1* gene was suppressed [26]. The application of in vitro synthesized dsRNA complementary to the *SlIAA9* and *SlAGL6* genes of tomato (*S. lycopersicum* cv UC82) resulted in the decreased expression of these genes and increased ovary size during flowering [34]. Treating grapevines with specific dsRNA targeting the *LBDIf7* gene resulted in the silencing of the *VviLBDIf7* gene and led to a decrease in infection and sporulation of *Plasmopara viticola*, a pathogenic organism affecting grapevines [23]. We have also previously shown that exogenous dsRNA targeting the *AtCHS* gene, as well as two important genes encoding transcriptional repressors of anthocyanin biosynthesis in Arabidopsis, AtMYBL2 and AtANAC032, can effectively influence gene expression [4]. Thus, the potential of using exogenous treatment of plants with targeted dsRNAs to modify plant phenotypes and biochemical responses has been demonstrated. This includes altering flower morphology, enhancing resistance to fungi and increasing tolerance to drought stress.

It is well established that the efficiency of gene silencing via dsRNA depends on various factors. These include the specific plant species involved, the selection of the dsRNA target, the design of the dsRNA itself, the method and timing of delivery to the plant surface and the prevailing environmental conditions. Each of these factors can significantly influence the success and effectiveness of the RNA interference (RNAi) process in plants [29]. While bioinformatics offers numerous tools for the design, analysis and evaluation of small RNA (sRNA) agents, the impact of dsRNA design on its processing and the efficiency of target gene silencing remains poorly understood. Further research is needed to elucidate how specific design parameters influence the effectiveness of dsRNA in achieving desired gene silencing outcomes. [35]. Höfle et al. [35] demonstrated that the reduction in SIGS-mediated disease resistance in *Fusarium graminearum* was significantly correlated with the length of the dsRNA construct that was sprayed. Longer dsRNA constructs were more effective in conferring disease resistance. It has also been shown that using high concentrations of dsRNA targeting the EGFP and NPTII genes of *A. thaliana* causes a stronger silencing effect [38]. As of now, there is a lack of studies specifically assessing the influence of the region of the transcript targeted by dsRNA on the efficacy of RNA interference in plants. Research in this area could provide valuable insights into whether certain regions of the transcript are more conducive to effective gene silencing and could potentially optimize dsRNA design strategies for enhanced RNAi efficiency.

This study was the first to examine the effects of four different dsRNAs targeting various regions of the *SlTRY* gene, an important negative regulator of anthocyanin biosynthesis. Although anthocyanins are present in a variety of plant species, their quantities are often limited or entirely absent in many plants due to restricted activity of the flavonoid biosynthesis pathway [39]. Our aim was to study the effect of dsRNA specific to different regions of immature and mature *SlTRY* transcripts on the silencing of this gene. We designed and produced dsRNA targeting the promoter part of the gene, the promoter and protein-coding region, the protein-coding region only and the intron. We found that the greatest activation of anthocyanin biosynthesis was achieved using dsRNA complementary to the protein-coding region of *SlTRY*. We also obtained interesting data indicating that dsRNA-Intron treatment increased the expression of the *SlTRY* gene compared to the control. However, this effect was observed only under standard conditions. We hypothesize that exogenous treatment with dsRNA-Intron may induce an increase in *SlTRY* gene expression through an intron-mediated enhancement mechanism [31]. It is possible that stress conditions could negate the regulatory effects of dsRNA-Intron enhancement, as abiotic stresses have been shown to affect the efficiency of dsRNA treatment [29].

Using the SIGS approach, we increased the total amount of anthocyanins in tomato leaves by 3.5 and 4.4 times under control and anthocyanin accumulation-inducing conditions, respectively, compared to water-treated plants. This convincingly demonstrates the role of the R3-MYB TRY protein as a competitive inhibitor of MBW complexes [14,40]. TFs are thought to exert their repressive function through competition, known as passive repression, for bHLH partners with R2R3-MYB factors that activate anthocyanin synthesis. This competitive interaction can limit the availability of bHLH partners for activator MYB proteins, thereby reducing the expression of genes involved in the anthocyanin biosynthesis pathway [8,14,41]. The most studied genes are *CAPRICE* (*CPC*) and *TRIPTYCHON* (*TRY*) in Arabidopsis, which mainly inhibit MBW complexes involved in the formation of trichomes and root hairs but also affect flavonoid production [17,18]. A CPC mutant in Arabidopsis has been reported to have elevated anthocyanin levels, suggesting that CPC may normally contribute to anthocyanin regulation [40].

Interestingly, we found that silencing the *SlTRY* gene not only increases the content of anthocyanins but also other substances in the phenylpropanoid pathway, such as caffeoyl putrescine, chlorogenic acid, ferulic acid glucoside, feruloyl quinic acid and rutin. These results align with a study demonstrating that the overexpression of *VvMTBC2L2*, a *Vitis vinifera* transcription factor, leads to a significant downregulation of genes involved in flavonoid production [42]. Nakatsuka et al. employed a chimeric RNAi construct to downregulate two genes, flavonol synthase (*FLS*) and flavonoid 3′-hydroxylase (*F3′H*), along with the expression of the gerbera dihydroflavonol 4-reductase (*DFR*) gene, resulting in abundant accumulation of flavonol derivatives in the form of colorless flavonoids and colored anthocyanins [43].

In summary, this study demonstrated the critical importance of selecting the appropriate gene target region for dsRNA action. Targeting the protein coding sequence of a gene maximally silenced its expression, highlighting dsRNA as a powerful tool for rapidly investigating gene function in plant research. Moreover, this approach holds promise in plant biotechnology for enhancing the production of bioactive compounds.

4. Materials and Methods

4.1. Plant Material and Growth Conditions

Seeds of the wild-type tomato cultivar Micro-Tom (*S. lycopersicum*) were sourced from the Laboratory of Biotechnology at the Federal Scientific Center of the East Asia Terrestrial Biodiversity, Vladivostok, Russia. The seeds underwent vapor-phase sterilization as described in [44] and were subsequently planted in pots (9 cm × 9 cm) containing 200 g of commercially available nutrient-rich soil, which was well irrigated from the bottom with filtered water. The plants were cultivated in a growth chamber (Sanyo MLR-352, Panasonic, Osaka, Japan) under a light intensity of approximately 120 $\mu mol \cdot m^{-2} \cdot s^{-1}$ with a 16 h daily light cycle at 22 °C for four weeks before receiving dsRNA treatments. Following the dsRNA treatments, the *S. lycopersicum* plants were further incubated for seven days either under control conditions (22 °C, 16 h light cycle) or under anthocyanin-inducing conditions (12 °C, 23 h light cycle) in a growth chamber (KS-200, Smolenskoye SKTB SPU, Smolensk, Russia), without additional irrigation, to promote anthocyanin accumulation.

4.2. *SlTRY* Gene Isolation and Sequencing

The promoter and protein-coding sequence of the *SlTRY* gene (XM_010328616) was amplified bia RT-PCR using RNA samples isolated from *S. lycopersicum* leaves. To obtain the sequence of the intron region of the *SlTRY* gene, we amplified the sequence from *S. lycopersicum* DNA. The primers used in this work are listed in Table S1. The RT-PCRs were conducted using a Bis-M1105 Thermal Cycler (Bis-N, Novosibirsk, Russia). All obtained RT-PCR products were cloned into pJET1.2/blunt vectors and sequenced as described in [45].

4.3. dsRNA Synthesis and Application

All dsRNAs used in this work were synthesized using the T7 RiboMAX™Express RNAi System (Promega, Madison, WI, USA). Four different dsRNA sequences were used in the work. The first dsRNA construct targeted only the promoter region of the *SlTRY* gene (dsRNA-Prom1, 308 bp). The second dsRNA construct targeted the promoter region and the initial part of the protein-coding sequence of the *SlTRY* gene (dsRNA-Prom2, 386 bp). The third dsRNA construct targeted the protein coding sequence of the *SlTRY* gene (dsRNA-TRY, 285 bp). The fourth dsRNA construct targeted the intron of the *SlTRY* gene (dsRNA-Intron, 483 bp).

All sequences were amplified through PCR to enable in vitro transcription and the production of dsRNA. A T7 promoter was incorporated at both the 5′ and 3′ termini of each sequence type using the PCRs and primers detailed in Table S1. The PCR reactions took place in the Bis-M1105 Thermal Cycler, with the program set according to the specifications of the T7 RiboMAX™ Express RNAi System. The resultant PCR products served as templates for transcription and dsRNA synthesis, following the manufacturer's guidelines. To evaluate their purity, integrity and concentration, the produced dsRNAs were analyzed through gel electrophoresis and spectrophotometry. The Prom1-, Prom2-, Intron- and TRY-dsRNAs were administered to individual five-week-old *S. lycopersicum* plants using soft brushes, as described in [38].

For each dsRNA application, 70 µg of the dsRNA was dissolved in 400 µL of nuclease-free water and then sprayed onto the foliage (both the upper and lower surfaces of all tomato leaves were treated for each condition). In a separate experiment, two *S. lycopersicum* plants were utilized for each treatment type, specifically, two received 400 µL of sterile filtered water, while two others were administered the dsRNA of each variant. Next, the treated plants were divided into two groups for incubation: one under control conditions (22 °C, 16 h light) and the other under anthocyanin-inducing conditions (12 °C, 23 h light), maintained for seven days (Figure 2). Each type of analysis was conducted across at least four independent experiments. In all experiments, the dsRNAs were applied to five-week-old *S. lycopersicum* plants during late evening (21:00–21:30) under low-soil-moisture conditions. These specific conditions (appropriate plant age, late-evening timing, and low soil moisture) were identified as crucial for effective gene silencing in *A. thaliana* based on our recent analysis [29].

4.4. RNA Isolation and Reverse Transcription Reaction

For RNA extraction, one leaf of tomato was collected from a plant either from a (1) water application or before dsRNA and (2) seven days post treatment in a separate experiment. Total RNA was extracted as described in [46]. cDNAs were synthesized using the MMLV RT Kit (Evrogen, Moscow, Russia). The reactions were conducted in 40 µL of the reaction mixture, which included the first strand buffer, 4 µL of dNTP mix (10 mM each), 1.5 µL of oligo-(dT)15 primer (100 µM), 4 µL of DTT (dithiothreitol, 20 mM) and 3.4 µL of MMLV reverse transcriptase (100 u/µL) at 37 °C for 80 min. The resulting products were then amplified by PCR and checked for DNA contamination using primers for the *Actin* gene (NM_001330119.1).

4.5. Gene Expression Analysis by qRT–PCR

The qRT-PCRs were conducted using SYBR Green I dye and a real-time PCR kit following the manufacturer's instructions (Evrogen, Moscow, Russia). The process was carried out in a thermocycler equipped with a real-time PCR detection system (DNA Technology, Moscow, Russia) as described in [47]. Two internal control genes were used to normalize the data (*SlActin* (NM_001330119.1) and *SlUBI* (NM_001366381.1)). The expression was calculated with the $2-\Delta\Delta CT$ method [48]. The average values of gene expression before dsRNA treatment were taken as a unit.

4.6. Analysis of Secondary Metabolites

For HPLC-MS analysis, 100 mg of tomato leaves were frozen at −20 °C and homogenized using a mortar and a pestle. Shredded tissue was weighed and extracted for one day at 4 °C in 2 mL of 1% (*v*/*v*) HCl in methanol. Then, the mixture was centrifuged at 12,500 rpm for 10 min. The samples were passed through a nylon membrane with a pore size of 0.25 µm for further analysis. Secondary metabolites were identified using a 1260 Infinity analytical HPLC system (Agilent Technologies, Santa Clara, CA, USA) connected to aBruker HCT ultra PTM Discovery System (Bruker Daltonik GmbH, Bremen, Germany). The data were acquired in a positive and negative ion mode under the operating conditions as described in [49]. The MS spectra were recorded across an m/z range of 100–1500. HPLC with diode array detection (HPLC–DAD) for the quantification of all compounds was performed using an HPLC LC-20AD XR analytical system (Shimadzu, Kyoto, Japan). DAD data were collected in the 200–600 nm range. For the quantitative determination of anthocyanins, chromatograms were acquired at 530 nm, and for the remaining analyzed compounds, at 325 nm. The chromatographic separation was performed on a Shim-pack GIST C18 column (Shimadzu, Kyoto, Japan). The compounds were separated in a gradient of 0.1% formic acid (A) and acetonitrile (B). The elution profile was as follows: 0 to 35 min 0% of B; 35 to 40 min 40% of B; 40 to 50 min 50% of B; 50 to 65 min 100% of B. In total, 3 µL of the sample was injected at a constant column temperature of 40 °C and a flow rate of 0.2 mL/min.

The contents of anthocyanins were determined with external standard methods using the four-point regression calibration curves built with the available standards. The analytical standards—cyanidin chloride, petunidin chloride, delphinidin chloride, malvidin chloride, rutin, chlorogenic acid, caffeic acid and ferulic acid—were obtained from Sigma-Aldrich (St. Louis, MO, USA). All solvents were of HPLC grade.

4.7. Statistical Analysis

The data, presented as mean ± standard error (SE), were subjected to a one-way analysis of variance (ANOVA) followed by Tukey's pairwise comparison test. The 0.05 level was selected as the point of minimal statistical significance. For each type of analysis, at least three independent experiments were performed, each with several technical replicates.

Supplementary Materials: The following supporting information can be downloaded at https://www.mdpi.com/article/10.3390/plants13172489/s1: Table S1: Primers used in RT-PCR and qRT-PCR; Table S2: The content of anthocyanins in the leaves of *S. lycopersicum* grown under the control and anthocyanin-inducing conditions.

Author Contributions: A.R.S. and K.V.K. carried out the research design, experiments and data analysis and participated in paper preparation. A.Y.M. performed the HPLC analysis. E.V.T. growing plants and conducting experiments. A.S.D. was responsible for project administration, funding acquisition, paper preparation and data analysis. All authors have read and agreed to the published version of the manuscript.

Funding: This work was supported by a grant from the Russian Science Foundation (grant number 23-26-00253, https://rscf.ru/project/23-26-00253/ (accessed on 6 August 2024)).

Data Availability Statement: The data presented in this study are available within the article and Supplementary Materials.

Conflicts of Interest: The authors declare no conflicts of interest.

References

1. Lin, T.; Zhu, G.; Zhang, J.; Xu, X.; Yu, Q.; Zheng, Z.; Zhang, Z.; Lun, Y.; Li, S.; Wang, X.; et al. Genomic Analyses Provide Insights into the History of Tomato Breeding. *Nat. Genet.* **2014**, *46*, 1220–1226. [CrossRef] [PubMed]
2. Gonzali, S.; Mazzucato, A.; Perata, P. Purple as a Tomato: Towards High Anthocyanin Tomatoes. *Trends Plant Sci.* **2009**, *14*, 237–241. [CrossRef] [PubMed]
3. Sun, C.; Deng, L.; Du, M.; Zhao, J.; Chen, Q.; Huang, T.; Jiang, H.; Li, C.-B.; Li, C. A Transcriptional Network Promotes Anthocyanin Biosynthesis in Tomato Flesh. *Mol. Plant* **2020**, *13*, 42–58. [CrossRef]
4. Suprun, A.R.; Kiselev, K.V.; Dubrovina, A.S. Exogenously Induced Silencing of Four MYB Transcription Repressor Genes and Activation of Anthocyanin Accumulation in *Solanum lycopersicum*. *Int. J. Mol. Sci.* **2023**, *24*, 9344. [CrossRef] [PubMed]
5. Tanaka, Y.; Sasaki, N.; Ohmiya, A. Biosynthesis of Plant Pigments: Anthocyanins, Betalains and Carotenoids. *Plant J.* **2008**, *54*, 733–749. [CrossRef]
6. Qiu, Z.; Wang, X.; Gao, J.; Guo, Y.; Huang, Z.; Du, Y. The Tomato Hoffman's Anthocyaninless Gene Encodes a bHLH Transcription Factor Involved in Anthocyanin Biosynthesis That Is Developmentally Regulated and Induced by Low Temperatures. *PLoS ONE* **2016**, *11*, e0151067. [CrossRef]
7. Meng, X.; Wang, J.-R.; Wang, G.-D.; Liang, X.-Q.; Li, X.-D.; Meng, Q.-W. An R2R3-MYB Gene, *LeAN2*, Positively Regulated the Thermo-Tolerance in Transgenic Tomato. *J. Plant Physiol.* **2015**, *175*, 1–8. [CrossRef]
8. Xu, Y.; Liu, X.; Huang, Y.; Xia, Z.; Lian, Z.; Qian, L.; Yan, S.; Cao, B.; Qiu, Z. Ethylene Inhibits Anthocyanin Biosynthesis by Repressing the R2R3-MYB Regulator SlAN2-like in Tomato. *Int. J. Mol. Sci.* **2022**, *23*, 7648. [CrossRef]
9. Lila, M.A. Anthocyanins and Human Health: An In Vitro Investigative Approach. *J. Biomed. Biotechnol.* **2004**, *2004*, 306–313. [CrossRef]
10. Gonçalves, A.C.; Nunes, A.R.; Falcão, A.; Alves, G.; Silva, L.R. Dietary Effects of Anthocyanins in Human Health: A Comprehensive Review. *Pharmaceuticals* **2021**, *14*, 690. [CrossRef]
11. Kang, S.-Y.; Seeram, N.P.; Nair, M.G.; Bourquin, L.D. Tart Cherry Anthocyanins Inhibit Tumor Development in ApcMin Mice and Reduce Proliferation of Human Colon Cancer Cells. *Cancer Lett.* **2003**, *194*, 13–19. [CrossRef]
12. Lynn, A.; Mathew, S.; Moore, C.T.; Russell, J.; Robinson, E.; Soumpasi, V.; Barker, M.E. Effect of a Tart Cherry Juice Supplement on Arterial Stiffness and Inflammation in Healthy Adults: A Randomised Controlled Trial. *Plant Foods Hum. Nutr.* **2014**, *69*, 122–127. [CrossRef] [PubMed]
13. Chaves-Silva, S.; Dos Santos, A.L.; Chalfun-Júnior, A.; Zhao, J.; Peres, L.E.P.; Benedito, V.A. Understanding the Genetic Regulation of Anthocyanin Biosynthesis in Plants—Tools for Breeding Purple Varieties of Fruits and Vegetables. *Phytochemistry* **2018**, *153*, 11–27. [CrossRef] [PubMed]
14. Albert, N.W.; Davies, K.M.; Lewis, D.H.; Zhang, H.; Montefiori, M.; Brendolise, C.; Boase, M.R.; Ngo, H.; Jameson, P.E.; Schwinn, K.E. A Conserved Network of Transcriptional Activators and Repressors Regulates Anthocyanin Pigmentation in Eudicots. *Plant Cell* **2014**, *26*, 962–980. [CrossRef] [PubMed]
15. Li, Q.; Zhang, C.; Li, J.; Wang, L.; Ren, Z. Genome-Wide Identification and Characterization of R2R3MYB Family in Cucumis Sativus. *PLoS ONE* **2012**, *7*, e47576. [CrossRef]

16. Cao, X.; Qiu, Z.; Wang, X.; Van Giang, T.; Liu, X.; Wang, J.; Wang, X.; Gao, J.; Guo, Y.; Du, Y.; et al. A Putative R3 MYB Repressor Is the Candidate Gene Underlying Atroviolacium, a Locus for Anthocyanin Pigmentation in Tomato Fruit. *J. Exp. Bot.* **2017**, *68*, 5745–5758. [CrossRef]
17. Colanero, S.; Perata, P.; Gonzali, S. The Atroviolacea Gene Encodes an R3-MYB Protein Repressing Anthocyanin Synthesis in Tomato Plants. *Front. Plant Sci.* **2018**, *9*, 830. [CrossRef]
18. Tominaga-Wada, R.; Nukumizu, Y.; Wada, T. Tomato (*Solanum lycopersicum*) Homologs of TRIPTYCHON (SlTRY) and GLABRA3 (SlGL3) Are Involved in Anthocyanin Accumulation. *Plant Signal Behav.* **2013**, *8*, e24575. [CrossRef]
19. Dubrovina, A.; Kiselev, K. Exogenous RNAs for Gene Regulation and Plant Resistance. *Int. J. Mol. Sci.* **2019**, *20*, 2282. [CrossRef]
20. Sifuentes-Romero, I.; Milton, S.L.; García-Gasca, A. Post-Transcriptional Gene Silencing by RNA Interference in Non-Mammalian Vertebrate Systems: Where Do We Stand? *Mutat. Res. Rev. Mutat. Res.* **2011**, *728*, 158–171. [CrossRef]
21. Kamthan, A.; Chaudhuri, A.; Kamthan, M.; Datta, A. Small RNAs in Plants: Recent Development and Application for Crop Improvement. *Front. Plant Sci.* **2015**, *6*, 208. [CrossRef]
22. Bilir, Ö.; Göl, D.; Hong, Y.; McDowell, J.M.; Tör, M. Small RNA-Based Plant Protection against Diseases. *Front. Plant Sci.* **2022**, *13*, 951097. [CrossRef] [PubMed]
23. Marcianò, D.; Ricciardi, V.; Marone Fassolo, E.; Passera, A.; Bianco, P.A.; Failla, O.; Casati, P.; Maddalena, G.; De Lorenzis, G.; Toffolatti, S.L. RNAi of a Putative Grapevine Susceptibility Gene as a Possible Downy Mildew Control Strategy. *Front. Plant Sci.* **2021**, *12*, 667319. [CrossRef] [PubMed]
24. Samarskaya, V.O.; Spechenkova, N.; Markin, N.; Suprunova, T.P.; Zavriev, S.K.; Love, A.J.; Kalinina, N.O.; Taliansky, M. Impact of Exogenous Application of Potato Virus Y-Specific dsRNA on RNA Interference, Pattern-Triggered Immunity and Poly(ADP-Ribose) Metabolism. *Int. J. Mol. Sci.* **2022**, *23*, 7915. [CrossRef] [PubMed]
25. Sammons, R.D.; Ivashuta, S.; Liu, H.; Wang, D.; Feng, P.C.C.; Kouranov, A.Y.; Andersen, S.E. Polynucleotide Molecules for Gene Regulation in Plants. US20110296556A1, 1 December 2011.
26. Lau, S.E.; Schwarzacher, T.; Othman, R.Y.; Harikrishna, J.A. dsRNA Silencing of an R2R3-MYB Transcription Factor Affects Flower Cell Shape in a *Dendrobium hybrid*. *BMC Plant Biol.* **2015**, *15*, 194. [CrossRef]
27. Li, H.; Guan, R.; Guo, H.; Miao, X. New Insights into an RNAi Approach for Plant Defence against Piercing-Sucking and Stem-Borer Insect Pests. *Plant Cell Environ.* **2015**, *38*, 2277–2285. [CrossRef]
28. Nerva, L.; Guaschino, M.; Pagliarani, C.; De Rosso, M.; Lovisolo, C.; Chitarra, W. Spray-Induced Gene Silencing Targeting a Glutathione S-Transferase Gene Improves Resilience to Drought in Grapevine. *Plant Cell Environ.* **2022**, *45*, 347–361. [CrossRef]
29. Kiselev, K.V.; Suprun, A.R.; Aleynova, O.A.; Ogneva, Z.V.; Dubrovina, A.S. Physiological Conditions and dsRNA Application Approaches for Exogenously Induced RNA Interference in *Arabidopsis thaliana*. *Plants* **2021**, *10*, 264. [CrossRef]
30. Rigano, M.M.; Raiola, A.; Docimo, T.; Ruggieri, V.; Calafiore, R.; Vitaglione, P.; Ferracane, R.; Frusciante, L.; Barone, A. Metabolic and Molecular Changes of the Phenylpropanoid Pathway in Tomato (*Solanum lycopersicum*) Lines Carrying Different *Solanum pennellii* Wild Chromosomal Regions. *Front. Plant Sci.* **2016**, *7*, 1484. [CrossRef]
31. Gallegos, J.E.; Rose, A.B. The Enduring Mystery of Intron-Mediated Enhancement. *Plant Sci.* **2015**, *237*, 8–15. [CrossRef]
32. Mohan, R.; Spells, S.; Wang, D.; Fu, Z. Caffeoylputrescine-Hexenal-Mediated Nonhost Resistance against Leafhoppers. *Trends Plant Sci.* **2022**, *27*, 837–839. [CrossRef] [PubMed]
33. Killiny, N.; Gonzalez-Blanco, P.; Gowda, S.; Martini, X.; Etxeberria, E. Plant Functional Genomics in A Few Days: Laser-Assisted Delivery of Double-Stranded RNA to Higher Plants. *Plants* **2021**, *10*, 93. [CrossRef] [PubMed]
34. Molesini, B.; Pennisi, F.; Cressoni, C.; Vitulo, N.; Dusi, V.; Speghini, A.; Pandolfini, T. Nanovector-Mediated Exogenous Delivery of dsRNA Induces Silencing of Target Genes in Very Young Tomato Flower Buds. *Nanoscale Adv.* **2022**, *4*, 4542–4553. [CrossRef] [PubMed]
35. Höfle, L.; Biedenkopf, D.; Werner, B.T.; Shrestha, A.; Jelonek, L.; Koch, A. Study on the Efficiency of dsRNAs with Increasing Length in RNA-Based Silencing of the Fusarium CYP51 Genes. *RNA Biol.* **2020**, *17*, 463–473. [CrossRef]
36. Yoon, J.; Fang, M.; Lee, D.; Park, M.; Kim, K.-H.; Shin, C. Double-Stranded RNA Confers Resistance to Pepper Mottle Virus in *Nicotiana benthamiana*. *Appl. Biol. Chem.* **2021**, *64*, 1. [CrossRef]
37. Zhang, H.; Li, H.; Guan, R.; Miao, X. Lepidopteran Insect Species-Specific, Broad-Spectrum, and Systemic RNA Interference by Spraying dsRNA on Larvae. *Entomol. Exp. Et Appl.* **2015**, *155*, 218–228. [CrossRef]
38. Dubrovina, A.S.; Aleynova, O.A.; Kalachev, A.V.; Suprun, A.R.; Ogneva, Z.V.; Kiselev, K.V. Induction of Transgene Suppression in Plants via External Application of Synthetic dsRNA. *Int. J. Mol. Sci.* **2019**, *20*, 1585. [CrossRef]
39. Sakuta, M.; Tanaka, A.; Iwase, K.; Miyasaka, M.; Ichiki, S.; Hatai, M.; Inoue, Y.T.; Yamagami, A.; Nakano, T.; Yoshida, K.; et al. Anthocyanin Synthesis Potential in Betalain-Producing Caryophyllales Plants. *J. Plant Res.* **2021**, *134*, 1335–1349. [CrossRef]
40. Zhu, H.-F.; Fitzsimmons, K.; Khandelwal, A.; Kranz, R.G. CPC, a Single-Repeat R3 MYB, Is a Negative Regulator of Anthocyanin Biosynthesis in *Arabidopsis*. *Mol. Plant* **2009**, *2*, 790–802. [CrossRef]
41. De Kroon, H.; Huber, H.; Stuefer, J.F.; Van Groenendael, J.M. A Modular Concept of Phenotypic Plasticity in Plants. *New Phytol.* **2005**, *166*, 73–82. [CrossRef]
42. Zhu, Z.; Li, G.; Liu, L.; Zhang, Q.; Han, Z.; Chen, X.; Li, B. A R2R3-MYB Transcription Factor, VvMYBC2L2, Functions as a Transcriptional Repressor of Anthocyanin Biosynthesis in Grapevine (*Vitis vinifera* L.). *Molecules* **2018**, *24*, 92. [CrossRef]
43. Nakatsuka, T.; Abe, Y.; Kakizaki, Y.; Yamamura, S.; Nishihara, M. Production of Red-Flowered Plants by Genetic Engineering of Multiple Flavonoid Biosynthetic Genes. *Plant Cell Rep.* **2007**, *26*, 1951–1959. [CrossRef] [PubMed]

44. Dubrovina, A.S.; Aleynova, O.A.; Suprun, A.R.; Ogneva, Z.V.; Kiselev, K.V. Transgene Suppression in Plants by Foliar Application of in Vitro-Synthesized Small Interfering RNAs. *Appl. Microbiol. Biotechnol.* **2020**, *104*, 2125–2135. [CrossRef]
45. Dubrovina, A.S.; Aleynova, O.A.; Ogneva, Z.V.; Suprun, A.R.; Ananev, A.A.; Kiselev, K.V. The Effect of Abiotic Stress Conditions on Expression of Calmodulin (CaM) and Calmodulin-like (CML) Genes in Wild-Growing Grapevine Vitis Amurensis. *Plants* **2019**, *8*, 602. [CrossRef]
46. Kiselev, K.V.; Dubrovina, A.S.; Shumakova, O.A.; Karetin, Y.A.; Manyakhin, A.Y. Structure and Expression Profiling of a Novel Calcium-Dependent Protein Kinase Gene, CDPK3a, in Leaves, Stems, Grapes, and Cell Cultures of Wild-Growing Grapevine *Vitis amurensis* Rupr. *Plant Cell Rep.* **2013**, *32*, 431–442. [CrossRef]
47. Kiselev, K.V.; Dubrovina, A.S. A New Method for Analyzing Gene Expression Based on Frequency Analysis of RT-PCR Products Obtained with Degenerate Primers. *Acta Physiol. Plant* **2010**, *32*, 495–502. [CrossRef]
48. Livak, K.J.; Schmittgen, T.D. Analysis of Relative Gene Expression Data Using Real-Time Quantitative PCR and the 2(-Delta Delta C(T)) Method. *Methods* **2001**, *25*, 402–408. [CrossRef] [PubMed]
49. Kiselev, K.V.; Grigorchuk, V.P.; Ogneva, Z.V.; Suprun, A.R.; Dubrovina, A.S. Stilbene Biosynthesis in the Needles of Spruce *Picea jezoensis*. *Phytochemistry* **2016**, *131*, 57–67. [CrossRef]

plants

MDPI

Article

Impact of Exogenous dsRNA on miRNA Composition in *Arabidopsis thaliana*

Nikolay N. Nityagovsky , Konstantin V. Kiselev , Andrey R. Suprun and Alexandra S. Dubrovina *

Laboratory of Biotechnology, Federal Scientific Center of the East Asia Terrestrial Biodiversity, Far Eastern Branch of the Russian Academy of Sciences, Vladivostok 690022, Russia; nityagovskii@biosoil.ru (N.N.N.); suprun@biosoil.ru (A.R.S.)

* Correspondence: dubrovina@biosoil.ru; Tel.: +7-423-2310410; Fax: +7-423-2310193

Abstract: The application of double-stranded RNAs (dsRNAs) to plant surfaces has emerged as a promising tool for manipulating gene expression in plants and pathogens, offering new opportunities for crop improvement. While research has shown the capability of exogenous dsRNAs to silence genes, the full spectrum of their impact, particularly on the intricate network of microRNAs (miRNAs), remains largely unexplored. Here, we show that the exogenous application of chalcone synthase (*CHS*)-encoding dsRNA to the rosette leaves of *Arabidopsis thaliana* induced extensive alterations in the miRNA profile, while non-specific bacterial neomycin phosphotransferase II (*NPTII*) dsRNA had a minimal effect. Two days after treatment, we detected 60 differentially expressed miRNAs among the 428 miRNAs found in the *A. thaliana* genome. A total of 59 miRNAs were significantly changed after *AtCHS*-dsRNA treatment compared with water and *NPTII*-dsRNA, and 1 miRNA was significantly changed after *AtCHS*-dsRNA and *NPTII*-dsRNA compared with the water control. A comprehensive functional enrichment analysis revealed 17 major GO categories enriched among the genes potentially targeted by the up- and downregulated miRNAs. These categories included processes such as aromatic compound biosynthesis (a pathway directly related to *CHS* activity), heterocycle biosynthesis, RNA metabolism and biosynthesis, DNA transcription, and plant development. Several predicted targets of upregulated and downregulated miRNAs, including *APETALA2*, *SCL27*, *SOD1*, *GRF1*, *AGO2*, *PHB*, and *PHV*, were verified by qRT-PCR. The analysis showed a negative correlation between the expression of miRNAs and the expression of their predicted targets. Thus, exogenous plant gene-specific dsRNAs induce substantial changes in the plant miRNA composition, ultimately affecting the expression of a wide range of genes. These findings have profound implications for our understanding of the effects of exogenously induced RNA interference, which can have broader effects beyond targeted mRNA degradation, affecting the expression of other genes through miRNA regulation.

Keywords: exogenous dsRNA; miRNA; gene silencing; RNA interference; chalcone synthase; plant gene regulation

check for updates

Citation: Nityagovsky, N.N.; Kiselev, K.V.; Suprun, A.R.; Dubrovina, A.S. Impact of Exogenous dsRNA on miRNA Composition in *Arabidopsis thaliana*. *Plants* **2024**, *13*, 2335. https://doi.org/10.3390/plants13162335

Academic Editors: Jian Zhang and Zhiyong Li

Received: 10 July 2024
Revised: 11 August 2024
Accepted: 18 August 2024
Published: 22 August 2024

1. Introduction

Spray-induced gene silencing (SIGS), also known as exogenous RNA interference (exo-RNAi), is a novel approach in plant science that utilizes externally applied RNA molecules to manipulate plant traits [1–5]. In this method, plant surfaces are sprayed with solutions containing double-stranded RNA (dsRNA) or short-interfering RNA (siRNA). These RNA molecules act like tiny blueprints that instruct the plant's RNA interference (RNAi) machinery to degrade specific mRNA molecules, effectively silencing the targeted genes.

The most promising application of SIGS lies in plant fungal and virus protection [5–7]. By delivering dsRNAs designed to match the virulence genes in attacking pathogens or plant viruses, researchers can achieve targeted silencing, essentially disarming the invaders.

This approach offers a significant advantage over traditional methods such as chemical pesticides, as it is specific and avoids harming beneficial organisms or leaving harmful substances. SIGS is considered a potential alternative to genetically modified (GM) plants, addressing public concerns surrounding GMOs [5,8].

Beyond disease control, SIGS holds immense potential for tailoring plant traits. Researchers have successfully used exogenous dsRNAs to silence specific plant genes and observed the resulting biochemical and phenotypic changes in the plant [9–15]. This approach has provided opportunities to understand the role of individual plant genes in plant development [9], metabolism [10–13], and the stress response [14,15]. While research in this area is still emerging, several studies have shown promise, revealing that SIGS can be used to have an effect on plants and change their properties over a certain period of time. These studies applied dsRNAs targeting specific genes in Arabidopsis, tobacco, amaranth, orchid, tomato, and grapevine. The results showed a decrease in the targeted gene's mRNA levels, indicating successful gene silencing. The exogenously induced silencing of plant endogenous genes translated into desirable changes in the plants, such as altered flower morphology [9], improved resistance to downy mildew [14], enhanced tolerance to drought stress [15], and anthocyanin biosynthesis [11–13]. Researchers are also exploring novel delivery methods for dsRNAs to target plant endogenes, such as nanoparticle carriers [16,17] and laser-assisted treatments [18], to improve the dsRNA uptake and efficacy within plant cells.

In our recent study, we demonstrated that the silencing of genes involved in anthocyanin production was demonstrated in *Arabidopsis thaliana* [11,13] and *Solanum lycopersicum* [12]. Derived from the phenylpropanoid biosynthetic pathway, anthocyanins represent a category of pigmented secondary metabolites responsible for the rich colors of fruits, flowers, and vegetables. They are also recognized for their beneficial properties, including antioxidant and anti-inflammatory activities, and applications in the food industry and horticulture [19,20]. By applying dsRNAs targeting the chalcone synthase *AtCHS*, a specific anthocyanin biosynthesis gene, or negative regulatory genes directly to the plant surface, we observed a significant decrease or increase in the anthocyanin pigment production in *A. thaliana*. This highlights the potential of exogenous exo-RNAi for manipulating plant pigment production and potentially other plant traits.

In addition to the degradation of mRNA, RNAi has also been found to trigger epigenetic modifications, such as alterations in DNA methylation, histone modifications, or microRNA (miRNA)-mediated processes [21,22]. These modifications can persist for a certain period of time after the initial RNAi event, potentially leading to unintended consequences. Consideration of potential unintended consequences and the specificity of gene silencing is crucial. Studies have demonstrated the ability of external dsRNA to silence genes; however, the comprehensive effects, especially on the complex system of miRNAs, have yet to be fully investigated. miRNAs are short, noncoding RNAs that are encoded in the genome and bind to target mRNAs to regulate gene expression by controlling mRNA translation and decay [23,24]. miRNAs can also be regulated by epigenetic modifications, creating a feedback loop between miRNAs and the epigenetic pathway.

In this study, we aimed to investigate the effect of external *AtCHS*-specific dsRNAs on the miRNA transcriptome in *A. thaliana*. We found that the exogenous application of *AtCHS*-dsRNA to the *A. thaliana* leaves induced significant changes in the miRNA transcriptome. These findings have profound implications for our understanding of RNAi-mediated gene silencing in plants and highlight the need to analyze the potential epigenetic effects of external dsRNA application.

2. Results

2.1. The Analysis of Small RNA Fraction in dsRNA-Treated Arabidopsis Plants

To investigate the effects of exogenous dsRNA on the miRNA population, we analyzed six RNA-sequencing (RNA-seq) libraries generated in our previous study using the Illumina NovaSeq 6000 platform [25]. These libraries included two from water-treated plants, two from plants treated with dsRNA specific to the chalcone synthase *AtCHS* gene of *A. thaliana* (ds*CHS*), and two from plants treated with dsRNA specific to *NPTII*, a non-related bacterial neomycin phosphotransferase II gene (ds*NPTII*). The *NPTII*-dsRNA was synthesized as a control to determine whether any observed effect of exogenous dsRNA on the miRNA profiles or other effects were sequence-specific. For each plant, 100 µL of dsRNAs (equivalent to 35 µg) or 100 µL of sterile water was spread onto the leaf surfaces of four-week-old *A. thaliana* rosettes, ensuring coverage of both the upper and lower leaf areas with sterile brushes.

Following the dsRNA treatments, the *A. thaliana* rosettes were incubated under conditions that promote *AtCHS* expression and anthocyanin production (7 °C with 23 h of light exposure) for a duration of two days. Two days after the dsRNA treatments, RNA fractions were isolated from the experimental plants. Low-molecular-weight RNA was used for the small RNA sequencing using Illumina technology [25]. High-molecular-weight RNA was used for the cDNA synthesis and quantitative real-time PCR (qRT-PCR) analysis of the miRNA target gene expression.

After processing and filtering the raw sequencing reads to remove low-quality and irrelevant sequences (adapters, reads shorter than 20 and longer than 24 nucleotides, nongenomic reads (except for *NPTII*)), we obtained 4.4–12.8 million clean reads (Table S1). These reads were aligned to the *A. thaliana* genome to identify and quantify individual miRNAs. We analyzed the proportion and composition of *A. thaliana* miRNAs across different read lengths ranging from 20 to 24 nucleotides and also examined the small RNAs aligned to the *CHS* and *NPTII* sequences (Figure 1).

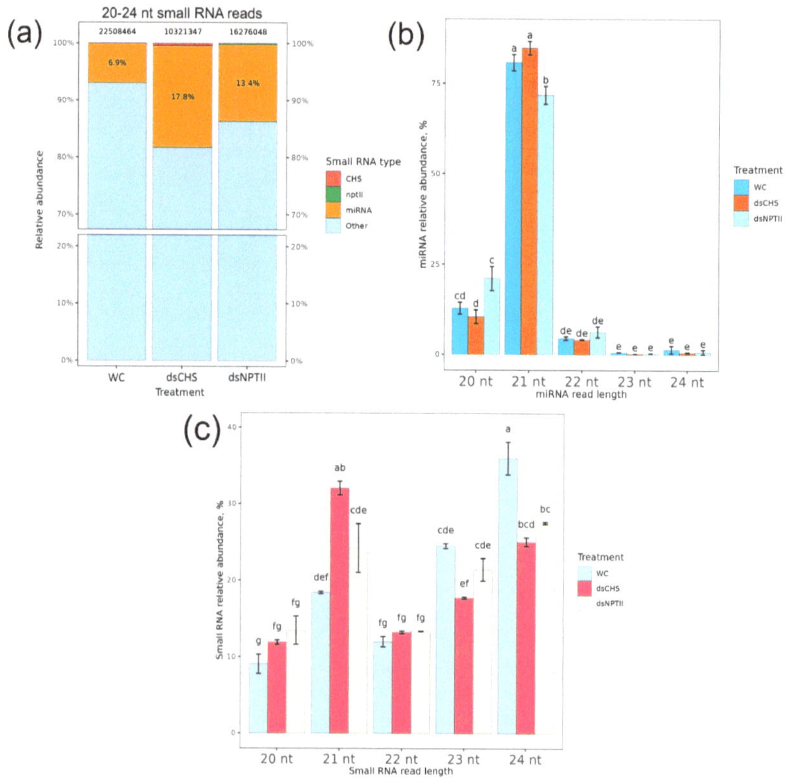

Figure 1. Effect of exogenous dsRNA on miRNA abundance and length distribution. (**a**) The ratio of small RNAs of 20–24 nucleotides in length in plants after water or dsRNA treatments. (**b**) Length

distribution of miRNAs of 20–24 nucleotides after dsRNA and water treatments (100%—total read number of all 20–24 nt miRNAs under each treatment). (**c**) Length distribution and relative abundance of all small RNAs of 20–24 nt in length. WC—*Arabidopsis thaliana* plants treated with sterile water (two plants); ds*CHS*—*A. thaliana* treated with *AtCHS*-coding dsRNA (two plants); ds*NPTII*—*A. thaliana* treated with *NPTII*-coding dsRNA (two plants). The main classes of analyzed small RNAs are miRNAs, *AtCHS*-encoding small RNAs (*CHS*), *NPTII*-encoding small RNAs (*nptII*), and other small RNAs. Means followed by the same letter were not different using one-way analysis of variance (ANOVA), followed by the Tukey HSD multiple comparison test ($p < 0.05$).

The analysis of the *A. thaliana* genome identified a total of 428 miRNAs (Table S2). For the miRNA profile analysis, we examined whether exogenous *AtCHS*-dsRNA and *NPTII*-dsRNA influenced the total 20–24 nt miRNA levels within the small RNA pool (Figures 1a and S1). The data revealed that the proportion of miRNAs was higher after application of the *AtCHS*-dsRNA than after treatment with the water and *NPTII* controls, reaching 17.8% in relative abundance (Figure 1a). The miRNA length size distribution analysis revealed that the prevalent miRNA lengths were 20, 21, and 22 nt miRNAs, with 21 nt miRNAs being the dominant size class, reaching 72–81% of all miRNAs (Figure 1b). The data also showed that the ds*CHS* and ds*NPTII* treatments considerably downregulated all 24 nt small RNAs compared to the water treatment (Figure 1c). It is important to note that the 24 nt small RNAs constitute the dominant small RNA class in the water control probes, mostly excluding miRNAs (Figure 1c,b). Therefore, the decrease in the 24 nt small RNA class (Figure 1c) after the dsRNA treatments contributed to the elevated proportion of miRNAs up to 17.8% and 13.4% (Figure 1a). The analysis also showed that 0.81% of 20–24 nt reads obtained after preprocessing from ds*CHS*-treated *A. thaliana* contained *AtCHS*-specific small RNAs, and 0.41% contained *NPTII*-specific small RNAs in the ds*NPTII*-treated *A. thaliana* among all the obtained 20–24 nt reads after preprocessing (Figure 1a). In contrast, no such sequences were found in the reads from the water-treated control plants.

2.2. The Differential Expression Profile of miRNAs in Response to Exogenous dsRNAs

We used the following analysis to determine the effect of the exogenous dsRNAs on the miRNA population two days after treatment with dsRNA or water. Using a cutoff threshold of an adjusted $p < 0.05$, we identified 60 miRNAs (14% of all detected miRNAs) that exhibited significant differential expression for ds*CHS*-treated plants or both types of dsRNA treatments vs. the water control (Figure 2; Tables S3 and S4). These differentially expressed miRNAs can be categorized into two primary groups. In the first group, 35 miRNAs were downregulated upon the *AtCHS*-dsRNA application in comparison to the water or *NPTII*-dsRNA applications (Figure 2b, Table S3). In the second group, 24 miRNAs were, in turn, upregulated upon the *AtCHS*-dsRNA application in comparison to the water control and ds*NPTII*-treated plants (Figure 2b, Table S4). We also identified ath-miR396a-5p, the level of which was significantly upregulated ($p < 0.05$) in both the ds*CHS* and ds*NPTII* treatments compared to the water treatment (Table S5). Thus, the total number of upregulated miRNAs reached 25 miRNAs. The activation of ath-miRNA396a-5p under the dsRNA treatments indicates the non-sequence-specific dsRNA effect.

A principal component analysis (PCA) of the miRNA profiles indicated that exogenous *AtCHS*-dsRNA was the primary factor affecting the overall miRNA expression in the leaves of *A. thaliana* (Figure 2a). Principal component 1 (PC1), accounting for 60% of the total variance, distinctly differentiated ds*CHS*-treated plants from the water and ds*NPTII* treatments (Figure 2a). While typically miRNAs downregulate target gene transcription via the control of mRNA translation or mRNA decay [23,24], our analysis revealed a higher number of underexpressed miRNAs after the *AtCHS*-dsRNA application. This suggests that the predominant secondary effect of gene-specific dsRNA application may be the alleviation of the silencing of the expression of certain target plant genes.

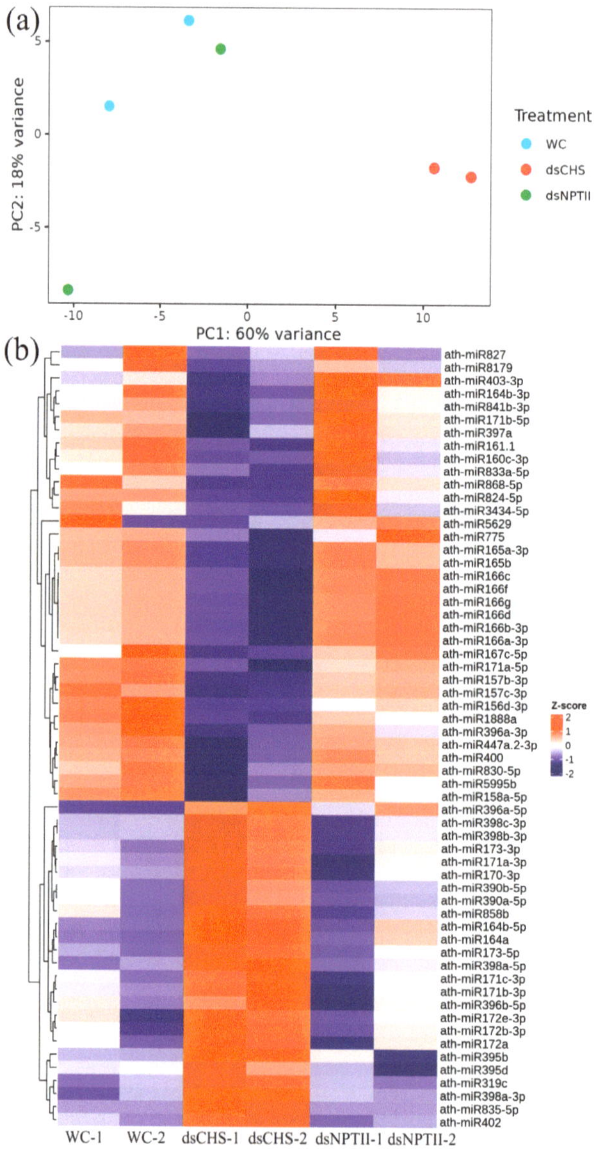

Figure 2. Effect of exogenous dsRNAs on miRNA profile in the *Arabidopsis thaliana* plants. (**a**) Principal component analysis (PCA) of miRNAs differentially expressed in *A. thaliana*. PCA shows the variance of two replicates per treatment x three treatments, including the water, *AtCHS*-dsRNA, and *NPTII*-dsRNA treatments. Each dot represents a sample, and each color represents a treatment. The percentage on each axis indicates the degree of variation explained by the principal components. (**b**) Heatmap showing Z-score-transformed expression levels of miRNAs that are differentially expressed in treated plants. Each row represents an miRNA, and each column represents a sample. WC-1, WC-2—two *A. thaliana* plants treated with sterile water; ds*CHS*-1, ds*CHS*-2—two *A. thaliana* plants treated with *AtCHS*-dsRNA; ds*NPTII*-1, ds*NPTII*—two *A. thaliana* plants treated with *NPTII*-dsRNA. The color intensity from red to blue indicates up- and downregulation, respectively.

2.3. Target Prediction for dsRNA-Affected miRNAs and Gene Enrichment Analysis

miRNAs play a crucial role in post-transcriptional gene regulation [23,24], and their ability to target multiple transcripts simultaneously, or, in turn, the ability of a single transcript to bear several miRNA target sites, highlight the complexity of miRNA-mediated regulation. To elucidate the potential functions of miRNAs affected in response to exogenous dsRNAs, computational target prediction and gene enrichment analysis could provide valuable insights.

Candidate mRNA targets for the differentially expressed miRNAs observed in *A. thaliana* under exogenous dsRNA treatments were predicted with the TarDB database using sequence complementarity (Tables S6 and S7). The analysis revealed that the 35 miRNAs underexpressed only after the *AtCHS*-dsRNA application (Table S3) had 282 predicted mRNA targets (Table S6), while the 24 upregulated miRNAs (Table S4) exhibited 582 predicted mRNA targets (Table S7). These target gene mRNAs represent potential candidates for miRNA-mediated regulation in response to dsRNA treatment. We also identified targets of ath-miR396a-5p, the expression of which was significantly upregulated ($p < 0.05$) in both the ds*CHS* and ds*NPTII* treatments compared to the water treatment (Table S8).

Gene Ontology (GO) analysis is a widely used technique to uncover the putative biological functions of genes [26]. By categorizing target genes based on their functional annotations, GO analysis could aid in understanding the overall impact of miRNA regulation in response to external dsRNA. Enrichment analysis of miRNA targets is a common approach to uncovering the hierarchical activities of miRNAs in gene regulation networks [27]. Using the R package gprofiler2 (version 0.2.3), GO-based enrichment analysis was performed on the mRNA targets of the differentially expressed miRNAs (Figure 3a,b). Gene list functional enrichment analysis of the targets of the 25 dsRNA-upregulated miRNAs (Figure 3a) showed the enrichment of processes associated with aromatic and heterocycle biosynthetic processes, the regulation of primary metabolism, DNA-templated transcription, and RNA biosynthesis and RNA metabolism; for a complete list of terms, see Tables S9 and S11. Notably, these processes were inferred to be downregulated by the miRNA-mediated regulation. This finding suggests that the 25 upregulated miRNAs may contribute to the suppression of these biological pathways.

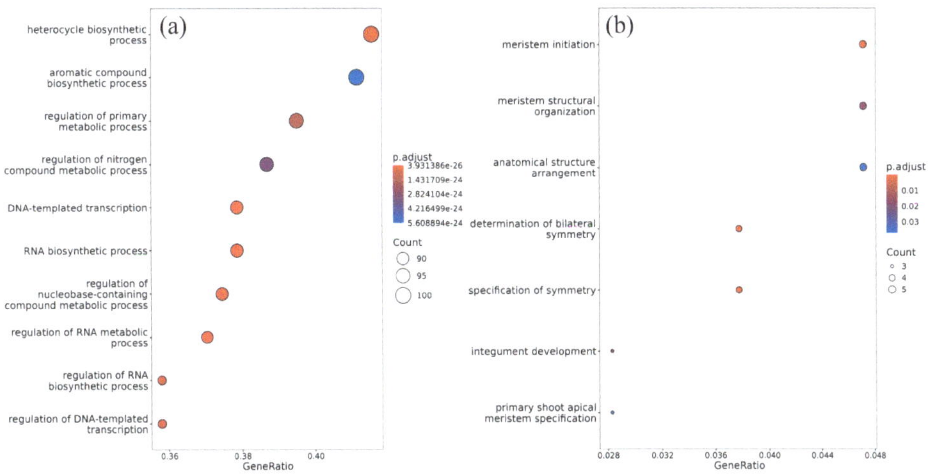

Figure 3. Gene set functional enrichment analysis performed with predicted targets of differentially expressed miRNAs. (**a**) Target genes of the upregulated miRNAs after the *AtCHS*-dsRNA treatment;

(**b**) Target genes of the downregulated miRNAs after the *AtCHS*-dsRNA treatment. The vertical axis shows the biological processes and molecular functions of the top 10 enriched GO terms. The horizontal axis represents the gene ratio, which is defined as the ratio of the targets of differentially expressed miRNAs in a pathway to the total number of genes. The different bubble sizes represent the number of genes for that pathway. The different colors of the bubbles represent the *p*-values. The dot plots were generated using the enrichplot R package. GO analysis was performed using the gprofiler2 R package.

At the same time, gene list functional enrichment analysis of the genes targeted by the 35 ds*CHS*-downregulated miRNAs showed the enrichment of development-related processes involved in meristem initiation, organization, and specification (Figure 3b; Table S10). It is possible that these processes were upregulated in response to the *AtCHS*-dsRNA treatment. However, the support of the gene overrepresentation of the downregulated miRNAs was considerably lower than that of the upregulated miRNAs.

2.4. Validation of miRNA Target Gene Data by qRT-PCR

By qRT-PCR analysis, we confirmed that the application of *AtCHS*-dsRNA to the *A. thaliana* leaves resulted in a significant 4-fold reduction in the *AtCHS* mRNA levels compared to the water treatment (Figure 4). Conversely, the unrelated *NPTII*-dsRNA showed no significant effect on the *AtCHS* expression. Then, the investigation turned to validating several predicted targets of seven distinct miRNA families either ds*CHS*-upregulated (ath-miR171, ath-miR172, ath-miR398, ath-miR396) or ds*CHS*-downregulated (ath-miR166, ath-miR165, ath-miR403) in comparison with the water control and ds*NPTII* application. The selected miRNAs were chosen due to their significant differential expression between the ds*CHS*-treated plants vs. the water and ds*NPTII* treatments (Figure 2; Tables S3 and S4) and the established roles of their corresponding mRNA targets [28–32]. Their respective mRNA targets, including *APETALA2*, *SCL27*, *SOD1*, *GRF1*, *AGO2*, *PHB*, and *PHV*, are associated with diverse biological functions ranging from plant development to stress responses.

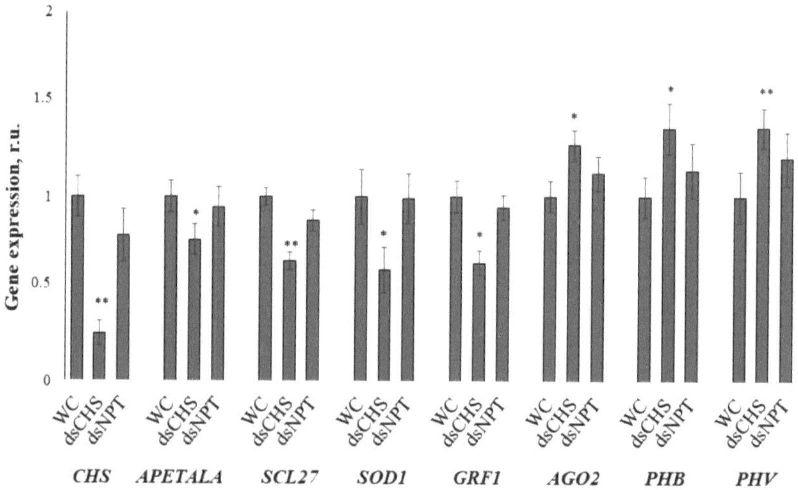

Figure 4. RT-qPCR validation of differentially expressed microRNA target genes under dsRNA treatments and control samples. The effects of external *AtCHS*- and *NPTII*-encoding dsRNAs on miRNA target genes and *AtCHS* mRNA level in *Arabidopsis thaliana* were analyzed by quantitative real-time PCR. *A. thaliana* plants were grown under anthocyanin-inducing (7 °C, 23 h light) conditions for two days after treatment with sterile water or synthetic dsRNA. qRT-PCR data are presented as the mean ± SE. *, **—significantly different from WC at $p \leq 0.01$ according to Student's *t*-test.

To verify the miRNA-target interactions, we measured the expression levels of the chosen target genes in the treated *A. thaliana* by qRT-PCR (Figure 4). It is known that miRNAs negatively regulate gene expression either by inhibiting mRNA translation or promoting RNA degradation [23,24]. Therefore, the expected outcome is a negative correlation between the expression of a specific miRNA and its corresponding target mRNA. In other words, downregulation of an miRNA should lead to the increased expression of its target gene, while upregulation of an miRNA should result in the decreased expression of its target gene.

The qRT-PCR analysis reveals a strong negative correlation in the expression of the pairs of miRNAs and their predicted targets (Figures 2 and 4). For example, three closely related miR172 family members, ath-miR172a, ath-miR172b-3p, and ath-miR172e-3p, were upregulated after treatment with *AtCHS*-dsRNA (Figure 2), while the expression of their target—a floral homeotic *APETALA2* gene—decreased in comparison with the water treatment and *NPTII* controls (Figure 4). Similarly, closely related family members ath-miR171a-3p/ath-miR171b-3p/ath-miR171c-3p, ath-miR398b-3p/ath-miR398c-3p, and ath-miR396b-5p were upregulated after the *AtCHS*-dsRNA treatment (Figure 2), while the expression of their respective targets—GRAS family transcription factor SCL27, the superoxide dismutase *SOD1* gene, and growth regulating factor *GRF1* genes—was downregulated (Figure 4). Conversely, while the ath-miR403-3p and seven ath-miR166 family members were downregulated by *AtCHS*-dsRNA, their target genes *AGO2*, *PHB*, and *PHV* were activated (Figure 4).

3. Discussion

The potential of using exogenous dsRNA to manipulate gene expression in plants and pathogens is gaining significant attention. This innovative approach, known as exo-RNAi or SIGS, offers a promising alternative to traditional crop management methods, providing environmentally friendly and low-risk solutions for crop trait tailoring [1–5]. Recently, we showed that applying gene-specific dsRNAs and siRNAs to the leaves of *A. thaliana* considerably diminished the expression of the target genes implicated in anthocyanin biosynthesis regulation, including *AtCHS* and five repressors of anthocyanin biosynthesis [11,13]. This reduction in *AtCHS* expression, a key enzyme in the anthocyanin pathway, was accompanied by the emergence of specific small RNAs against *AtCHS*, indicating the involvement of RNAi in the plant response to the exogenous dsRNA [25]. Furthermore, a study revealed that exogenous *AtCHS*-dsRNA not only penetrates individual plant cells but also travels through the vascular system, reaching various parts of the plant. This systemic movement suggests the wide-reaching potential for dsRNA-based treatments [11].

Several other studies have shown that exogenous dsRNAs applied to plant leaves are capable of regulating plant endogenes [9,10,12,14,15], yet little is known about the exogenous plant dsRNA treatment impacts on plant epigenetics, including miRNA profiles. In this study, we found that the exogenous application of *AtCHS*-dsRNA to the leaves of *A. thaliana* induced extensive changes in the miRNA transcriptome, considerably affecting the expression of 59 miRNAs, while the non-sequence-specific bacterial *NPTII*-dsRNA had a low effect, affecting the expression of only 1 miRNA. The miRNA profile after the *AtCHS*-dsRNA application was considerably different from that after the control treatments. First of all, this finding revealed that gene-specific dsRNAs induce specific and massive cellular responses at the epigenetic level after exogenous application. Thus, *AtCHS*-dsRNA induced not only specific *AtCHS* gene downregulation but also extensive changes in the epigenetic environment after treatment. Recently, exogenous dsRNA targeting a plant promoter sequence was shown to induce the de novo methylation of promoter sequences in tobacco plants [33]. To the best of our knowledge, no other reports on the epigenetic environment after dsRNA plant treatments have been published. The dsRNA-induced miRNA alterations that are presumably the result of the inhibited *AtCHS* gene expression suggest that the *AtCHS* gene is tightly epigenetically regulated by miRNAs. It is possible that our treatment disrupts this miRNA-mediated feedback.

The qRT-PCR analysis revealed a strong negative correlation between the expression of miRNAs and their predicted targets. This compelling evidence solidifies the validity of these predicted interactions, providing valuable insight into the intricate regulatory networks orchestrated by these miRNA families. For example, ath-miR403 was downregulated after the *AtCHS*-dsRNA application, and verification confirmed the increased gene expression of its target *AGO2*, which is the key effector component of the RNAi machinery [34]. This indicates an activated RNAi system in response to exogenous gene-specific dsRNA application.

The miR171-SCL and miR396 modules are involved in the regulation of chlorophyll biosynthesis and leaf growth in *A. thaliana* [29,30]. This study showed that three closely related miR171 family members (ath-miR171a-3p, ath-miR171b-3p, and ath-miR171c-3p) were upregulated after the *AtCHS*-dsRNA application, while the expression of the target gene *SCL27* was inhibited. The data suggest that the ds*CHS* treatment had a negative effect on chlorophyll biosynthesis and new organ formation in the *A. thaliana*. SOD1 is implicated in the regulation of plant stress resistance and immune reactions [32]. Activation of ath-miR398b-3p, ath-miR398b-3p, and ath-miR398c-3 was associated with the downregulation of the *SOD1* target gene. Thus, it is possible that the exogenous dsRNA compromises plant stress resistance mechanisms. Downregulated miR165 and miR166 levels correlated with the increased expression of the *PHB* and *PHV* genes, suggesting the positive effect of exogenous *AtCHS*-dsRNA on shoot apical meristem and floral development, which are regulated by miR166/165. Thus, this study indicates that exogenous dsRNA activates RNAi-mediated processes, meristem formation, and flowering, and it downregulates new plant organ formation and stress tolerance.

4. Methods and Materials

4.1. Plant Material and Growth Conditions

The seeds of wild-type *Arabidopsis thaliana* from the Columbia (Col-0) ecotype underwent a sterilization process using vapor-phase sterilization. Then, the seeds were sown on solid ½ Murashige and Skoog (MS) medium as described [25]. One week later, the germinated seedlings were planted in 7 cm × 7 cm pots filled with 100 g of commercially available, rich soil. The plants were cultivated in a controlled environmental chamber at 22 °C under plastic wrap for an additional three weeks without extra watering [25]. To induce anthocyanin production and allow for the evaluation of the gene silencing efficiency after the dsRNA treatment, the four-week-old plants were incubated without further irrigation for an additional two days in a growth chamber (KS-200, Smolenskoe SKTB SPU, Smolensk, Russia) with a temperature of 7 °C and daily light period of 23 h before RNA isolation.

4.2. dsRNA Synthesis

The T7 RiboMAX™ Express RNAi System (Promega, Madison, WI, USA) was employed as described to generate *AtCHS*-dsRNA and *NPTII*-dsRNA [25]. For this purpose, a large cDNA fragment of *AtCHS* (AT5G13930.1, 736 bp out of 1188 bp) was amplified via PCR for in vitro transcription and dsRNA synthesis. Additionally, a significant portion of *NPTII* (GenBank AJ414108, 599 bp out of 798 bp) was amplified using the pZP-RCS2-nptII plasmid [35]. In a single PCR for each gene, the T7 promoter sequence was added to both the 5′ and 3′ ends of the amplified *AtCHS* or *NPTII* as detailed [25]. These PCR fragments were then run on a 1% agarose gel for separation and subsequently purified. The purified PCR products served as templates for the in vitro transcription and dsRNA synthesis according to the manufacturer's instructions. The resulting dsRNAs were assessed through gel electrophoresis and spectrophotometry to determine their purity, integrity, and quantity.

4.3. Plant dsRNA Treatment

For the dsRNA application, four-week-old wild-type *A. thaliana* rosettes were treated by applying *AtCHS*- and *NPTII*-dsRNAs to the leaf surfaces with individual sterilized soft brushes made from natural pony hair, which had been sterilized in an autoclave [11,36]. Each treatment involved diluting 35 μg of the dsRNAs in 100 μL of nuclease-free water and covering all rosette leaves on both the upper and lower surfaces with the mixture [25]. Two rosettes per treatment condition were treated, with 35 μg of dsRNA administered to each rosette. Plant dsRNA treatments were performed during late evening hours (21:00–21:30) under conditions of low soil moisture [25]. This method of dsRNA application has been shown to be effective in recent studies [11,36].

4.4. RNA Isolation

To extract RNA from samples, we collected two whole four-week-old rosettes of *A. thaliana* two days post-treatments: water treatment (rosettes WC-1, WC-2), *AtCHS*-dsRNA treatment (rosettes ds*CHS*-1, ds*CHS*-2), and *NPTII*-dsRNA treatment (rosettes ds*NPTII*-1, ds*NPTII*-2). The RNA isolation process involved separating high-molecular-weight and low-molecular-weight RNA fractions by a modified protocol based on cetyltrimethylammonium bromide (CTAB) and lithium chloride (LiCl) application as described [25]. Initially, total RNA extraction was initiated using the CTAB–polyvinylpyrrolidone K30 (PVP) extraction buffer prepared as detailed [37]. Subsequently, high-molecular-weight RNAs were isolated by treating the pellets as described [25]. The pellets after air drying were resuspended in sterile, filtered water and utilized for complementary DNA (cDNA) synthesis. To obtain the low-molecular-weight RNA fraction, after precipitating with LiCl, 400 μL of the subsequent upper aqueous layer was moved to a new, clean 1.5 mL collection tube and precipitated [25].

4.5. Reverse Transcription and qRT-PCRs

Synthesis of cDNAs was performed with 1.5 μg of total RNA according to the procedure described [37]. To ensure the absence of DNA contamination, 1 μL aliquots of reverse transcription products were subjected to PCR amplification with specific primers for the *AtCHS* gene (Table S9). Reverse transcription and the qRT-PCRs were conducted using SYBR Green I Real-Time PCR dye and a real-time PCR kit (Evrogen, Moscow, Russia), as described [38], with *UBQ* and *GAPDH* serving as internal controls. These genes were previously identified as appropriate reference genes for qRT-PCRs in Arabidopsis [39]. A DTprime real-time thermocycler (DNA Technology, Moscow, Russia) was used for amplification with an initial step of 2 min at 95 °C, followed by 50 cycles at 95 °C for 10 s and at 62 °C for 25 s. The calculation of the expression was performed using the $2^{-\Delta\Delta CT}$ method [40]. The data analysis was carried out utilizing RealTime_PCR version 7.3 (DNA Technology, Russia). Table S12 contains all gene identification numbers along with the primers employed.

4.6. Sequencing of Small RNAs

The RNA fractions with low molecular weights, isolated through ethanol precipitation, were sent to Evrogen in Moscow, Russia, for high-throughput sequencing utilizing Illumina technology [25]. Quality control of the incoming samples, RNA preparation, and small RNA sequencing were conducted as previously described [25]. After performing quality control and measuring the DNA concentration, the library collection was sequenced using an Illumina NovaSeq 6000 system with single-read sequencing at 100 bp according to the established protocol [25]. This process yielded a total of 526,174,120 reads. An overview of the read numbers is provided in Table S1. The sequences from the sRNA library have been deposited in the National Center for Biotechnology Information (NCBI) with Accession Number PRJNA827691 and are also accessible in the database of the Laboratory of Biotechnology at the Federal Scientific Center of East Asia Terrestrial Biodiversity, Far Eastern Branch of the Russian Academy of Sciences, Russia (https://biosoil.ru/downloads/biotech/RNAseq/Arabidopsis/2021-03-20012551-data1(Our-RNAseq-1(2))/ (accessed on 4 July 2024)).

4.7. Small RNA Identification and Analysis

Adaptor sequences (5′AACTGTAGGCACCATCAAT, 5′AGATCGGAAGAGCACACGT) were cut off, and reads of poor quality as well as those shorter than 20 nucleotides and longer than 24 nucleotides were filtered out from the obtained high-throughput sequencing data using the BBDuk program (version 39.06) [41]. Using the Bowtie program (version 1.3.1), allowing 0 mismatches [42], reads aligned to the *A. thaliana* TAIR10.1 genome [43] (NCBI Accession Number GCF_000001735.4) with the added *NPTII* gene and unaligned reads were excluded from analysis. The obtained SAM format files were converted into BAM files and indexed using SAMtools (version 1.20) [44]. The processing of the data was performed using the R programming language. Using the GenomicAlignments R package (version 1.38.0) [45], we analyzed the small RNAs aligned to *AtCHS* and *NPTII* sequences and miRNAs obtained from the TAIR10.1 genome (Table S1) with miRBase accession IDs [46]. *AtCHS*, *NPTII*, and miRNA reads were counted using the summarizeOverlaps function from the GenomicAlignments R package. The bar plots of the relative abundance of small RNA fractions were plotted using the ggplot2 (version 3.5.0) [47] and ggbreak (version 0.1.2) [48] R packages based on 20–24 nt reads.

4.8. Differential Expression Analysis of miRNAs

Using DESeq2 (version 1.44.0) [49], we analyzed differentially expressed miRNAs for the ds*CHS* treatment compared to the water and ds*NPTII* controls, and for the ds*CHS* and ds*NPTII* treatments compared to the water control. We identified differentially expressed miRNAs according to the standard algorithm [50–53]. For the differential miRNA expression analysis, we used counts of reads aligned to Arabidopsis miRNAs as input for the DESeq2 tool [49]. MiRNA count data were transformed using the variance-stabilizing transformation method and visualized on a PCA ordination plot using the vst and plotPCA functions of the DESeq2 R package, respectively. Differentially expressed miRNA counts were visualized on a heatmap using the ComplexHeatmap R package (version 2.20.0) [54].

4.9. Target Gene Prediction and Functional Annotation

The target genes of differentially expressed miRNAs were obtained using the TarDB database [55]. Gene set enrichment analysis (GSEA) of target genes was performed using the gProfiler2 R package (version 0.2.3) [56] across the Gene Ontology–Biological Process (GO–BP) database [26], and the results were visualized using the enrichplot R package (version 1.24.2) [57].

4.10. Statistical Analysis

The distribution data of miRNAs, which are 20–24 nucleotides long, are shown as the mean $\pm$ standard error (SE) and were analyzed using a one-way ANOVA, followed by the Tukey HSD test for multiple comparisons. In the DESeq2 analysis, *p*-values were corrected for multiple testing using the Benjamini–Hochberg procedure. In the GSEA, the default *p*-value correction method g:SCS was used. The RT-qPCR validation data are presented as the mean $\pm$ SE and were evaluated using a paired Student's *t*-test. A *p*-value of less than 0.05 was determined as the threshold for statistical significance in all the tests. For each treatment type, two whole rosettes of *A. thaliana*, four weeks old, were collected two days post-treatment (yielding two biological replicates for each type of treatment).

5. Conclusions

This study found that the external *AtCHS*-specific dsRNAs had a profound effect on the miRNA transcriptome in *A. thaliana*, inducing significant changes in the miRNA composition and affecting the expression of the miRNA target genes. These findings provide new insights into the complex interplay between dsRNA treatments and miRNA expression in plants. Further research is needed to determine whether external dsRNAs can trigger epigenetic alterations that persist over time, and to assess the potential risks and benefits of using exo-RNAi approaches. In addition, this study sheds light on the delicate interplay between miRNAs and their target genes, highlighting their crucial role in shaping the complex biological processes in *A. thaliana*.

Supplementary Materials: The following supporting information can be downloaded at https://www.mdpi.com/article/10.3390/plants13162335/s1, Figure S1. 10–100 nt RNAs obtained after high-throughput sequencing and aligned on the Arabidopsis genome. Table S1. Samples analyzed by sRNA-seq and read numbers of 20–24 nt small RNAs obtained after high-throughput sequencing (Illumina NovaSeq 6000 instrument). Table S2. MicroRNAs in *A. thaliana* TAIR10.1 genome. Table S3. Significantly downregulated microRNAs after dsCHS treatment compared to water and dsNPTII treatment (adjusted $p < 0.05$). Table S4. Significantly upregulated microRNAs after dsCHS treatment compared to water and dsNPTII treatment (adjusted $p < 0.05$). Table S5. Significantly upregulated microRNAs after both dsCHS and dsNPTII dsRNA treatments compared to water treatment (adjusted $p < 0.05$). Table S6. Targets of significantly downregulated microRNAs after dsCHS treatment compared to water and dsNPTII treatment according to the TarDB database. Table S7. Targets of significantly upregulated microRNAs after dsCHS treatment compared to water and dsNPTII treatment according to the TarDB database. Table S8. Targets of significantly upregulated microRNAs after both dsCHS and dsNPTII dsRNA treatments compared to water treatment according to the TarDB database. Table S9. List of significantly enriched functions in predicted mRNA targets of dsCHS-upregulated miRNAs. Table S10. List of significantly enriched functions in predicted mRNA targets of dsCHS-downregulated miRNAs. Table S11. Targets of significantly upregulated microRNAs after both dsCHS and dsNPTII dsRNA treatments compared to water treatment according to the TarDB database. Table S12. Primers used in RT-PCR and qRT-PCRs.

Author Contributions: N.N.N. performed the bioinformatic analysis and qRT-PCRs. K.V.K. performed the dsRNA treatments, RNA isolation, and data interpretation. A.R.S. was involved in the plant management. A.S.D. was responsible for the research design, data analysis, and paper preparation. All authors have read and agreed to the published version of the manuscript.

Funding: This work was supported by a grant from the Russian Science Foundation (grant number 23-26-00253, https://rscf.ru/project/23-26-00253/ assessed on 10 July 2024).

Data Availability Statement: The data presented in this study are available within the article and Supplementary Materials.

Conflicts of Interest: The authors declare no conflicts of interest.

References

1. Wang, M.; Jin, H. Spray-induced gene silencing: A powerful innovative strategy for crop protection. *Trends Microbiol.* **2017**, *25*, 4–6. [CrossRef] [PubMed]
2. Dubrovina, A.S.; Kiselev, K.V. Exogenous RNAs for gene regulation and plant resistance. *Int. J. Mol. Sci.* **2019**, *20*, 2282. [CrossRef]
3. Dalakouras, A.; Wassenegger, M.; Dadami, E.; Ganopoulos, I.; Pappas, M.L.; Papadopoulou, K. Genetically modified organism-free RNA interference: Exogenous application of RNA molecules in plants. *Plant Physiol.* **2020**, *182*, 38–50. [CrossRef] [PubMed]
4. Hoang, B.T.L.; Fletcher, S.J.; Brosnan, C.A.; Ghodke, A.B.; Manzie, N.; Mitter, N. RNAi as a foliar spray: Efficiency and challenges to field applications. *Int. J. Mol. Sci.* **2022**, *23*, 6639. [CrossRef] [PubMed]
5. Vatanparast, M.; Merkel, L.; Amari, K. Exogenous application of dsRNA in plant protection: Efficiency, safety concerns and risk assessment. *Int. J. Mol. Sci.* **2024**, *25*, 6530. [CrossRef]
6. Mitter, N.; Worrall, E.A.; Robinson, K.E.; Xu, Z.P.; Carroll, B.J. Induction of virus resistance by exogenous application of double-stranded RNA. *Curr. Opin. Virol.* **2017**, *26*, 49–55. [CrossRef]
7. Morozov, S.Y.; Solovyev, A.G.; Kalinina, N.O.; Taliansky, M.E. Double-stranded RNAs in plant protection against pathogenic organisms and viruses in agriculture. *Acta Nat.* **2019**, *11*, 13–21. [CrossRef]
8. Arpaia, S.; Christiaens, O.; Giddings, K.; Jones, H.; Mezzetti, B.; Moronta-Barrios, F.; Perry, J.N.; Sweet, J.B.; Taning, C.N.T.; Smagghe, G.; et al. Biosafety of GM crop plants expressing dsRNA: Data requirements and EU regulatory considerations. *Front. Plant Sci.* **2020**, *11*, 940. [CrossRef]
9. Lau, S.E.; Schwarzacher, T.; Othman, R.Y.; Harikrishna, J.A. dsRNA silencing of an R2R3-MYB transcription factor affects flower cell shape in a Dendrobium hybrid. *BMC Plant Biol.* **2015**, *15*, 194. [CrossRef]
10. Warnock, N.D.; Wilson, L.; Canet-Perez, J.V.; Fleming, T.; Fleming, C.C.; Maule, A.G.; Dalzell, J.J. Exogenous RNA interference exposes contrasting roles for sugar exudation in host-finding by plant pathogens. *Int. J. Parasitol.* **2016**, *46*, 473–477. [CrossRef]
11. Kiselev, K.V.; Suprun, A.R.; Aleynova, O.A.; Ogneva, Z.V.; Kalachev, A.V.; Dubrovina, A.S. External dsRNA downregulates anthocyanin biosynthesis-related genes and affects anthocyanin accumulation in *Arabidopsis thaliana*. *Int. J. Mol. Sci.* **2021**, *22*, 6749. [CrossRef]

12. Suprun, A.R.; Kiselev, K.V.; Dubrovina, A.S. Exogenously induced silencing of four MYB transcription repressor genes and activation of anthocyanin accumulation in *Solanum lycopersicum*. *Int. J. Mol. Sci.* **2023**, *24*, 9344. [CrossRef]
13. Kiselev, K.V.; Suprun, A.R.; Aleynova, O.A.; Ogneva, Z.V.; Dubrovina, A.S. Simultaneous Application of several exogenous dsRNAs for the regulation of anthocyanin biosynthesis in *Arabidopsis thaliana*. *Plants* **2024**, *13*, 541. [CrossRef] [PubMed]
14. Marcianò, D.; Ricciardi, V.; Fassolo, E.M.; Passera, A.; BIANCO, P.A.; Failla, O.; Casati, P.; Maddalena, G.; De Lorenzis, G.; Toffolatti, S.L. RNAi of a putative grapevine susceptibility gene as a possible downy mildew control strategy. *Front. Plant Sci.* **2021**, *12*, 667319. [CrossRef] [PubMed]
15. Nerva, L.; Guaschino, M.; Pagliarani, C.; De Rosso, M.; Lovisolo, C.; Chitarra, W. Spray-induced gene silencing targeting a glutathione S-transferase gene improves resilience to drought in grapevine. *Plant Cell Environ.* **2022**, *45*, 347–361. [CrossRef] [PubMed]
16. Jiang, L.; Ding, L.; He, B.; Shen, J.; Xu, Z.; Yin, M.; Zhang, X. Systemic gene silencing in plants triggered by fluorescent nanoparticle-delivered double-stranded RNA. *Nanoscale* **2014**, *6*, 9965–9969. [CrossRef]
17. Molesini, B.; Pennisi, F.; Cressoni, C.; Vitulo, N.; Dusi, V.; Speghini, A.; Pandolfini, T. Nanovector-mediated exogenous delivery of dsRNA induces silencing of target genes in very young tomato flower buds. *Nanoscale Adv.* **2022**, *4*, 4542–4553. [CrossRef]
18. Killiny, N.; Gonzalez-Blanco, P.; Gowda, S.; Martini, X.; Etxeberria, E. Plant functional genomics in a few days: Laser-assisted delivery of double-stranded RNA to higher plants. *Plants* **2021**, *10*, 93. [CrossRef] [PubMed]
19. Kong, J.M.; Chia, L.S.; Goh, N.K.; Chia, T.F.; Brouillard, R. Analysis and biological activities of anthocyanins. *Phytochemistry* **2003**, *64*, 923–933. [CrossRef]
20. Khoo, H.E.; Azlan, A.; Tang, S.T.; Lim, S.M. Anthocyanidins and anthocyanins: Colored pigments as food, pharmaceutical ingredients, and the potential health benefits. *Food Nutr. Res.* **2017**, *61*, 1361779. [CrossRef]
21. Simon, S.A.; Meyers, B.C. Small RNA-mediated epigenetic modifications in plants. *Curr. Opin. Plant Biol.* **2011**, *14*, 148–155. [CrossRef]
22. Holoch, D.; Moazed, D. RNA-mediated epigenetic regulation of gene expression. *Nat. Rev. Genet.* **2015**, *16*, 71–84. [CrossRef] [PubMed]
23. Dugas, D.V.; Bartel, B. MicroRNA regulation of gene expression in plants. *Curr. Opin. Plant Biol.* **2004**, *7*, 512–520. [CrossRef] [PubMed]
24. Liu, W.W.; Meng, J.; Cui, J.; Luan, Y.S. Characterization and Function of MicroRNAs in Plants. *Front. Plant Sci.* **2017**, *8*, 2200. [CrossRef] [PubMed]
25. Nityagovsky, N.N.; Kiselev, K.V.; Suprun, A.R.; Dubrovina, A.S. Exogenous dsRNA induces RNA interference of a chalcone synthase gene in *Arabidopsis thaliana*. *Int. J. Mol. Sci.* **2022**, *23*, 5325. [CrossRef]
26. Ashburner, M.; Ball, C.A.; Blake, J.A.; Botstein, D.; Butler, H.; Cherry, J.M.; Davis, A.P.; Dolinski, K.; Dwight, S.S.; Eppig, J.T.; et al. Gene Ontology: Tool for the Unification of Biology. *Nat. Genet.* **2000**, *25*, 25–29. [CrossRef]
27. Xu, J.; Wong, C.-W. Enrichment analysis of miRNA targets. *Methods Mol Biol.* **2013**, *936*, 91–103.
28. Chen, X. A microRNA as a translational repressor of APETALA2 in Arabidopsis flower development. *Science* **2004**, *303*, 2022–2025. [CrossRef]
29. Ma, Z.; Hu, X.; Cai, W.; Huang, W.; Zhou, X.; Luo, Q.; Yang, H.; Wang, J.; Huang, J. Arabidopsis miR171-targeted scarecrow-like proteins bind to GT cis-elements and mediate gibberellin-regulated chlorophyll biosynthesis under light conditions. *PLoS Genet.* **2014**, *10*, e1004519. [CrossRef]
30. Liu, D.; Song, Y.; Chen, Z.; Yu, D. Ectopic expression of miR396 suppresses GRF target gene expression and alters leaf growth in Arabidopsis. *Physiol. Plant.* **2009**, *136*, 223–236. [CrossRef]
31. Jung, J.H.; Park, C.M. MIR166/165 genes exhibit dynamic expression patterns in regulating shoot apical meristem and floral development in *Arabidopsis*. *Planta* **2007**, *225*, 1327–1338. [CrossRef]
32. Sunkar, R.; Kapoor, A.; Zhu, J.K. Posttranscriptional induction of two Cu/Zn superoxide dismutase genes in *Arabidopsis* is mediated by downregulation of miR398 and important for oxidative stress tolerance. *Plant Cell* **2006**, *18*, 2051–2065. [CrossRef] [PubMed]
33. Dalakouras, A.; Ganopoulos, I. Induction of promoter DNA methylation upon high-pressure spraying of double-stranded RNA in plants. *Agronomy* **2021**, *11*, 789. [CrossRef]
34. Liang, C.; Wang, X.; He, H.; Xu, C.; Cui, J. Beyond loading: Functions of plant ARGONAUTE proteins. *Int. J. Mol. Sci.* **2023**, *24*, 16054. [CrossRef] [PubMed]
35. Tzfira, T.; Tian, G.W.; Lacroix, B.; Vyas, S.; Li, J.; Leitner-Dagan, Y.; Krichevsky, A.; Taylor, T.; Vainstein, A.; Citovsky, V. pSAT vectors: A modular series of plasmids for autofluorescent protein tagging and expression of multiple genes in plants. *Plant Mol. Biol.* **2005**, *57*, 503–516. [CrossRef] [PubMed]
36. Kiselev, K.V.; Suprun, A.R.; Aleynova, O.A.; Ogneva, Z.V.; Dubrovina, A.S. Physiological conditions and dsRNA application approaches for exogenously induced RNA interference in *Arabidopsis thaliana*. *Plants* **2021**, *10*, 264. [CrossRef]
37. Aleynova, O.A.; Kiselev, K.V.; Ogneva, Z.V.; Dubrovina, A.S. The grapevine calmodulin-like protein gene *CML21* is regulated by alternative splicing and involved in abiotic stress response. *Int. J. Mol. Sci.* **2020**, *21*, 7939. [CrossRef]
38. Dubrovina, A.S.; Kiselev, K.V. The role of calcium-dependent protein kinase genes *VaCPK1* and *VaCPK26* in the response of *Vitis amurensis* (*in vitro*) and *Arabidopsis thaliana* (*in vivo*) to abiotic stresses. *Russ. J. Genet.* **2019**, *55*, 319–329. [CrossRef]

39. Czechowski, T.; Stitt, M.; Altmann, T.; Udvardi, M.K.; Scheible, W.R. Genome-wide identification and testing of superior reference genes for transcript normalization in *Arabidopsis*. *Plant Physiol.* **2005**, *139*, 5–17. [CrossRef]
40. Livak, K.J.; Schmittgen, T.D. Analysis of relative gene expression data using real-time quantitative PCR and the 2(-Delta Delta C(T)) method. *Methods* **2001**, *25*, 402–408. [CrossRef]
41. Bushnell, P. BBMap. 2015. Available online: https://sourceforge.net/projects/bbmap/ (accessed on 8 July 2024).
42. Langmead, B.; Trapnell, C.; Pop, M.; Salzberg, S.L. Ultrafast and Memory-Efficient Alignment of Short DNA Sequences to the Human Genome. *Genome Biol.* **2009**, *10*, R25. [CrossRef] [PubMed]
43. Reiser, L.; Bakker, E.; Subramaniam, S.; Chen, X.; Sawant, S.; Khosa, K.; Prithvi, T.; Berardini, T.Z. The Arabidopsis Information Resource in 2024. *Genetics* **2024**, *227*, iyae027. [CrossRef]
44. Li, H.; Handsaker, B.; Wysoker, A.; Fennell, T.; Ruan, J.; Homer, N.; Marth, G.; Abecasis, G.; Durbin, R. 1000 Genome Project Data Processing Subgroup The Sequence Alignment/Map Format and SAMtools. *Bioinformatics* **2009**, *25*, 2078–2079. [CrossRef]
45. Lawrence, M.; Huber, W.; Pagès, H.; Aboyoun, P.; Carlson, M.; Gentleman, R.; Morgan, M.T.; Carey, V.J. Software for Computing and Annotating Genomic Ranges. *PLoS Comput. Biol.* **2013**, *9*, e1003118. [CrossRef] [PubMed]
46. Kozomara, A.; Birgaoanu, M.; Griffiths-Jones, S. miRBase: From microRNA Sequences to Function. *Nucleic Acids Res.* **2019**, *47*, D155–D162. [CrossRef]
47. Wickham, H.; Averick, M.; Bryan, J.; Chang, W.; McGowan, L.D.; François, R.; Grolemund, G.; Hayes, A.; Henry, L.; Hester, J.; et al. Welcome to the Tidyverse. *J. Open Source Softw.* **2019**, *4*, 1686. [CrossRef]
48. Xu, S.; Chen, M.; Feng, T.; Zhan, L.; Zhou, L.; Yu, G. Use Ggbreak to Effectively Utilize Plotting Space to Deal with Large Datasets and Outliers. *Front. Genet.* **2021**, *12*, 774846. [CrossRef]
49. Love, M.I.; Huber, W.; Anders, S. Moderated Estimation of Fold Change and Dispersion for RNA-Seq Data with DESeq. *Genome Biol.* **2014**, *15*, 550. [CrossRef]
50. Sang, Q.; Fan, L.; Liu, T.; Qiu, Y.; Du, J.; Mo, B.; Chen, M.; Chen, X. MicroRNA156 conditions auxin sensitivity to enable growth plasticity in response to environmental changes in Arabidopsis. *Nat. Commun.* **2023**, *14*, 1449. [CrossRef]
51. Zhu, X.; He, S.; Fang, D.; Guo, L.; Zhou, X.; Guo, Y.; Gao, L.; Qiao, Y. High-Throughput Sequencing-Based Identification of Arabidopsis miRNAs Induced by *Phytophthora capsici* Infection. *Front. Microbiol.* **2020**, *11*, 1094. [CrossRef]
52. Wu, C.; Li, X.; Guo, S.; Wong, S.M. Analyses of RNA-Seq and sRNA-Seq data reveal a complex network of anti-viral defense in TCV-infected *Arabidopsis thaliana*. *Sci. Rep.* **2016**, *6*, 36007. [CrossRef] [PubMed]
53. Mehdi, S.M.M.; Krishnamoorthy, S.; Szczesniak, M.W.; Ludwików, A. Identification of Novel miRNAs and Their Target Genes in the Response to Abscisic Acid in Arabidopsis. *Int. J. Mol. Sci.* **2021**, *22*, 7153. [CrossRef] [PubMed]
54. Gu, Z.; Eils, R.; Schlesner, M. Complex Heatmaps Reveal Patterns and Correlations in Multidimensional Genomic Data. *Bioinformatics* **2016**, *32*, 2847–2849. [CrossRef]
55. Liu, J.; Liu, X.; Zhang, S.; Liang, S.; Luan, W.; Ma, X. TarDB: An Online Database for Plant miRNA Targets and miRNA-Triggered Phased siRNAs. *BMC Genom.* **2021**, *22*, 348. [CrossRef] [PubMed]
56. Kolberg, L.; Raudvere, U.; Kuzmin, I.; Vilo, J.; Peterson, H. Gprofiler2—An R Package for Gene List Functional Enrichment Analysis and Namespace Conversion Toolset g: Profiler. *F1000Research* **2020**, *9*, ELIXIR-709. [CrossRef]
57. Yu, G. Enrichplot: Visualization of Functional Enrichment Result. 2024. Available online: https://bioconductor.org/packages/release/bioc/html/enrichplot.html (accessed on 8 July 2024).

plants
MDPI

Communication

Fine Mapping of *qAL5.2* Controlling Anther Length in *Oryza sativa*

Xinyong Liu [1,2,†], Zixuan Yu [1,†], Xiaohong Tong [1], Longxue Chang [1], Jie Huang [1], Yifeng Wang [1], Jiezheng Ying [1], Xingwang Li [2], Shen Ni [1,*] and Jian Zhang [1,*]

1 State Key Laboratory of Rice Biology and Breeding, China National Rice Research Institute, Hangzhou 311400, China; liuxinyong1234@gmail.com (X.L.); wangyifeng@caas.cn (Y.W.)
2 College of Life Science and Technology, Huazhong Agricultural University, Wuhan 430070, China
* Correspondence: nishen@caas.cn (S.N.); zhangjian@caas.cn (J.Z.)
† These authors contributed equally to this article.

Abstract: Anther length is the critical floral trait determining hybrid rice seed production and is controlled by many quantitative trait loci (QTL). However, the cloning of genes specifically controlling anther size has yet to be reported. Here, we report the fine mapping of *qAL5.2* for anther size using backcross inbred lines (BILs) in the genetic background of *Oryza sativa indica* Huazhan (HZ). Gene chip analysis on the BC_4F_2 and BC_5F_1 population identified effective loci on Chr1, Chr5, and Chr8 and two genomic regions on Chr5, named *qAL5.1* and *qAL5.2*. *qAL5.2* was identified in both populations with LOD values of 17.54 and 10.19, which explained 35.73% and 25.1% of the phenotypic variances, respectively. Ultimately *qAL5.2* was localized to a 73 kb region between HK139 and HK140 on chromosome 5. And we constructed two near-isogenic lines (NILs) for RNA-seq analysis, named NIL-qAL5.2^{HZ} and NIL-qAL5.2^{KLY}, respectively. The result of the GO enrichment analysis revealed that differential genes were significantly enriched in the carbohydrate metabolic process, extracellular region, and nucleic acid binding transcription, and KEGG enrichment analysis revealed that alpha-linolenic acid metabolism was significantly enriched. Meanwhile, candidate genes of *qAL5.2* were analyzed in RNA-seq, and it was found that *ORF8* is differentially expressed between NIL-qAL5.2^{HZ} and NIL-qAL5.2^{KLY}. The fine mapping of *qAL5.2* conferring anther length will promote the breed improvement of the restorer line and understanding of the mechanisms driving crop mating patterns.

Keywords: anther length; outcrossing; *Oryza sativa*; fine mapping; hybrid rice

Citation: Liu, X.; Yu, Z.; Tong, X.; Chang, L.; Huang, J.; Wang, Y.; Ying, J.; Li, X.; Ni, S.; Zhang, J. Fine Mapping of *qAL5.2* Controlling Anther Length in *Oryza sativa*. *Plants* **2024**, *13*, 1130. https://doi.org/10.3390/plants13081130

Academic Editor: Jiangqi Wen

Received: 8 March 2024
Revised: 8 April 2024
Accepted: 16 April 2024
Published: 18 April 2024

1. Introduction

Rice is one of the most populous crops in the world, and hybrid rice technology is an important technique for improving rice yield. Hybrid rice technology breaks the limitations of self-pollinating crops, using outcrossing pollination between restoring and sterile lines to obtain offspring and exhibiting strong heterosis in F1 generation, which can significantly improve rice yield [1]. Hybrid rice technology mainly includes the "three-line system" and the "two-line system". The earliest hybrid varieties were obtained through the three-line system. Hybrid varieties developed with a three-line (CMS, maintainer, and restorer) system can improve 20–40% higher yields and are widely recommended since 1976 in China [1].

Three-line hybrid technology includes rice cytoplasmic male sterile lines, a rice cytoplasmic male sterile maintainer line, and a rice cytoplasmic male sterile restorer line [2,3]. *WA352* and *Rf4* were the first genes to be cloned from wild abortive cytoplasmic male sterility (CMS-WA) [4,5]. It consists of two main steps, first breeding the female sterile parent by crossing the sterile line with the maintainer line and then crossing the sterile line with the restorer line to produce the F1 hybrid seed. In subsequent research, scientists discovered the photoperiod/thermo-sensitive male sterile lines and achieved "two-line system" technology without the maintainer line, which effectively simplified the process of hybrid rice seed production [6,7]. Since rice is a self-pollinating crop and needs to be pollinated artificially, enough restorer lines should be planted to ensure that the sterile line can obtain sufficient pollen to complete fertilization, whether the three-line system or the two-line system [8]. However, the overplanting of restorer lines will reduce the area of sterile lines, causing a decrease in the yield of hybrid seed production. In recent years, the high labor cost and low yield in hybrid rice seed production led to high seed prices, severely limiting the promotion and application of hybrid rice [9,10].

Improving the anther size of the restoring line or the exsertion rate of the sterile line stigma can effectively increase the yield of hybrid seed production [11]. Some researchers have detected QTL for anther length by different populations. Tazib et al. detected four major QTL located on chromosomes 2, 3, 5, and 7 through backcross inbred lines derived from the backcrossing of the rice cultivars (Nipponbare × Kasalath) × Nipponbare [12]. Four QTL for anther length were also detected on chromosomes 2, 3, and 8 by populations derived from a cross between an indica (SPR1) and a common wild rice [13]. Recently, some new anther QTL were identified by the advanced backcross line of *O. longistaminata* accession W1508 and chromosomal segment substitution lines in the genetic background of *O. sativa* Taichung 65 and found anther size was regulated by cell elongation and cell proliferation in two different ways [14]. However, the region of these QTL is too rough for accurate mapping, making it challenging to apply in hybrid seed production.

In this study, we detected six QTL by backcross inbred lines derived from two rice cultivars HZ and Koliya (KLY). And *qAL5.2* was narrowed down to a 73 kb region by fine mapping. Regulatory pathways of anther size were analyzed by RNA-seq, and it was revealed that anther size might be regulated by pathways related to alpha-linolenic acid metabolism.

2. Results

2.1. Anther Morphology of NIL-HZ and NIL-KLY

To discover probable quantitative trait loci (QTL) controlling anther length, we investigated the parental phenotype of the restore line Huazhan (HZ, *Oryza sativa* ssp. *Indica*) and Koliya (KLY) from south-east Asia (Figure 1). The anther length of HZ and KLY were 1.96 ± 0.04 mm and 2.53 ± 0.09 mm, respectively. However, KLY exhibits much shorter grain length than HZ (Figure 1F). The grain length of KLY was only 6.5 mm, which is 23% shorter than that of HZ, resulting in a much higher anther/glume length ratio (Figure 1E,F). Subsequently, we constructed a mapping population and the near-isogenic line (NIL) using anther length as the trait for selection (Figure S1). At the same time, we investigated the phenotype of NIL-HZ and NIL-KLY derived from BC_4F_1 with the genetic background of HZ (Figure 2). The anther length of NIL-HZ and NIL-KLY were 2.00 ± 0.1 mm and 2.52 ± 0.05 mm, respectively (Figure 2C,D). At the same time, there were no noticeable significant differences in anther width, 1000-grain weight, grain length, or grain width but there were in plant height (Figure 2E–I).

Figure 1. The phenotypic characterization of Huazhan (HZ) and Koliya (KLY). (**A**,**B**) The gestational stage (**A**) and maturation stage (**B**) of HZ and KLY grown under natural field conditions, Bar = 25 cm. (**C**) The grain length of HZ (down) and KLY (up), Bar = 5 cm. (**D**) The anther of the spikelet of HZ (left) and KLY (right), Bar = 1 cm. (**E**–**H**) The statistics of the agronomic traits of HZ and KLY: anther length (**E**), grain length (**F**), grain width (**G**), and plant height (**H**). Values are means ± SD from three biological replicates. Asterisks indicate statistical significance as determined by Student's *t*-test (** $p < 0.01$).

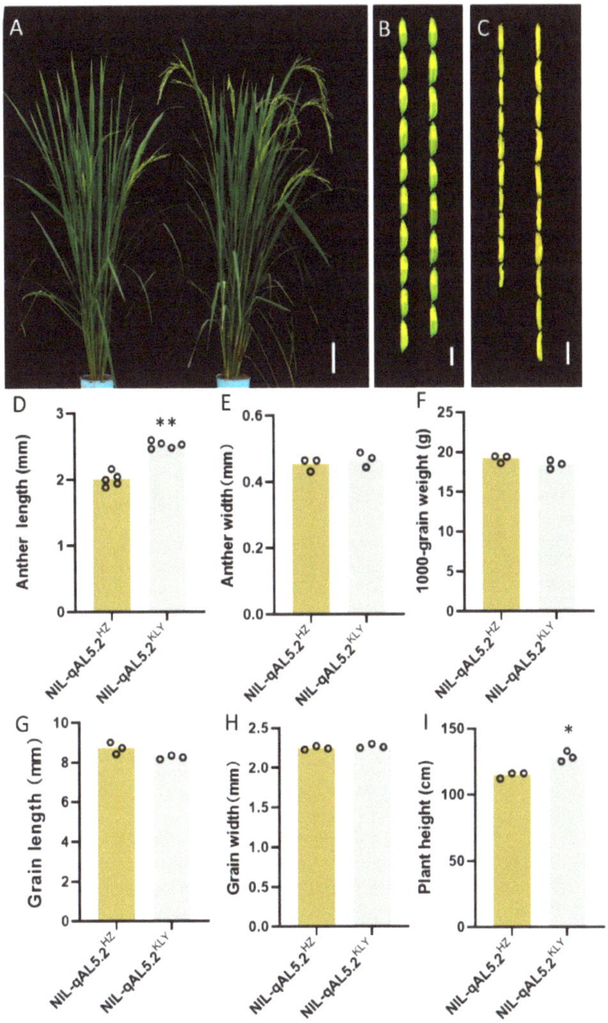

Figure 2. The phenotypic characterization of NIL-HZ and NIL-KLY. (**A**) The gestational stage of NIL-HZ and NIL-KLY grown under natural field conditions, Bar = 10 cm. (**B**,**C**) The length of the spikelet (**B**) and anther (**C**) of NIL-HZ (left) and NIL-KLY (right), Bar = 5 mm and 2 mm. (**D**–**I**) The statistics of the agronomic traits of NIL-HZ and NIL-KLY: anther length (**D**), anther width (**E**), the weight of 1000 grains (**F**), grain length (**G**), grain width (**H**), and plant height (**I**). Values are means ± SD from three biological replicates. Asterisks indicate statistical significance as determined by Student's *t*-test (* $p < 0.05$; ** $p < 0.01$).

2.2. Effects of *qAL5.2* for Anther Length

For mapping quantitative trait loci (QTL) of anther length, the BC_4F_2 population was constructed with the genetic background of HZ (Figure S1). BC_4F_2 individuals with extremely large or small anthers were pool sampled, respectively, for gene chip analysis (Figure 3A). In the BC_4F_2 population (n = 20), five loci on Chr1-1, Chr5, Chr6, Chr8-1, and Chr9 (Figure 3A) were detected. Subsequently, six loci were detected on Chr1-2, Chr3, Chr5, Chr8-2, Chr10, and Chr12 in the BC_5F_1 population (n = 10) (Figure 3B).

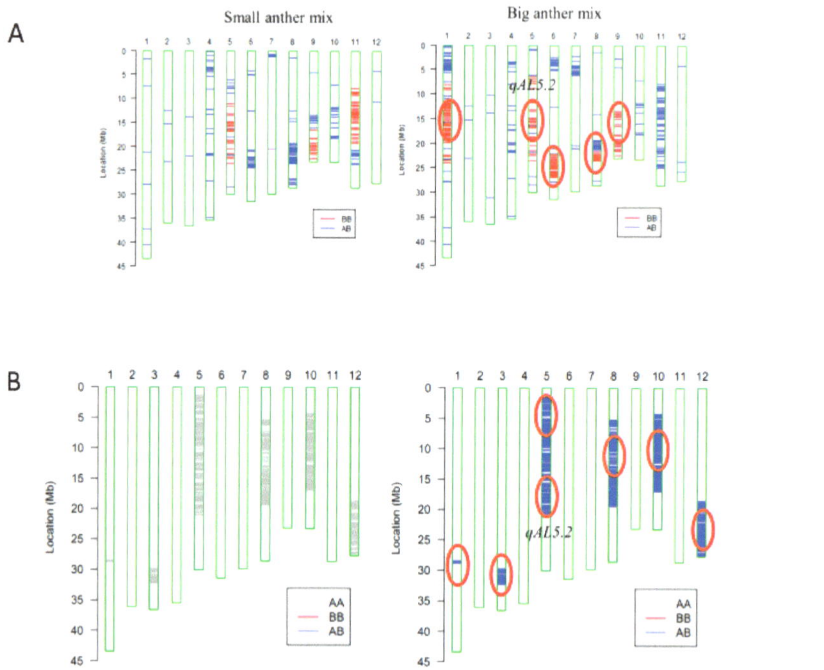

Figure 3. Primary mapping of QTL for anther length using gene chips containing 10 K molecular markers. (**A**) Analysis of plants with extremely short and long anther size in BC_4F_2 (n = 20). (**B**) Analysis in F1 of BC_5F_1 (n = 10). AA stands for HZ; BB stands for KLY. Red circles represent QTL locations.

The draft mapping results revealed that two genomic regions in Chr5 were repeatedly detected in the BC_4F_2 and BC_5F_1 populations and thereafter named as *qAL5.1* and *qAL5.2*, respectively. To fine map QTL of anther length, we constructed seven BC_5F_2 populations with the segregating regions covering the chip mark R0516540382GA- F0520537816TG in *qAL5.2*, and QTL analysis was performed (Figure 4). The result showed that *qAL5.2* was observed in LY-2 and LY-3 populations. In LY-2 and LY-3 populations, the peak LOD values were 17.54 and 10.19, and the additive effects were 0.12 mm and 0.08 mm, explaining 35.73% and 25.1% of the phenotypic variances, respectively (Figure S2). There were no significant QTL in other populations (Table 1). So *qAL5.2* might be localized to a 73 kb interval between HK139 and HK140 on chromosome 5.

Table 1. QTL detected for anther length in seven BC_5F_2 populations. A, the additive effect of replacing an HZ allele with a KLY allele; D, dominance effect; R^2, the proportion of phenotypic variance explained by the QTL effect; ns, no significance.

Population	Heterozygous Interval	LOD	A	D	R^2
LY-1	HK134–HK03	ns	-	-	-
LY-2	HK134–HK62	17.54	0.12	0.01	35.73
LY-3	HK03–HK62	10.19	0.08	-	25.10
LY-4	HK134–HK138	ns	-	-	-
LY-5	HK140–HK62	ns	-	-	-
LY-6	HK61–HK62	ns	-	-	-
LY-7	HK134–HK139	ns	-	-	-

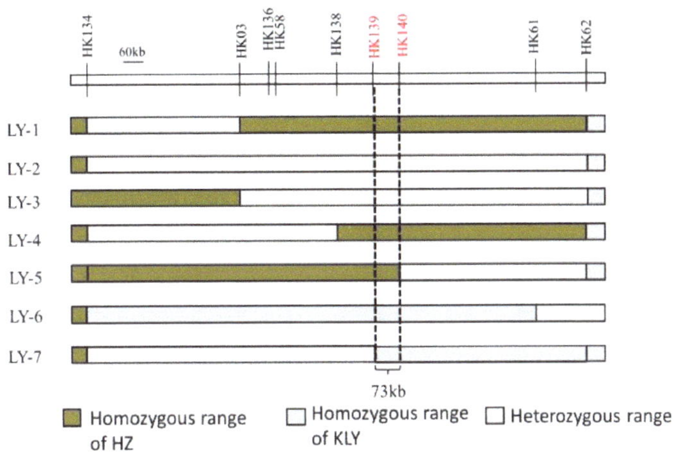

Figure 4. The genotypic compositions of populations in the segregating regions.

2.3. Enrichment Analysis of RNA-Seq

To analyze how *qAL5.2* affects the development of the anther, we conducted an RNA-sequence experiment using the anther samples in the S4-stage (pollen microspore metaphase stage) from the near-isogenic lines NIL-qAL5-2^{HZ} and NIL-qAL5-2^{KLY}, respectively. Compared with NIL-qAL5-2^{KLY}, there were 952 genes observably up-regulated and 541 genes significantly down-regulated in the S4-stage anther in differential expression gene (DEG) analysis (Figure 5A and Table S3). In order to verify the DEG results, three genes were randomly detected by qRT-PCR, and the results showed that the relative expression of the three genes was consistent with the DEG results (Figure 5B). So it could be used in the following analysis.

A

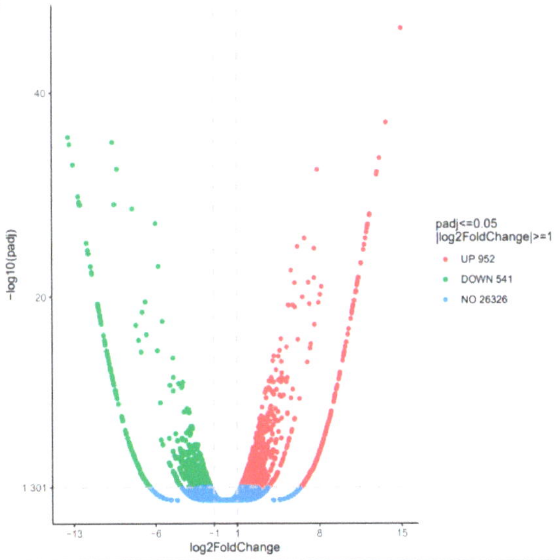

Figure 5. *Cont.*

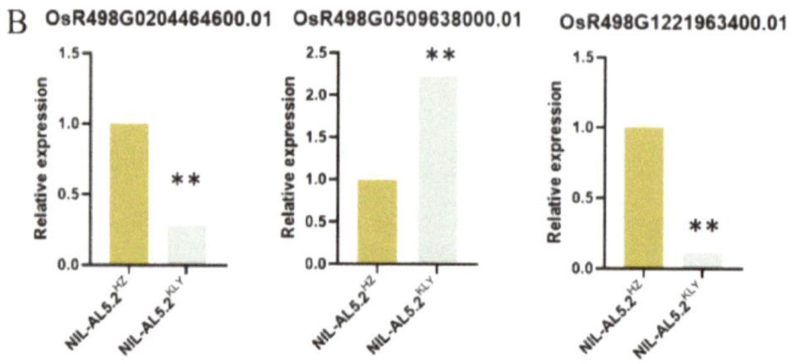

Figure 5. Differential expression analysis in RNA-seq. (**A**) Volcano plot of mRNA expression in S4 anther (pollen microspore metaphase stage). (**B**) Verification of DEG by qRT-PCR. Asterisks indicate statistical significance as determined by Student's *t*-test (** $p < 0.01$).

The result of the GO enrichment analysis showed that differential genes were mainly enriched in the carbohydrate metabolic process, extracellular region, and nucleic acid binding transcription (Figure 6A). The result of the KEGG enrichment analysis revealed that they were significantly ($p < 0.05$) enriched in alpha-linolenic acid metabolism, porphyrin and chlorophyll metabolism, and amino sugar and nucleotide sugar metabolism pathways (Figure 6B), indicating *qAL5.2* is involved in the regulation of the energy metabolism of anthers.

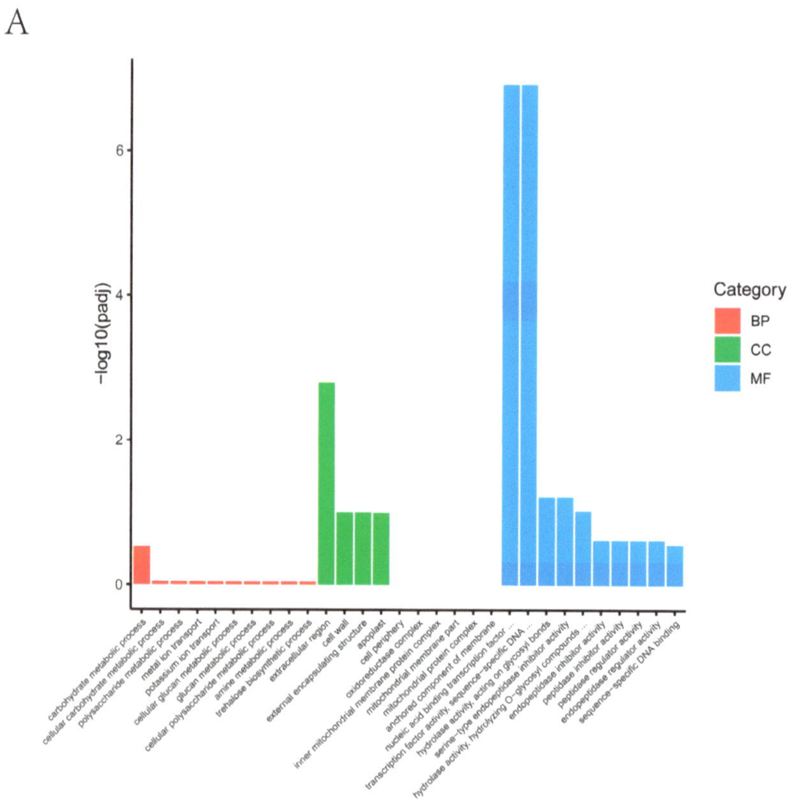

Figure 6. *Cont.*

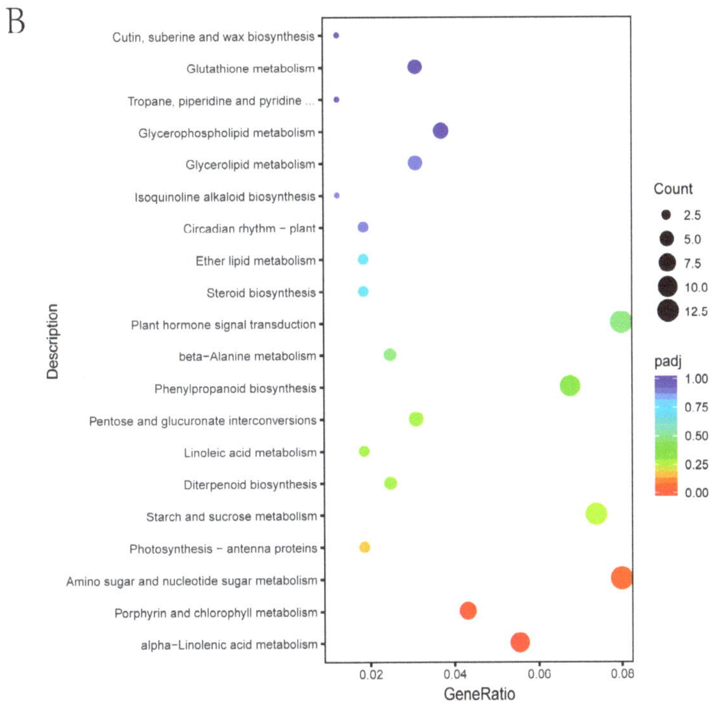

Figure 6. Candidate gene selection from RNA-seq analysis. (**A**) Enrichment of GOs for differentially expressed mRNAs. (**B**) Bubble plot of pathways for differentially expressed mRNAs. BP biological process; CC cellular component; MF molecular function.

2.4. Candidate Gene Analysis of *qAL5.2*

There were eight predicted reading frames (ORFs) covered by *qAL5.2*. We analyzed the expression of all eight candidate genes using the data from the Rice Genome Annotation Project. We found that only *ORF1*, *ORF3*, *ORF4*, and *ORF8* showed expression in the rice anther, and *ORF8* had a particularly high level (Table 2 and Figure S3). And only *ORF8* expression has a significant difference in RNA-Seq.

Table 2. Candidate genes in the target region of *qAL5.2*.

ORF	ID	Gene Product	The Highest Expression
ORF1	LOC_Os05g29900	expressed protein	Seed-S4
ORF2	LOC_Os05g29910	retrotransposon protein, putative, expressed	No
ORF3	LOC_Os05g29920	expressed protein	Seed-S5
ORF4	LOC_Os05g29930	late embryogenesis abundant protein, expressed	SAM
ORF5	LOC_Os05g29940	expressed protein	No
ORF6	LOC_Os05g29950	expressed protein	No
ORF7	LOC_Os05g29960	expressed protein	No
ORF8	LOC_Os05g29974	lipase, putative, expressed	Anther

According to the Sanger sequencing results, there was a nonsynonymous mutation with T to C in exon 8 of *ORF8*, resulting in one amino acid substitution form Ala to Val. *ORF3* also had a nonsynonymous SNP with G to C in exon and caused one acid substitution from Ser to Trp. (Figure 7A). A subcellular localization analysis of *ORF3* and 8 was performed. The result showed that ORF3-GFP localized in plasma membranes and cytosol, and ORF8-GFP localized in the nucleus, plasma membranes, and cytosol (Figure 7B).

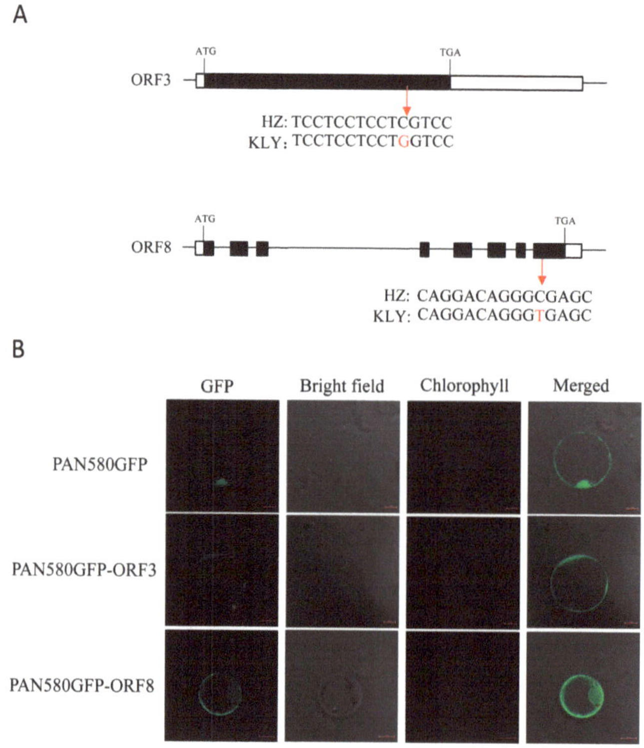

Figure 7. The identification of candidate genes for anther length. (**A**) A schematic diagram of the SNP of the candidate genes between HZ and KLY. An SNP occurred in *ORF3*, resulting in the residues being changed from Ser in HZ to Trp in KLY. An SNP occurred in *ORF8*, resulting in the residues being changed from Ala in HZ to Val in KLY. White boxes indicate UTR, black boxes indicate exons, and the lines between them represent introns. (**B**) Free green fluorescent protein (GFP) and ORF3/ORF8-GFP fusion protein were transiently expressed in rice protoplasts. Green fluorescence shows GFP.

3. Discussion

The anther length was an essential trait in improving the outcrossing yield in rice, and the majority of QTL controlling anther size could be used in marker-assisted selection breeding. In this research, we constructed advanced backcross populations of HZ and KLY, and eight QTL sites of anther length were obtained by gene chip analysis. *qAL5.2* was fine mapped to a 73 kb region between HK139 and HK140 on chromosome 5 by seven BC_5F_2 populations, and candidate genes were analyzed by RNA-sequence.

In the past decade, wild rice segregating populations were used for the QTL analysis of anther size, such as *O. longistaminata* and *O. rufipogon* [13,14]. Although there were significant differences in anthers between wild rice and cultivated rice, the fine mapping of the genes remains difficult, primarily due to the challenges in phenotyping. We used two cultivated rice HZ and KLY for a QTL analysis of anther size. The significant difference in the anther/glume ratio between KLY and HZ helped to rule out the influence of the grain length QTL on anther length (Figures 1 and 2). At the same time, it quickly generated the mapping populations, and we obtained seven BC_5F_2 populations for *qAL5.2* mapping. The genotypes of LY-1, LY-4, and LY-5 were HZ homozygous in *qAL5.2* (Figure 4), and the average anther length was 2.03 mm, 2.12 mm, and 2.15, respectively (Table S1). The anther length of LY-1 was similar to HZ; LY-4 and LY-5 were longer than HZ. We speculated that LY-4 and LY-5 populations closer to the heterozygous range were more prone to genetic recombination. The genotypes of LY-6 and LY-7 were KLY homozygous range in *qAL5.2*, named $qAL5.2^{KLY}$, and the average anther length of LY-6 was 2.26 mm, which was the longest anther in all BC_5F_2 populations. Notably, the anther of homozygous $qAL5.2^{KLY}$ is not identical to KLY, indicating that other QTL may co-regulate anther size.

This research showed that the genomic regions were located on Chr1, Chr3, Chr5, Chr6, Chr8, Chr9, Chr10, and Chr12 by DNA microarray mapping (Figure 3A,B). Previously, several other groups have attempted to map QTL for anther size. Eighteen QTL for anther traits were detected using five mapping populations, and the QTL sites of Chr1, Chr3, Chr5, Chr8, and Chr9 overlapped with the current study's locations [15]. It suggested that gene chip mapping was reliable for QTL preliminary mapping. There was no coincident QTL that was detected by all mapping populations in Uga et al., and it was similar to other research [14,16]. Therefore, we supposed that anther length was regulated by different minor QTL in different populations, and major QTL have yet to be touched. The *qAL5.2* site was close to the RM18569 marker, and a similar site was detected in multiple populations, including *O. longistaminata* (W1508), *O. rufipogon* (P16), and *indica* (IR24, T65, Aijiao Nante) [14–16]. *qAL5.2* should be a major QTL for anther length.

Although *qAL5* has been detected before, the candidate gene remains elusive. Based on the expression profile, there were eight predicted reading frames in the *qAL5.2* region, and only four genes showed expression in rice (Table 2). *ORF1* and *ORF3* were mainly expressed during seed development; *ORF4* and *ORF8* were mainly expressed in the panicle formation stage, and the expression of *ORF4* was also higher during the young seed stage (Table 2 and Figure S3). Only *ORF3* and *ORF8* had a nonsynonymous variation between HZ and KLY (Figure 7A), and *ORF8* encoded a lipase belonging to the alpha/beta-hydrolase (ABH) family, and have been reported in regulating plant development [17,18]. *SDP1* encodes lipase with a patatin-like acyl-hydrolase domain, mainly expressed in developing seeds, and *sdp1* exhibited a post-germinative growth arrest phenotype in Arabidopsis [19]. *RVMS* (Reversible Male Sterile) encodes a GDSL lipase/hydrolase protein predominantly expressed in anthers. The mutant of *rvms* is fertile at a low temperature (17 °C) but is male-sterile at normal temperature (24 °C), so lipase is also necessary for anther development [20,21]. And CSE, the Arabidopsis homolog protein of *ORF8*, has been reported to be involved in lignin synthesis, and *cse2* shows severe dwarfing and reduction in lignin content, so we thought *ORF8* was probably a candidate gene of *qAL5.2* [22,23].

The development of the anther is a complex process involving many regulatory pathways, such as the hormone pathway, phosphorylation pathway, and ubiquitination pathway [24,25]. In this research, we conducted an RNA-sequence experiment using an S4-stage anther of the near-isogenic line of *qAL5.2* to analyze possible regulatory pathways. The differentially expressed genes were highly enriched in the nucleic acid binding transcription and alpha-linolenic acid metabolism pathway, indicating that many enzymes or substrates related to lipid synthesis are transcribed. So we guess *qAL5.2* may be involved in regulating transcription factors and lipid metabolism in developing anthers.

In conclusion, we detected a new QTL and narrowed in a 73 kb region by fine mapping. At the same time, candidate genes of *qAL5.2* were analyzed, and RNA-sequence analysis showed that *qAL5.2* was involved in the regulation of the anther's energy metabolism.

4. Materials and Methods

4.1. Plant Materials and Growth Conditions

Oryza sativa L. ssp. *indica* Huazhan (HZ) was a recurrent parent. *Oryza sativa* L. ssp. *Japonica* Poliya (PLY) was the donor parent. F1 of a big anther was identified from the HZ/KLY population derived from the cross between HZ and KLY. The BC_4F_2 population was derived from F1 with the recurrent parent HZ for four consecutive generations. NIL-$qAL5.2^{HZ}$ and NIL-$qAL5.2^{KLY}$ were derived from BC_5F_3. All plants were grown in the nature field at Hangzhou, Zhejiang Province of China, from May to October and at Linshui, Hainan Province of China, from January to April.

4.2. Measurement of Traits

The agronomic traits, including plant height, anther length, grain length, grain width and 1000-grain weight, were measured with more than three replicates at the mature stage. The anther length, anther width, grain length, and grain width were measured by SC-G software (Wanshen Detection Technology Co., Ltd., Hangzhou, China). The phenotypic variance was calculated by SPSS 17.0 software.

4.3. Genotype Analysis

The genomic DNA of each plant was extracted by the CTAB method [26]. For PCR amplification, a 20 μL reaction mixture consisted of 0.2 μM primers, 10 μL 2 × Taq PCR MasterMix (LSC, Hangzhou, China), and ~15 ng of a genomic DNA sample. The PCR amplified profile was as follows: a pre-denaturation of 5 min at 94 °C, 30 cycles of 30 s at 94 °C, 30 s at 55 °C, and 30 s at 72 °C, and a final elongation at 72 °C for 5 min. Insertion and deletion markers were designed with NCBI and Primer5, and the primer sequences are listed in Table S2. The amplified products were electrophoresed in 3% agarose gel in 1× TAE buffer.

4.4. Genetic Mapping

The BC_4F_2 and BC_5F_1 population, individuals with extreme phenotypes of a big anther or small anther, were used for gene chip bulked segregation analysis by the company (Shuang Lv Yuang Bioinformatics Technology, Wuhan, China). Fine mapping was experimented with seven populations and nine markers by the method described previously. Briefly, genotype and phenotype analyses were conducted using Mapmaker/Exp 3.0, in which genetic distances between markers were presented in centiMorgans (cM) derived with the Kosambi function. QTL analysis was performed with the IM method by Windows QTL Cartographer 2.5, and an LOD value of 2.5 was taken as the threshold value [27].

4.5. Transcriptomics Analysis

About 1 g of the S4-stage (pollen microspore metaphase stage) anther of NIL-qAL5.2^{HZ} and NIL-qAL5.2^{KLY} was collected for Transcriptomics sequence. Different expression analyses, GO and KEGG, were performed by Tianjin Novogene Bioinformatics Technology. In brief, the PCR products were purified by AMPure XP system (Beckman Coulter, Pasadena, CA, USA), and the library quality was assessed using the Agilent Bioanalyzer 2100 system (Agilent, Santa Clara, CA, USA). After cluster generation, the library preparations were sequenced on the Illumina HiSeq platform (Illumina, San Diego, CA, USA) and 125 bp/150 bp paired-end reads were generated [28].

4.6. Subcellular Localization of qAL5

CDS of *ORF3* and *ORF8* without the stop codon were cloned into a transient expression vector PAN580-GFP to determine the subcellular localization. The fusions of GFP were transformed into protoplasts that were extracted from 15 d old HZ seedlings based on the $CaCl_2$-PEG4000 method [29]. An empty PAN580-GFP vector was the control.

Supplementary Materials: The following supporting information can be downloaded at https://www.mdpi.com/article/10.3390/plants13081130/s1, Figure S1: Development of advanced backcross lines; Figure S2. The LOD score plots by WinQTLCart. (A) The LOD score of LY2 population. (B) The LOD score of LY3 population; Figure S3. The expression profiles of the candidate genes in different tissues; Table S1: Anther length of populations; Table S2: Prime list used in this study; Table S3: Differential expression genes list.

Author Contributions: Conceptualization, J.Z., S.N. and X.L. (Xinwang Li); methodology, X.L. (Xinyong Liu), S.N. and Z.Y.; software, X.L. (Xinyong Liu) and Z.Y.; validation, X.L. (Xinyong Liu),

Z.Y. and L.C.; formal analysis, X.L. (Xinyong Liu) and Z.Y.; investigation, X.L. (Xinyong Liu) and Z.Y.; resources, J.H., Y.W., X.T., J.Y. and X.L. (Xinwang Li); data curation, X.L. (Xinyong Liu) and Z.Y.; writing—original draft preparation, X.L. (Xinyong Liu); writing—review and editing, J.Z. and J.Y.; visualization, X.L. (Xinyong Liu) and Z.Y.; supervision, J.Z. and S.N.; project administration, J.Z.; funding acquisition, J.Z. All authors have read and agreed to the published version of the manuscript.

Funding: This work was supported by National Key R&D Program of China (2022YFE0125600), Special support plan for high-level talents in Zhejiang (Grant No. 2022R52020), National Natural Science Foundation of China (Grant No. U22A20456), and Natural Science Foundation of Zhejiang (Grant No. LD24C130001).

Data Availability Statement: The data presented in this study are available in the article or the Supplementary Materials.

Conflicts of Interest: The authors declare no competing interests.

References

1. Qian, Q.; Zhang, F.; Xin, Y. Yuan Longping and Hybrid Rice Research. *Rice* **2021**, *14*, 101. [CrossRef]
2. Liao, C.; Yan, W.; Chen, Z.; Xie, G.; Deng, X.W.; Tang, X. Innovation and development of the third-generation hybrid rice technology. *Crop J.* **2021**, *9*, 693–701. [CrossRef]
3. Cheng, S.H.; Zhuang, J.Y.; Fan, Y.Y.; Du, J.H.; Cao, L.Y. Progress in Research and Development on Hybrid Rice: A Super-domesticate in China. *Ann. Bot.* **2007**, *100*, 959–966. [CrossRef]
4. Luo, D.; Xu, H.; Liu, Z.; Guo, J.; Li, H.; Chen, L.; Fang, C.; Zhang, Q.; Bai, M.; Yao, N.; et al. A detrimental mitochondrial-nuclear interaction causes cytoplasmic male sterility in rice. *Nat. Genet.* **2013**, *45*, 573–577. [CrossRef]
5. Tang, H.; Luo, D.; Zhou, D.; Zhang, Q.; Tian, D.; Zheng, X.; Chen, L.; Liu, Y.-G. The Rice Restorer Rf4 for Wild-Abortive Cytoplasmic Male Sterility Encodes a Mitochondrial-Localized PPR Protein that Functions in Reduction of WA352 Transcripts. *Mol. Plant* **2014**, *7*, 1497–1500. [CrossRef]
6. Kim, Y.-J.; Zhang, D. Molecular Control of Male Fertility for Crop Hybrid Breeding. *Trends Plant Sci.* **2018**, *23*, 53–65. [CrossRef]
7. Fujii, S.; Toriyama, K. Suppressed expression of RETROGRADE-REGULATED MALE STERILITY restores pollen fertility in cytoplasmic male sterile rice plants. *Proc. Natl. Acad. Sci. USA* **2009**, *106*, 9513–9518. [CrossRef]
8. Malik, T.U.H.; Baba, A. Techniques hybrid seed production in rice. *J. Pharmacogn. Phytochem.* **2018**, *7*, 962–967.
9. Li, J.; Luo, X.; Zhou, K. Research and development of hybrid rice in China. *Plant Breed.* **2023**, *143*, 96–104. [CrossRef]
10. Ying, J.; Qin, Y.; Zhang, F.; Duan, L.; Cheng, P.; Yin, M.; Wang, Y.; Tong, X.; Huang, J.; Li, Z.; et al. A weak allele of TGW5 enables greater seed propagation and efficient size-based seed sorting for hybrid rice production. *Plant Commun.* **2024**, *5*, 100811. [CrossRef]
11. Rubiyo; Satoto; Widyastuti, Y.; Indrawanto, C. Hybrid rice development in Indonesia: Constraints and opportunities. *E3S Web Conf.* **2021**, *306*, 01047. [CrossRef]
12. Tazib, T.; Kobayashi, Y.; Koyama, H.; Matsui, T. QTL analyses for anther length and dehiscence at flowering as traits for the tolerance of extreme temperatures in rice (*Oryza sativa* L.). *Euphytica* **2014**, *203*, 629–642. [CrossRef]
13. Khumto, S.; Pusadee, T.; Olsen, K.M.; Jamjod, S. Genetic relationships between anther and stigma traits revealed by QTL analysis in two rice advanced-generation backcross populations. *Euphytica* **2017**, *214*, 5. [CrossRef]
14. Ogami, T.; Yasui, H.; Yoshimura, A.; Yamagata, Y. Identification of Anther Length QTL and Construction of Chromosome Segment Substitution Lines of Oryza longistaminata. *Plants* **2019**, *8*, 388. [CrossRef]
15. Uga, Y.; Siangliw, M.; Nagamine, T.; Ohsawa, R.; Fujimura, T.; Fukuta, Y. Comparative mapping of QTLs determining glume, pistil and stamen sizes in cultivated rice (*Oryza sativa* L.). *Plant Breed.* **2010**, *129*, 657–669. [CrossRef]
16. Xiong, L.Z.; Liu, K.D.; Dai, X.K.; Xu, C.G.; Zhang, Q. Identification of genetic factors controlling domestication-related traits of rice using an F2 population of a cross between *Oryza sativa* and *O. rufipogon*. *Theor. Appl. Genet.* **1999**, *98*, 243–251. [CrossRef]
17. Shen, G.; Sun, W.; Chen, Z.; Shi, L.; Hong, J.; Shi, J. Plant GDSL Esterases/Lipases: Evolutionary, Physiological and Molecular Functions in Plant Development. *Plants* **2022**, *11*, 468. [CrossRef]
18. Nardini, M.; Dijkstra, B.W. Alpha/beta hydrolase fold enzymes: The family keeps growing. *Curr. Opin. Struct. Biol.* **1999**, *9*, 732–737. [CrossRef]
19. Eastmond, P.J. SUGAR-DEPENDENT1 Encodes a Patatin Domain Triacylglycerol Lipase That Initiates Storage Oil Breakdown in Germinating Arabidopsis Seeds. *Plant Cell* **2006**, *18*, 665–675. [CrossRef]
20. Zhu, J.; Lou, Y.; Shi, Q.-S.; Zhang, S.; Zhou, W.-T.; Yang, J.; Zhang, C.; Yao, X.-Z.; Xu, T.; Liu, J.-L.; et al. Slowing development restores the fertility of thermo-sensitive male-sterile plant lines. *Nat. Plants* **2020**, *6*, 360–367. [CrossRef]
21. Kim, H.U.; Hsieh, K.; Ratnayake, C.; Huang, A.H.C. A Novel Group of Oleosins Is Present Inside the Pollen of Arabidopsis. *J. Biol. Chem.* **2002**, *277*, 22677–22684. [CrossRef]
22. Ha, C.M.; Escamilla-Trevino, L.; Yarce, J.C.S.; Kim, H.; Ralph, J.; Chen, F.; Dixon, R.A. An essential role of caffeoyl shikimate esterase in monolignol biosynthesis in *Medicago truncatula*. *Plant J.* **2016**, *86*, 363–375. [CrossRef] [PubMed]

23. Barros, J.; Escamilla-Trevino, L.; Song, L.; Rao, X.; Serrani-Yarce, J.C.; Palacios, M.D.; Engle, N.; Choudhury, F.K.; Tschaplinski, T.J.; Venables, B.J.; et al. 4-Coumarate 3-hydroxylase in the lignin biosynthesis pathway is a cytosolic ascorbate peroxidase. *Nat. Commun.* **2019**, *10*, 1994. [CrossRef]
24. Aya, K.; Ueguchi-Tanaka, M.; Kondo, M.; Hamada, K.; Yano, K.; Nishimura, M.; Matsuoka, M. Gibberellin Modulates Anther Development in Rice via the Transcriptional Regulation of GAMYB. *Plant Cell* **2009**, *21*, 1453–1472. [CrossRef]
25. Marchant, D.B.; Walbot, V. Anther development—The long road to making pollen. *Plant Cell* **2022**, *34*, 4677–4695. [CrossRef] [PubMed]
26. Allen, G.C.; Flores-Vergara, M.A.; Krasynanski, S.; Kumar, S.; Thompson, W.F. A modified protocol for rapid DNA isolation from plant tissues using cetyltrimethylammonium bromide. *Nat. Protoc.* **2006**, *1*, 2320–2325. [CrossRef]
27. Cheng, Y.C.; Li, G.; Yin, M.; Adegoke, T.V.; Wang, Y.F.; Tong, X.H.; Zhang, J.; Ying, J.Z. Verification and dissection of one quantitative trait locus for grain size and weight on chromosome 1 in rice. *Sci. Rep.* **2021**, *11*, 18252. [CrossRef]
28. Zhou, Q.; Hu, X. Systemic Stress and Recovery Patterns of Rice Roots in Response to Graphene Oxide Nanosheets. *Environ. Sci. Technol.* **2017**, *51*, 2022–2030. [CrossRef]
29. Li, Z.; Wei, X.; Tong, X.; Zhao, J.; Liu, X.; Wang, H.; Tang, L.; Shu, Y.; Li, G.; Wang, Y.; et al. The OsNAC23-Tre6P-SnRK1a feed-forward loop regulates sugar homeostasis and grain yield in rice. *Mol. Plant* **2022**, *15*, 706–722. [CrossRef]

plants MDPI

Article

Dissecting the Genetic Basis of Yield Traits and Validation of a Novel Quantitative Trait Locus for Grain Width and Weight in Rice

Man Yin, Xiaohong Tong, Jinyu Yang, Yichen Cheng, Panpan Zhou, Guan Li, Yifeng Wang and Jiezheng Ying *

State Key Laboratory of Rice Biology and Breeding, China National Rice Research Institute, Hangzhou 311400, China; 13545663721@163.com (M.Y.); jinyu.yang@syngentagroup.cn (J.Y.); 13872209530@163.com (Y.C.); zhoupanpan202206@163.com (P.Z.); li@ricescience.org (G.L.); wangyifeng@caas.cn (Y.W.)
* Correspondence: yingjiezheng@caas.cn; Tel.: +86-571-6337-0206

Abstract: Grain yield in rice is a complex trait and it is controlled by a number of quantitative trait loci (QTL). To dissect the genetic basis of rice yield, QTL analysis for nine yield traits was performed using an F_2 population containing 190 plants, which was developed from a cross between Youyidao (YYD) and Sanfenhe (SFH), and each plant in the population evaluated with respect to nine yield traits. In this study, the correlations among the nine yield traits were analyzed. The grain yield per plant positively correlated with six yield traits, except for grain length and grain width, and showed the highest correlation coefficient of 0.98 with the number of filled grains per plant. A genetic map containing 133 DNA markers was constructed and it spanned 1831.7 cM throughout 12 chromosomes. A total of 36 QTLs for the yield traits were detected on nine chromosomes, except for the remaining chromosomes 5, 8, and 9. The phenotypic variation was explained by a single QTL that ranged from 6.19% to 36.01%. Furthermore, a major QTL for grain width and weight, *qGW2-1*, was confirmed to be newly identified and was narrowed down to a relatively smaller interval of about ~2.94-Mb. Collectively, we detected a total of 36 QTLs for yield traits and a major QTL, *qGW2-1*, was confirmed to control grain weight and width, which laid the foundation for further map-based cloning and molecular design breeding in rice.

Keywords: rice; yield trait; grain width; grain weight; quantitative trait locus

Citation: Yin, M.; Tong, X.; Yang, J.; Cheng, Y.; Zhou, P.; Li, G.; Wang, Y.; Ying, J. Dissecting the Genetic Basis of Yield Traits and Validation of a Novel Quantitative Trait Locus for Grain Width and Weight in Rice. *Plants* **2024**, *13*, 770. https://doi.org/10.3390/plants13060770

Academic Editor: Igor G. Loskutov

Received: 28 December 2023
Revised: 29 February 2024
Accepted: 6 March 2024
Published: 8 March 2024

1. Introduction

Adequate food is extremely vital for people all over the world, and it can satisfy people's basic sense of security, achieve social stability, and ensure national security. The agricultural crisis, especially the grain price crisis from 2008 to 2011, triggered global unrest [1]. In 2017, approximately 11% of the world's population suffered from hunger, and it is expected that by 2050, climate change will put an additional 77 million people at risk of hunger [2]. In 2019 alone, approximately 340 million children worldwide suffered from micronutrient deficiency, and 144 million children under the age of 5 had developmental delays (http://www.fao.org/worldfoodsituation/csdb/en/, accessed on 21 February 2024). In the face of natural disasters, including rising global temperatures and unstable climate events, cultivating crops with high yield potential is crucial for ensuring food security [3]. Rice is one of the most important cereal crops and serves as a stable food for more than half the world's population [4]. Rice is grown in over 100 countries around the world [5], and increasing rice production can eliminate hunger and ensure food security. However, rice production faces many challenges. The increase in carbon dioxide concentration in the air will reduce the nutritional quality levels of B vitamins, proteins, iron, and zinc in rice [6]. In rice variety improvement, it is often found that one yield trait has improved while another yield trait has deteriorated, resulting in the phenomenon of trade-offs [7,8]. To enhance rice production, more attention is directed to deeply exploring the genetic mechanisms of rice yield.

As a complex agronomic trait, grain yield per plant (GYPP) is mainly determined by three key components: number of panicles per plant (NPPP), number of filled grains per panicle (NFGPP), and grain weight [9]. The three component traits and other related yield traits are typical quantitative traits and are controlled by quantitative trait loci (QTL). During the last 30 years, large numbers of QTLs for the yield traits have been identified by the approaches of classical QTL mapping and genome-wide association analysis (GWAS) with biparental populations and the natural populations, respectively. Biparental genetic populations have been widely used in the identification of QTLs for rice yield traits with PCR-based or genome-resequenced linkage maps. In comparison with biparental genetic populations, natural populations with a large scale of rice varieties contain more genetic diversity, and their corresponding GWAS can improve map resolution. However, rare alleles are difficult to identify in GWAS studies. Plentiful recombination events and allele differences play an important role in detecting the QTLs underlying rice yield traits. In recent decades, great advances in rice functional genomics have provided the foundation to increase rice production. The elucidation of the genetic mechanisms underlying each yield component will certainly benefit the improvement of rice variety.

The NPPP is closely related to tillering ability and is easily affected by environment, with low heritability and regulation by multiple genes [10]. Tillering is a fundamental trait in rice that ensures the number of panicles and sink size [11]. Rice tillers include primary, secondary, and tertiary tillers, of which the secondary and tertiary tillers in the late tillering stage contribute little to yield. Rice tillering involves two processes: one is the formation of the axillary meristem, and the other is the outgrowth of the tiller bud [12]. *MOC1*, *MOC2*, and *MOC3* are three vital genes related to rice tillering, and they affect the growth of rice tiller buds and the formation of axillary meristems, respectively. The *MOC1* plant mutant almost completely loses its tillering ability and only produces one main culm, while the wild-type plant has multiple tillers. *MOC1* encodes a putative GRAS family nuclear protein, which is mainly expressed in axillary buds and functions to initiate axillary bud formation and promote its outgrowth [13]. *MOC2* mutant plants show a significant decrease in tiller number, reduced growth rate, and dwarfing characteristics. The *MOC2* mutation leads to a deficiency in FBPase activity, resulting in an insufficient supply of sucrose and possibly ultimately inhibiting the growth of tillering buds [14]. *MOC3* mutant plants only have a monoculm and are sterile. Morphological and histological studies have shown that the disruption of tillering bud formation in *MOC3* leads to a monoculm phenotype in the plant [15]. Outgrowth of the rice tiller bud is mainly regulated by phytohormones. As one of the direct factors affecting rice yield, the NPPP has been widely studied in recent years. In a previous study, a set of rice populations were used to genotype 700,000 single nucleotide polymorphisms (SNPs). A total of 15 novel QTL loci related to rice tillering were identified by the GWAS. The DNA sequences of the above 15 novel QTL loci were analyzed using extreme tiller number phenotypic accessions, and a total of five candidate genes were identified [16].

The NFGPP, directly determining grain yield, is affected by the number of primary branches (NPB) and the number of secondary branches (NSB) [17]. There are extensive variations in the NFGPP and its correlation among different accessions. A large number of QTL mappings have been carried out by using biparental genetic populations [18–20]. The QTLs controlling the NFGPP are distributed on 12 chromosomes, with the most on chromosome 1, and more than 180 QTLs have been located by incomplete statistics (www.grarene.org, accessed on 1 April 2023). Abundant QTLs controlling the NFGPP and its directly related panicle traits have been identified by GWAS [21,22]. *Gn1a*, encoding the enzyme osckx2 that degrades cytokinin, was the first QTL to be cloned in rice to control the NFGPP. *Gn1a* delected leads to the accumulation of cytokinin and increases the NFGPP [23]. *GNP1* encodes an essential enzyme GA20 oxidase in the gibberellin synthesis pathway, which affects the NFGPP by affecting the number of secondary branches [24].

Grain weight is generally estimated by 1000-grain weight (TGW), which is mainly related to grain size and the filling process [25]. The heritability of grain size is relatively high, and so is affected by genetic factors, while the filling degree is easily influenced by the environment [26]. Grain size is composed of grain length (GL), grain width (GW), and grain thickness (GT), of which the first two are essential factors affecting yield [27,28]. Up to now, 21 QTLs controlling rice grain weight and grain size have been cloned and distributed on 8 chromosomes, except for chromosomes 1, 10, 11, and 12. Except for *GLW7* [28] and *GW5* [29] which were identified by GWAS, *OsLG3* [30] was identified by linkage analysis and association analysis, and the remaining eighteen QTLs were all identified by map-based cloning. Among these QTLs, 13 were found to mainly control GL and GW, including *GS2* [31] on chromosome 2; *GS3* [32], *SG3* [33], *OsLG3* [30], *qLG3* [34], *GL3.1* [35], *TGW3* [36] and *GSA1* [37] located on chromosome 3; *GL4* [38] on chromosome 4; *TGW6* [39], *GW6a* [40], and *GL6* [41] located on chromosome 6; and *GLW7* [28] on chromosome 7. *GL7* [42] on chromosome 7 and *GS9* [43] on chromosome 9 only control grain size, without affecting grain weight. The remaining six QTLs mainly control GW and TGW, including *GW2* [44] and *TGW2* [45] on chromosome 2; *GS5* [46] and *GW5* [29] on chromosome 5; and *GW6* [47] and *GW8* [48] on chromosomes 6 and 8, respectively.

In this study, the correlations among nine yield traits, GYPP, NPPP, number of total grains per plant (TNSP), NFGPP, number of grains per panicle (NSPP), seed-setting rate (SSR), GL, GW, and TGW, were analyzed, and a linkage map covering the whole genome was constructed through genotyping each plant with the selected polymorphic markers using a YYD/SFH F_2 population. Subsequently, QTLs controlling yield traits were identified using the combined information of genotype and phenotype. Furthermore, a QTL for GW and TGW, *qGW2-1*, was validated and narrowed down to smaller marker intervals with a set of near-isogenic line (NIL) population and the populations derived from the residual heterozygotes.

2. Results

2.1. Phenotypic Variation

To dissect the genetic basis of rice yield, two indica varieties, Youyidao (YYD) and Sanfenhe (SFH), that showed significant differences inn yield traits were selected as parents to develop an F_2 population (Figure 1a,b).

Figure 1. Comparison of grain size and plant type characters of YYD and SFH. (**a**) Grain phenotypes of two rice parents, YYD and SFH, bar: 5.00 mm. (**b**) Plant phenotypes of two rice parents, YYD and SFH, bar: 25 cm.

Nine yield traits were investigated in the parental varieties and the corresponding F_2 population. The female parent YYD showed 284.6%, 512.2%, 611.6%, and 62.7% more NPPP, TNSP, NFGPP, and NSPP; as well as 17.54% and 264.8% higher SSR and GYPP than the male parent SFH. However, the grains of YYD were 31.7% shorter, 21.0% narrower, and 48.5% lighter than those of SFH (Figure 2).

In the F_2 population, we found that wide variations existed in the phenotypic distribution of the nine yield traits (Figure 3). Significant transgressive segregations were observed in the distribution of the NFGPP, NSPP, SSR, and GYPP. All the traits showed continuous segregation, suggesting that they were controlled by multiple genes and were suitable for QTL analysis.

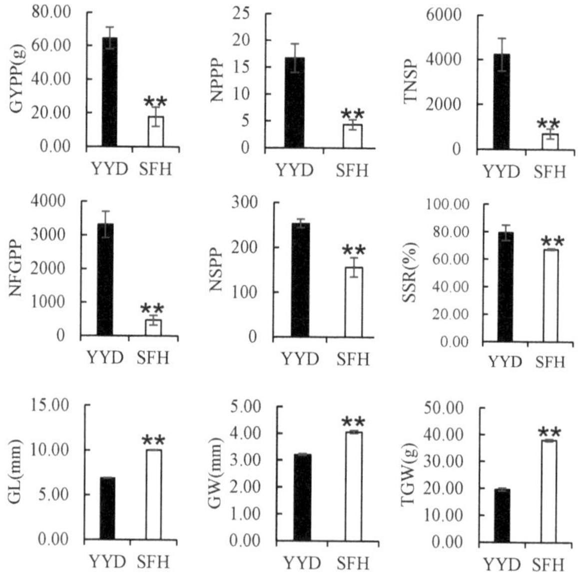

Figure 2. Comparison of 9 yield traits of YYD and SFH. Significant α = 0.01 **.

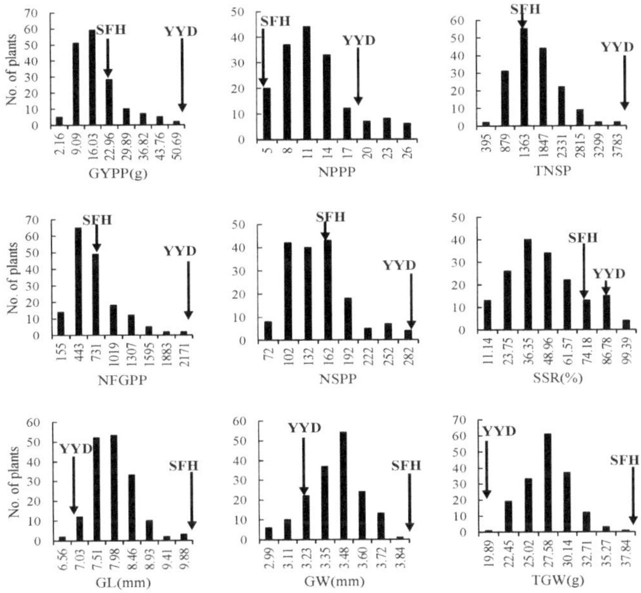

Figure 3. Frequency distribution of yield trait in the YYD/SFH F_2 population. GYPP, grain yield per plant; NPPP, number of panicles per plant; TNSP, total number of spikelets per plant; NFGPP, number of filled grains per plant; NSPP, number of spikelets per panicle; SSR, seed-setting rate; GL, grain length; GW, grain width, and TGW, 1000-grain weight. Trait values of the parental materials are indicated by black arrows.

2.2. Correlations between Yield Traits

The correlation coefficients between yield traits are exhibited in Figure 4. GYPP positively correlated with all six of the yield traits except for the grain size traits. No correlation was observed between GYPP, GL, and GW. NPPP showed positive correlations with TNSP, NFGPP, GYPP, and negative correlations with NSPP, SSR, and GW. For the relationship among the four grain/spikelet number traits including TNSP, NFGPP, NSPP, and SSR, significantly positive correlations were observed among all the traits, except that TNSP showed a negative correlation with SSR. For the relationships among the three grain size/weight traits including GL, GW, and TGW, significantly positive correlations were found among them. Negative or no correlations were observed between the three grain size/weight traits and the four grain/spikelet number traits, except that GW showed a slight positive correlation with SSR.

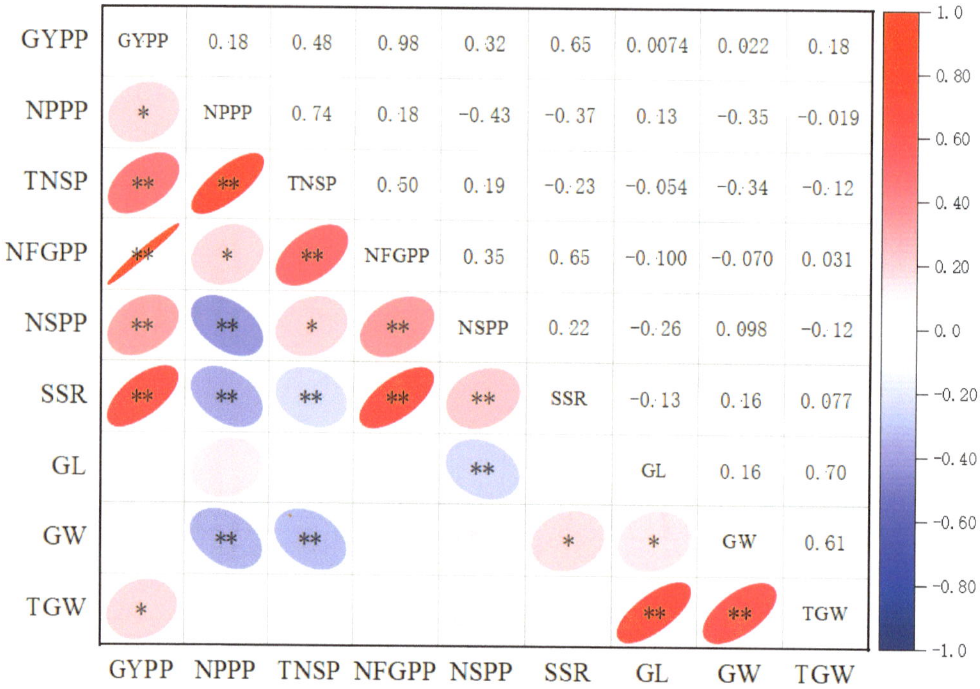

Figure 4. Correlation coefficients among 9 traits in the F_2 population. The upper panel contains correlation coefficients, and the lower panel contains significance analysis. * and ** represent significance levels at 0.05 and 0.01, respectively.

2.3. Creation of a Linkage Map

Based on genotyping 190 individuals with 133 DNA markers (82 SSR, 51 InDel), a linkage map was constructed with 12 linkage groups corresponding to 12 rice chromosomes (Figure 5). The linkage maps spanned 1831.7 cM. The genetic distances between neighboring markers ranged from 0.5 to 36.4 cM, with an average interval of 15.1 cM.

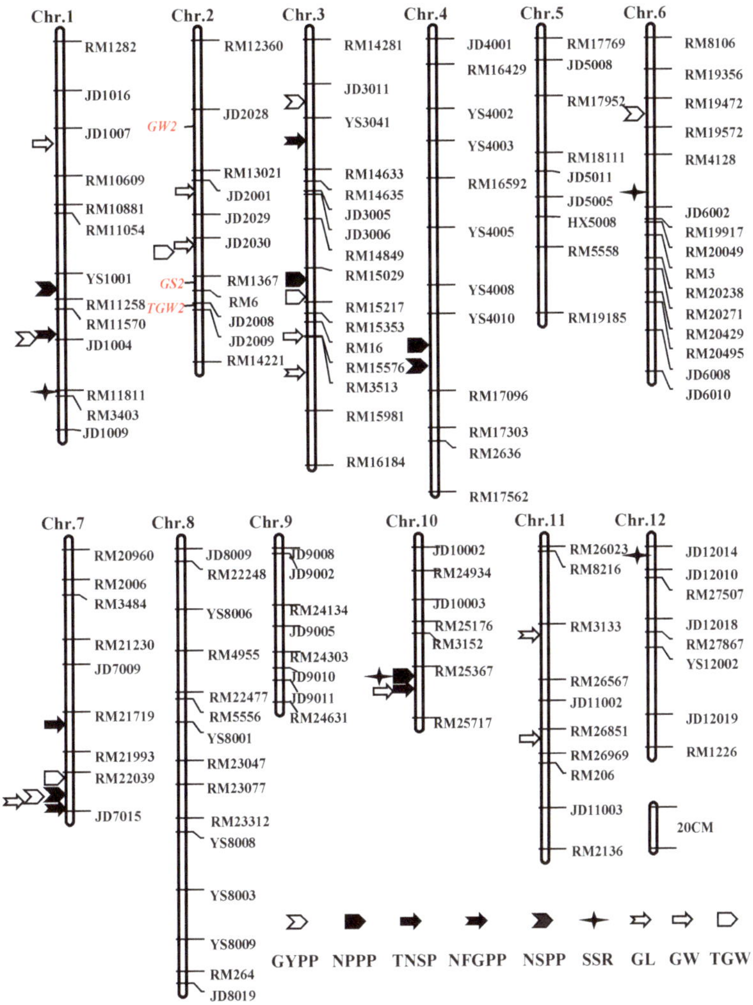

Figure 5. Genetic linkage map showing QTL positions detected in the F_2 population. GYPP, grain yield per plant; NPPP, number of panicles per plant; TNSP, total number of spikelets per plant; NFGPP, number of filled grains per plant; NSPP, number of spikelets per panicle; SSR, seed-setting rate; GL, grain length; GW, grain width, and TGW, 1000-grain weight.

2.4. QTLs for Rice Yield Traits

2.4.1. QTLs for GYPP and NPPP

Using the YYD/SFH F_2 population, three QTLs for GYPP with relatively small effects were identified on chromosome 3, and the LOD peak values ranged from 4.34 to 5.86 (Figure 5, Table 1). All the three QTLs explained 31.34% of the GYPP's variations, with the additive effects for enhancing the GYPP contributed by the YYD allele. No major QTL for the GYPP were detected. Four QTLs controlling NPPP were mapped on chromosomes 3, 4, 10, and 12. These QTLs collectively accounted for 47.01% of the total phenotypic variation. One major QTL with a peak LOD value of 7.61, *qNPPP3*, was mapped on chromosome 3 and was responsible for 17.85% of the NPPP's variation. The additive and dominance effects showed opposite directions. Although the male parent SFH only had 3 to 5 tillers, the SFH allele at *qNPPP3* increased the NPPP with a positive additive effect of 5.5 and showed a negative dominant effect of −4.1.

Table 1. QTLs for yield traits detected in the F_2 population.

Traits	QTL	Chr.	Marker Interval	Position	LOD	A	D	R^2 (%)
GYPP	*qGYPP3*	3	JD3011-YS3041	22.34	5.86	−5.37	1.12	12.35
	qGYPP6	6	RM19472-RM19572	30.43	4.43	−4.29	−1.11	8.95
	qGYPP7	7	RM22039-JD7015	111.44	4.34	−3.83	−3.11	10.04
NPPP	*qNPPP3*	3	RM14849-RM15029	89.92	7.61	5.5	−4.1	17.85
	qNPPP4	4	RM17096-RM17303	176.06	5.83	−3.0	0.3	10.82
	qNPPP10	10	RM25367-RM25717	59.62	5.35	−3.3	−0.3	10.76
	qNPPP12	12	JD12010-RM27507	11.64	3.76	1.6	2.0	7.58
TNSP	*qTNSP1*	1	JD1016-JD1007	24.55	3.37	−243.3	27.6	6.19
	qTNSP3	3	RM15029-RM15217	114.19	3.61	492.0	−462.3	11.30
	qTNSP7	7	RM21719-RM21993	80.99	4.18	−300.3	17.1	8.56
	qTNSP10	10	RM25367-RM25717	59.70	3.75	−330.9	−170.3	13.15
	qTNSP12	12	JD12010-RM27507	11.64	4.05	341.9	−5.8	8.30
NFGPP	*qNFGPP3*	3	JD3011-YS3041	22.34	5.44	−200.6	46.1	11.36
	qNFGPP6	6	RM19472-RM19572	30.43	3.86	−146.1	−53.7	7.71
	qNFGPP7	7	RM22039-JD7015	111.44	6.19	−179.1	−148.1	16.54
	qNFGPP12	12	YS12002-JD12019	50.69	3.20	7.9	218.0	8.16
NSPP	*qNSPP1*	1	RM11570-JD1004	133.83	5.03	−18.8	−15.4	10.40
	qNSPP3	3	RM15029-RM15217	114.15	4.47	−26.7	16.3	7.19
	qNSPP4	4	RM17096-RM17303	176.00	3.39	16.9	−5.4	6.76
	qNSPP7	7	RM22039-JD7015	111.44	11.92	−31.0	−11.4	25.11
SSR	*qSSR1*	1	RM11811-RM3403	174.36	4.55	−8.82	−5.22	8.70
	qSSR3	3	JD3011-YS3041	22.32	3.86	−10.07	−1.27	9.16
	qSSR6	6	RM19472-RM19572	30.53	5.00	−5.18	−12.31	11.88
	qSSR10	10	RM25367-RM25717	59.66	4.00	13.88	−5.66	10.64
	qSSR12	12	JD12014-JD12010	0.04	3.35	−0.84	−13.08	8.08
GL	*qGL3*	3	RM3513-RM15981	148.73	9.40	0.50	−0.41	36.01
	qGL7	7	RM22039-JD7015	111.46	3.75	0.20	0.18	9.71
	qGL11	11	RM3133-RM26567	38.72	3.49	0.28	0.05	12.06
GW	*qGW1*	1	JD1007-RM10609	43.43	3.47	−0.01	0.10	8.26
	qGW2-1	2	JD2001-JD2029	70.21	9.02	0.12	0.02	23.58
	qGW2-2	2	JD2030-RM1367	99.12	9.85	0.13	−0.01	25.89
	qGW3	3	RM16-RM15576	140.87	4.34	0.11	−0.04	10.81
	qGW10	10	RM25367-RM25717	59.69	3.89	0.06	0.10	17.81
TGW	*qTGW2*	2	JD2030-RM1367	99.14	4.40	1.62	−0.43	11.51
	qTGW3	3	RM15029-RM15353	114.32	8.01	2.68	−0.73	20.67
	qTGW7	7	RM22039-JD7015	111.32	3.94	0.97	1.35	10.07

A: additive effect, positive addictive effect means SFH allele increasing trait values; D, dominance effect; R^2, proportion of phenotype variance explained by the QTL. GYPP, grain yield per plant; NPPP, number of panicles per plant; TNSP, total number of spikelets per plant; NFGPP, number of filled grains per plant; NSPP, number of spikelets per panicle; SSR, seed-setting rate; GL, grain length; GW, grain width and TGW, 1000-grain weight.

2.4.2. QTLs for Grain Number Traits

Four grain number traits including TNSP, NFGPP, NSPP, and SSR were evaluated to perform QTL analysis in this study (Figure 5, Table 1). For TNSP, five minor QTLs were detected on chromosomes 1, 3, 7, 10, and 12, and accounted for 6.19% to 13.15% of the phenotypic variation. For NFGPP, four QTLs were identified on chromosomes 3, 6, 7, and 12, with the LOD peak values ranging from 3.20 to 6.19. Together, these QTLs were responsible for 43.77% of the total phenotypic variation. One major QTL located on chromosome 7, *qNFGPP7*, scored a peak LOD value of 6.19 and showed the additive effect of −179.1 with the YYD allele increasing the filled grain number. For NSPP, four QTLs were mapped on chromosomes 1, 3, 4, and 7, and showed peak LOD values ranging from 3.39 to 11.92. All four of the QTLs collectively explained 49.96% of the observed phenotypic variation. Noticeably, one major QTL for NSPP, *qNSPP7*, was co-located with *qNFGPP7* in the same marker interval RM22039-JD7015 and showed an additive effect of −31.0 and a dominant effect of −11.4. Similarly, the female parent YYD allele increased the spikelets. Five minor QTLs regulating the SSR were identified on chromosomes 1, 3, 6, 10, and 12, respectively. These QTLs together accounted for 48.46% of the SSR's variation and scored peak LOD values from 3.35 to 5.00.

2.4.3. QTLs for Grain Size and Weight

Three grain size and weight traits, including the GL, GW, and TGW, were measured to perform QTL analysis in the YYD/SFH F_2 population (Figure 5, Table 1). For the GL, three QTLs were detected on chromosomes 3, 7, and 11, which collectively accounted for 57.78% of phenotypic variation. One major QTL with an LOD value of 9.40, *qGL3*, was detected on chromosome 3 and was responsible for 36.01% of the total phenotypic variation, with the increasing GL contributed by the SFH allele. For GW, five QTLs were identified on chromosomes 1, 2, 3, and 10, which explained 81.23% of the phenotypic variation. On chromosome 2, two major QTLs with LOD values of 9.02 and 9.85, *qGW2-1* and *qGW2-2*, were identified to be responsible for 23.58% and 25.89% of the GW's variations. For TGW, three QTLs were detected on chromosomes 2, 3, and 7, which explained 11.51%, 20.67%, and 10.07% of the total phenotypic variation, respectively. One major QTL, *qTGW3*, was identified in the marker interval RM15029-RM15353. The SFH allele underlying *qTGW3* showed an additive effect and increased the TGW.

2.5. Validation of *qGW2-1*

Three RH-derived populations in F_7 with sequential heterozygous segments covering the target interval JD2001-JD2029, named YM1, YM2, and YM3, were developed to confirm the genetic effect and location of *qGW2-1*. The GL, GW, and TGW were continuously distributed in the three F_7 populations (Figure S1). Three segmental linkage maps were constructed for the YM1, YM2, and YM3 populations (Figure 6). Based on the combination of genotype and phenotype information of each individual in the three populations, *qGW2-1* for the GW and TGW were identified in each population. No significant QTL effects for the GL were found in the three populations. The enhancing alleles for the GW and TGW contributed to the positive additive effects and were derived from the male parent SFH in all the three populations (Table 2). The variations explained in the GW were 31.54%, 28.28%, and 45.79% in the YM1, YM2, and YM3 populations. Based on the above results, we concluded that *qGW2-1* was located at the common segregating interval YS2005-YS2006 of the three populations, and the two flanking cross-over intervals YS2004-YS2005 in YM2 and YS2006-YS2027 in YM1. Therefore, *qGW2-1* was delimitated to the marker interval YS2004-YS2027, that is, it corresponded to ~2.94-Mb region in the Nipponbare genome (Figure 6).

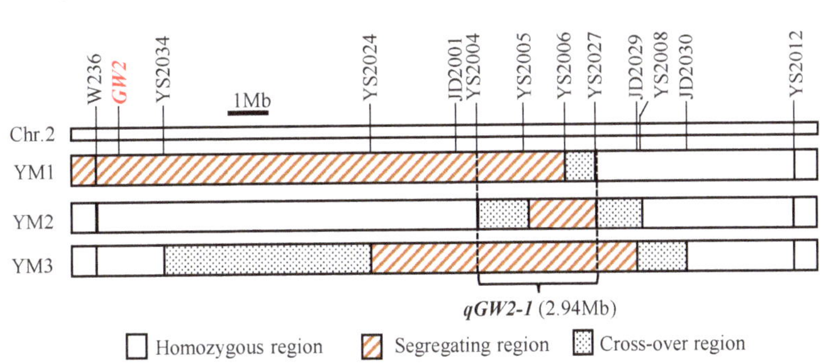

Figure 6. Genotype compositions of three F_2 populations. Bar: 1 Mb.

Table 2. QTLs of grain size in three F_7 populations.

Population	Trait	Interval	Sample	LOD	A	D	R^2 (%)
YM1	GL	W236-YS2006	128	ns	ns	ns	ns
	GW	W236-YS2006	128	8.14	0.09	−0.01	31.54
	TGW	W236-YS2006	128	4.10	1.21	−0.55	19.07
YM2	GL	YS2024-JD2029	140	ns	ns	ns	ns
	GW	YS2024-JD2029	140	13.07	0.05	−0.01	45.79
	TGW	YS2024-JD2029	140	3.56	0.57	−0.04	14.70
YM3	GL	W236-YS2010	179	ns	ns	ns	ns
	GW	W236-YS2010	179	3.33	0.06	0.01	11.17
	TGW	W236-YS2010	179	4.24	0.97	−0.18	17.77

A: additive effect, positive addictive effect means SFH allele increasing trait values; D, dominance effect; R^2, proportion of phenotype variance explained by the QTL. ns, no significance.

For further validation of *qGW2-1*, one set of NIL plants in F_8 that were produced through selecting from the YM3 population were planted during the growth season in Hangzhou. The phenotypic values of GL, GW, and TGW showed continuous distributions (Figure 7a). Obvious differentiations in the GW and TGW between the YYD and SFH homozygous genotypes were observed in the NIL population (Figure 7a–d). The YYD and SFH homozygous lines were in the low- and high-value areas of the GW and TGW, which suggests that *qGW2-1* is responsible for GW and TGW (Figure 7a).

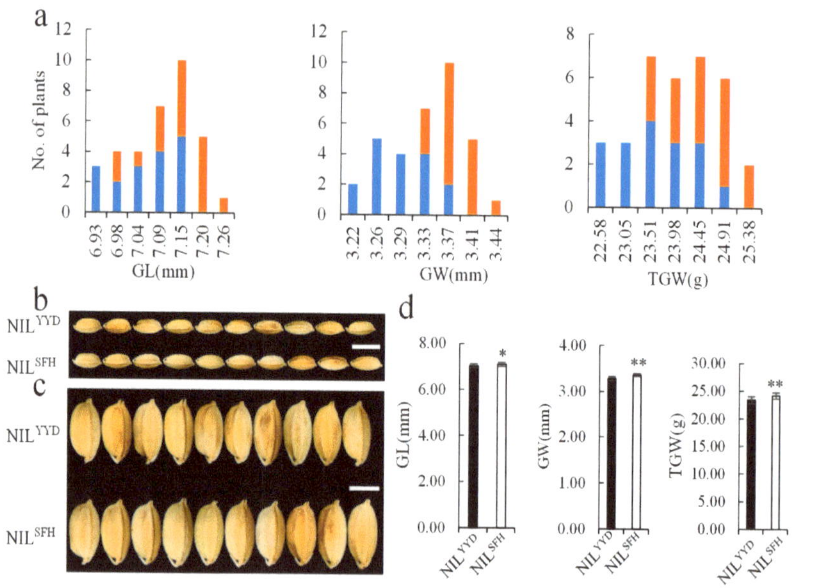

Figure 7. Comparative analysis of grain traits between NIL^{YYD} and NIL^{SFH} for *qGW2-1*. (**a**) Distributions of GL, GW, and TGW in the NIL population. Blue column: the plants carrying the homozygous YYD allele; orange column: the plants carrying the homozygous SFH alleles. (**b**) GL, scale bar: 10 mm. (**c**) GW, scale bar: 5 mm. (**d**) Phenotypes of NIL^{YYD} and NIL^{SFH} in GL, GW, and TGW. * and ** represent significance levels at 0.05 and 0.01, respectively.

3. Discussion

3.1. QTLs Dissect the Genetic Basis of Rice Yield Traits

To dissect the genetic basis of rice yield traits, we designed an F_2 population that was derived from a cross of YYD and SFH with contrasting yield traits. As we expected, wide variations in the nine yield traits were observed in the population. After genotyping, phenotyping, and QTL mapping, we created a linkage map covering the rice genome and identified a total of 36 QTLs for the nine yield traits. The identified QTLs were distributed throughout the whole genome, except for chromosomes 5, 8, and 9 and explained from 6.19% to 36.01% of the total phenotypic variation. Except for the GYPP, TNSP, and SSR, one or two major QTLs that explained > 15% of the phenotypic variation were identified for the remaining six yield traits. For each yield trait, the number of QTLs that were detected in the YYD/SFH F_2 population ranged from three to five. Similar to the earlier studies, the smallest number of QTLs was observed for the most complex trait, the GYPP, and no QTL could explain > 15% of the phenotypic variation. Through QTL analysis, it was found that there are QTLs that control different traits within the same molecular marker interval. In the molecular marker interval RM22039-JD7015 on chromosome 7, there are QTLs that control the GYPP, NFGPP, TNSP, GL, and TGW, respectively. It is believed that there may be a single factor for multiple effects or closely linked QTLs within a molecular marker interval.

3.2. Construction of Genetic Linkage Maps Is Important to Identify QTLs

Genetic linkage maps are the basis of QTL mapping, and appropriate molecular marker density is the key to the accuracy of QTL mapping. In this study, we constructed a low-density PCR-based genetic map containing 133 markers, of which the distance between some marker intervals was large, which may have caused some minor QTLs not to be detected. In order to confirm the reliability of the results of this map, we compared the QTL of yield-related traits found in this study with the previous mapping results. In this study, *qTNSP12* for the TNSP was identified in similar regions to the previous *qnspp12.1* and *qfgn12.1* controlling the TNSP and NFGPP, respectively [49]. *qGYPP3* and *qNFGPP3* for the GYPP and NFGPP were located in the intervals consistent with *qGY_S-3-1* controlling the GYPP and *qFG_S-3-1* controlling the NFGPP, which were detected by a high-density genetic map, respectively [50]. In our study, among the 11 QTLs for grain size and grain weight, *qGL3*, *qGW2-2*, and *qTGW3* were found adjacent to the cloned genes *GSA1* [37], *GS2* [31], and *SG3* [33], respectively.

3.3. Complex Correlations among Rice Yield and Yield Traits

Complex correlations were observed among rice yield and yield traits. Rice yield could be enhanced via the improvement of yield traits since the GYPP positively correlated with the yield traits (Figure 4). To improve the efficiency of rice breeding, elucidation of the relationship between yield and yield traits is very important and has been investigated in many studies [51,52]. Similar to our result, the GYPP was also found to positively correlate with the NPPP, NSPP, NFGPP, TGW, and SSR in the Pusa1266/Jaya recombinant inbred line population [53]. In the rice varieties released in China from 1978 to 2017, rice yield positively correlated with the NFGPP, TGW, NSPP, and SSR in the indica ecotype, whereas it positively correlated with panicle number per unit area in the japonica ecotype [26]. Previous studies have found that the colocation of QTLs responsible for different traits resulted in complex correlations [54–56]. In the present study, eight marker intervals each had at least two QTLs. All three of the QTLs for the GYPP, *qGYPP3*, *qGYPP6*, and *qGYPP7*, were all co-localized in the same genomic regions with the three QTLs for the NFGPP, *qNFGPP3*, *qNFGPP6*, and *qNFGPP7*, and their additive effects acted in the same direction, which might have resulted in the significant positive correlation between the GYPP and NFGPP, with the highest coefficient being 0.98.

3.4. *qGW2-1* Is Confirmed to Be a Novel QTL for Grain Width and Weight

The genetic background of the population used for the initial mapping is generally complex, which may cause certain deviations in QTL interval. Before conducting QTL fine mapping and marker-assisted selection in variety breeding, it is necessary to verify the genetic effects and determine the region of the target QTL. In this study, one major QTL for GW and TGW, *qGW2-1*, was detected on chromosome 2, and was further validated and delimited to the target interval YS2004-YS2027 (~2.94 Mb). For grain size and weight, three major QTLs, including *GW2*, *GS2* and *TGW2*, were map-based cloned, and many minor QTLs were detected on chromosome 2. *GW2* negatively regulates GW and encodes a RING-type protein with E3 ubiquitin ligase activity [44]. *TGW2*, a semi-dominant QTL for GW and TGW, was mapped on the bottom of chromosome 2 and encodes CELL NUMBER REGULATOR 1 [45]. Another semi-dominant QTL, *GS2/OsGRF4*, encodes Growth-Regulation Factor 4 and increases GL, GW, and TGW through elevating its expression [31]. Although *GW2*, *GS2*, and *TGW2* are all located on chromosome 2, their locations are completely different from *qGW2-1*. Furthermore, many QTLs controlling grain size and weight are mapped on chromosome 2 and most of them are collected in the Gramene and Q-TARO databases (http://qtaro.abr.affrc.go.jp and http://qtaro.abr.affrc.go.jp, accessed on 8 April 2023). After comparing the QTL locations, no QTL for GW and TGW showed any overlapping interval with *qGW2-1*. Therefore, *qGW2-1* which was detected in this study is a novel QTL for GW and TGW.

4. Materials and Methods

4.1. Plant Materials and Field Trials

Two indica varieties that show very significant differences in rice yield traits, including NPPP, NGPP, NSPP, SSR, GYPP, grain size and weight, YYD, and SFH, were selected to construct an F_2 mapping population containing 190 individuals (Figure 1a). Three plants with heterozygous chromosomal segments covering *qGW2-1* were selected and self-crossed three times to produce three F_7 populations by marker screening.

The field trials were carried out in two locations. We planted 190 plants of the F_2 population and the two parents in Lingshui (110.0° E, 18.5° N) in Hainan Province, China. Three F_7 populations and the parents, with each population containing of 190 plants, were grown in a paddy field in Hangzhou (120.2° E, 30.3° N) in Zhejiang Province, China. All the rice materials in this study were planted with a planting density of 16.7 cm $\times$ 26.5 cm during the rice-growing season. All kinds of field management were performed as in normal agriculture practice.

4.2. Phenotypic Measurement

After maturity, the panicles of each plant in the populations were harvested, respectively. The NPPP was counted manually. All the grains were threshed from the panicle, and the unfilled grains were separated from the filled grains and calculated artificiality. The filled grains were sun-dried and stored at room temperature for about three months. Then, the NFGPP and GYPP were measured by SC-A seed counting and grain weighting device (Wanshen Ltd., Hangzhou, China). Fully filled grains were separated by 3.5 mol/L NaCl solution and dried to measure the TGW, GL, and GW following the procedure reported by Zhang et al. [57]. Finally, the NGPP, NSPP, NSP, SSR, and GYPP were calculated.

4.3. DNA Extraction and DNA Marker Analysis

Total DNA was extracted from the young leaves of each plant in the populations according to the method of Zheng et al. [58]. PCR amplification was performed, and the products were visualized with 2.5% agarose gels with GelRed staining (Biotium, Fremont, CA, USA). The DNA markers used in this study included SSR markers and InDel markers. The SSR markers with the prefixes "RM" were selected from the public database (www.gramene.org, accessed on 9 April 2023). The InDel markers with the prefixes "JD" were taken from our previous study [36], and the InDel markers with the prefixes "YS" were newly designed using the online tool of Primer3.0 based on the 30$\times$ genome re-sequence of the parents YYD and SFH (Table S1).

4.4. Construction of the Linkage Map

Based on the genotyping of the 190 plants of the F_2 population with 133 SSR and InDel markers, twelve linkage groups corresponding to 12 chromosomes were established by using the computer program MAP-MAKER/EXP 3.0 [59]. The recombination frequencies among the markers in the same linkage group were converted into genetic distances using the kosambi function. An LOD score of 3.0 was used to determine the order of markers and the linkage group.

4.5. QTL Analysis and Statistical Analysis

Composite interval mapping using the Windows QTL Cartographer 2.5 was conducted to detect the QTLs for rice yield traits and further validate *qGW2-1* [60]. QTL analysis was carried out with 1000 permutations at the 0.05 probability level. An LOD score of 3.0 was fixed as the thresh value based on the 500 permutation tests for each trait. The correlation coefficients among different traits in the F_2 population were evaluated by using Correlation Plot in the Origin2022 software.

5. Conclusions

The correlations among nine yield traits were analyzed using an F_2 population derived from a cross of YYD and SFH with significant phenotypic differences. The GYPP positively correlated with six yield traits, except for GL and GW, and showed the highest correlation coefficient of 0.98 with the NFGPP. A genetic map was constructed and spanned 1831.7 cM throughout 12 chromosomes, with an average interval of 15.1 cM. A total of 36 QTLs for yield traits were detected on nine chromosomes, except for the remaining chromosomes 5, 8, and 9. The phenotypic variation that could be explained by a single QTL ranged from 6.19% to 36.01%. Furthermore, a major QTL, *qGW2-1*, was confirmed to control the GW and was narrowed down to relatively smaller intervals, defined in the marker interval of about ~2.94-Mb using three RH-derived populations in F_7 and one set of NIL plants in F_8. These results lay the foundation for further map-based cloning and molecular design breeding in rice.

Supplementary Materials: The following supporting information can be downloaded at: https://www.mdpi.com/article/10.3390/plants13060770/s1, Figure S1: Frequency distribution of GL, GW and TGW in the three F_7 populations. a: YM1. b:YM2. c: YM3, GL: grain length, GW: grain width; TGW: 1000-grain weight; Table S1: Sequence and physical location of the markers.

Author Contributions: Conceptualization, J.Y. (Jiezheng Ying), Y.W. and X.T.; Data curation, M.Y., J.Y. (Jinyu Yang), Y.C., P.Z. and G.L.; Formal analysis, M.Y.; Investigation, M.Y., J.Y. (Jinyu Yang) and Y.C.; Supervision, J.Y. (Jiezheng Ying); Writing—original draft, M.Y.; Writing—review and editing, Y.W. and X.T. All authors have read and agreed to the published version of the manuscript.

Funding: This work was supported by intergovernment science and technology innovation collaboration project, China national key R&D program (Grant No. 2022YFE0125600), National Natural Science Foundation of China (Grant No. 32072050, U22A20456 and U20A2030). Key R&D Program of Zhejiang (Grant No. 2022C02032), Natural Science Foundation of Zhejiang (Grant No. LD24C130001), and ASTIP program of CAAS.

Data Availability Statement: Data are contained within the article.

Conflicts of Interest: The authors declare no conflicts of interest.

References

1. Verneau, F.; Amato, M.; La Barbera, F. Edible Insects and Global Food Security. *Insects* **2021**, *12*, 472. [CrossRef]
2. IFPRI. *2019 Global Food Policy Report*; IFPRI: Washington, DC, USA, 2019.
3. Sade, N.; Peleg, Z. Future challenges for global food security under climate change. *Plant Sci.* **2020**, *295*, 110467. [CrossRef]
4. Kuroha, T.; Ashikari, M. Molecular mechanisms and future improvement of submergence tolerance in rice. *Mol. Breed.* **2020**, *40*, 41. [CrossRef]
5. Fukagawa, N.K.; Ziska, L.H. Rice: Importance for Global Nutrition. *J. Nutr. Sci. Vitaminol.* **2019**, *65*, S2–S3. [CrossRef] [PubMed]
6. Zhu, C.; Kobayashi, K.; Loladze, I.; Zhu, J.; Jiang, Q.; Xu, X.; Liu, G.; Seneweera, S.; Ebi, K.L.; Drewnowski, A.; et al. Carbon dioxide (CO(2)) levels this century will alter the protein, micronutrients, and vitamin content of rice grains with potential health consequences for the poorest rice-dependent countries. *Sci. Adv.* **2018**, *4*, eaaq1012. [CrossRef] [PubMed]
7. Li, M.; Li, X.; Zhou, Z.; Wu, P.; Fang, M.; Pan, X.; Lin, Q.; Luo, W.; Wu, G.; Li, H. Reassessment of the Four Yield-related Genes Gn1a, DEP1, GS3, and IPA1 in Rice Using a CRISPR/Cas9 System. *Front. Plant Sci.* **2016**, *7*, 377. [CrossRef] [PubMed]

8. Zeng, D.; Tian, Z.; Rao, Y.; Dong, G.; Yang, Y.; Huang, L.; Leng, Y.; Xu, J.; Sun, C.; Zhang, G.; et al. Rational design of high-yield and superior-quality rice. *Nat. Plants* **2017**, *3*, 17031. [CrossRef]
9. Xing, Y.; Zhang, Q. Genetic and molecular bases of rice yield. *Annu. Rev. Plant Biol.* **2010**, *61*, 421–442. [CrossRef] [PubMed]
10. Liu, G.; Zhu, H.; Liu, S.; Zeng, R.; Zhang, Z.; Li, W.; Ding, X.; Zhao, F.; Zhang, G. Unconditional and conditional QTL mapping for the developmental behavior of tiller number in rice (*Oryza sativa* L.). *Genetica* **2010**, *138*, 885–893. [CrossRef] [PubMed]
11. Takai, T. Potential of rice tillering for sustainable food production. *J. Exp. Bot.* **2024**, *75*, 708–720. [CrossRef]
12. Wang, Y.; Li, J. Branching in rice. *Curr. Opin. Plant Biol.* **2011**, *14*, 94–99. [CrossRef] [PubMed]
13. Li, X.; Qian, Q.; Fu, Z.; Wang, Y.; Xiong, G.; Zeng, D.; Wang, X.; Liu, X.; Teng, S.; Hiroshi, F.; et al. Control of tillering in rice. *Nature* **2003**, *422*, 618–621. [CrossRef] [PubMed]
14. Koumoto, T.; Shimada, H.; Kusano, H.; She, K.-C.; Iwamoto, M.; Takano, M. Rice monoculm mutation moc2, which inhibits outgrowth of the second tillers, is ascribed to lack of a fructose-1,6-bisphosphatase. *Plant Biotechnol.* **2013**, *30*, 47–56. [CrossRef]
15. Lu, Z.; Shao, G.; Xiong, J.; Jiao, Y.; Wang, J.; Liu, G.; Meng, X.; Liang, Y.; Xiong, G.; Wang, Y.; et al. MONOCULM 3, an ortholog of WUSCHEL in rice, is required for tiller bud formation. *J. Genet. Genom.* **2015**, *42*, 71–78. [CrossRef] [PubMed]
16. Jiang, S.; Wang, D.; Yan, S.; Liu, S.; Liu, B.; Kang, H.; Wang, G.L. Dissection of the Genetic Architecture of Rice Tillering using a Genome-wide Association Study. *Rice* **2019**, *12*, 43. [CrossRef] [PubMed]
17. Liang, Y.; Yan, C.; Zheng, J.; Nan, W.; Qin, X.; Zhang, H. Locating QTL associated with spike traits of Dongxiang wild rice (*Oryza rufipogon* Griff.). *Euphytica* **2019**, *215*, 26. [CrossRef]
18. Cui, K.H.; Peng, S.B.; Xing, Y.Z.; Yu, S.B.; Xu, C.G. Genetic analysis of the panicle traits related to yield sink size of rice. *Yi Chuan Xue Bao Acta. Genet. Sin.* **2002**, *29*, 144–152.
19. Zhang, X.; Jia, H.; Li, T.; Wu, J.; Nagarajan, R.; Lei, L.; Powers, C.; Kan, C.C.; Hua, W.; Liu, Z.; et al. TaCol-B5 modifies spike architecture and enhances grain yield in wheat. *Science* **2022**, *376*, 180–183. [CrossRef]
20. Yang, Y.; Zhang, Y.; Li, J.; Xu, P.; Wu, Z.; Deng, X.; Pu, Q.; Lv, Y.; Elgamal, W.; Maniruzzaman, S.; et al. Three QTL from Oryza meridionalis Could Improve Panicle Architecture in Asian Cultivated Rice. *Rice* **2023**, *16*, 22. [CrossRef]
21. Zhao, K.; Tung, C.W.; Eizenga, G.C.; Wright, M.H.; Ali, M.L.; Price, A.H.; Norton, G.J.; Islam, M.R.; Reynolds, A.; Mezey, J.; et al. Genome-wide association mapping reveals a rich genetic architecture of complex traits in *Oryza sativa*. *Nat. Commun.* **2011**, *2*, 467. [CrossRef]
22. Kadam, N.N.; Struik, P.C.; Rebolledo, M.C.; Yin, X.; Jagadish, S.V.K. Genome-wide association reveals novel genomic loci controlling rice grain yield and its component traits under water-deficit stress during the reproductive stage. *J. Exp. Bot.* **2018**, *69*, 4017–4032. [CrossRef]
23. Ashikari, M.; Sakakibara, H.; Lin, S.; Yamamoto, T.; Takashi, T.; Nishimura, A.; Angeles, E.R.; Qian, Q.; Kitano, H.; Matsuoka, M. Cytokinin oxidase regulates rice grain production. *Science* **2005**, *309*, 741–745. [CrossRef]
24. Wu, Y.; Wang, Y.; Mi, X.F.; Shan, J.X.; Li, X.M.; Xu, J.L.; Lin, H.X. The QTL GNP1 Encodes GA20ox1, Which Increases Grain Number and Yield by Increasing Cytokinin Activity in Rice Panicle Meristems. *PLoS Genet.* **2016**, *12*, e1006386. [CrossRef]
25. Huang, R.; Jiang, L.; Zheng, J.; Wang, T.; Wang, H.; Huang, Y.; Hong, Z. Genetic bases of rice grain shape: So many genes, so little known. *Trends Plant Sci.* **2013**, *18*, 218–226. [CrossRef]
26. Li, R.; Li, M.; Ashraf, U.; Liu, S.; Zhang, J. Exploring the Relationships Between Yield and Yield-Related Traits for Rice Varieties Released in China From 1978 to 2017. *Front. Plant Sci.* **2019**, *10*, 543. [CrossRef] [PubMed]
27. Zhang, X.; Wang, J.; Huang, J.; Lan, H.; Wang, C.; Yin, C.; Wu, Y.; Tang, H.; Qian, Q.; Li, J.; et al. Rare allele of OsPPKL1 associated with grain length causes extra-large grain and a significant yield increase in rice. *Proc. Natl. Acad. Sci. USA* **2012**, *109*, 21534–21539. [CrossRef]
28. Si, L.; Chen, J.; Huang, X.; Gong, H.; Luo, J.; Hou, Q.; Zhou, T.; Lu, T.; Zhu, J.; Shangguan, Y.; et al. OsSPL13 controls grain size in cultivated rice. *Nat. Genet.* **2016**, *48*, 447–456. [CrossRef] [PubMed]
29. Weng, J.; Gu, S.; Wan, X.; Gao, H.; Guo, T.; Su, N.; Lei, C.; Zhang, X.; Cheng, Z.; Guo, X.; et al. Isolation and initial characterization of GW5, a major QTL associated with rice grain width and weight. *Cell Res.* **2008**, *18*, 1199–1209. [CrossRef]
30. Yu, J.; Xiong, H.; Zhu, X.; Zhang, H.; Li, H.; Miao, J.; Wang, W.; Tang, Z.; Zhang, Z.; Yao, G.; et al. OsLG3 contributing to rice grain length and yield was mined by Ho-LAMap. *BMC Biol.* **2017**, *15*, 28. [CrossRef]
31. Hu, J.; Wang, Y.; Fang, Y.; Zeng, L.; Xu, J.; Yu, H.; Shi, Z.; Pan, J.; Zhang, D.; Kang, S.; et al. A Rare Allele of GS2 Enhances Grain Size and Grain Yield in Rice. *Mol. Plant* **2015**, *8*, 1455–1465. [CrossRef] [PubMed]
32. Fan, C.; Xing, Y.; Mao, H.; Lu, T.; Han, B.; Xu, C.; Li, X.; Zhang, Q. GS3, a major QTL for grain length and weight and minor QTL for grain width and thickness in rice, encodes a putative transmembrane protein. *Theor. Appl. Genet.* **2006**, *112*, 1164–1171. [CrossRef]
33. Li, Q.; Lu, L.; Liu, H.; Bai, X.; Zhou, X.; Wu, B.; Yuan, M.; Yang, L.; Xing, Y. A minor QTL, SG3, encoding an R2R3-MYB protein, negatively controls grain length in rice. *Theor. Appl. Genet.* **2020**, *133*, 2387–2399. [CrossRef]
34. Liu, Q.; Han, R.; Wu, K.; Zhang, J.; Ye, Y.; Wang, S.; Chen, J.; Pan, Y.; Li, Q.; Xu, X.; et al. G-protein $\beta\gamma$ subunits determine grain size through interaction with MADS-domain transcription factors in rice. *Nat. Commun.* **2018**, *9*, 852. [CrossRef] [PubMed]
35. Qi, P.; Lin, Y.S.; Song, X.J.; Shen, J.B.; Huang, W.; Shan, J.X.; Zhu, M.Z.; Jiang, L.; Gao, J.P.; Lin, H.X. The novel quantitative trait locus GL3.1 controls rice grain size and yield by regulating Cyclin-T1;3. *Cell Res.* **2012**, *22*, 1666–1680. [CrossRef] [PubMed]
36. Ying, J.Z.; Ma, M.; Bai, C.; Huang, X.H.; Liu, J.L.; Fan, Y.Y.; Song, X.J. TGW3, a Major QTL that Negatively Modulates Grain Length and Weight in Rice. *Mol. Plant* **2018**, *11*, 750–753. [CrossRef] [PubMed]

37. Dong, N.Q.; Sun, Y.; Guo, T.; Shi, C.L.; Zhang, Y.M.; Kan, Y.; Xiang, Y.H.; Zhang, H.; Yang, Y.B.; Li, Y.C.; et al. UDP-glucosyltransferase regulates grain size and abiotic stress tolerance associated with metabolic flux redirection in rice. *Nat. Commun.* **2020**, *11*, 2629. [CrossRef] [PubMed]
38. Wu, W.; Liu, X.; Wang, M.; Meyer, R.S.; Luo, X.; Ndjiondjop, M.N.; Tan, L.; Zhang, J.; Wu, J.; Cai, H.; et al. A single-nucleotide polymorphism causes smaller grain size and loss of seed shattering during African rice domestication. *Nat. Plants* **2017**, *3*, 17064. [CrossRef] [PubMed]
39. Ishimaru, K.; Hirotsu, N.; Madoka, Y.; Murakami, N.; Hara, N.; Onodera, H.; Kashiwagi, T.; Ujiie, K.; Shimizu, B.; Onishi, A.; et al. Loss of function of the IAA-glucose hydrolase gene TGW6 enhances rice grain weight and increases yield. *Nat. Genet.* **2013**, *45*, 707–711. [CrossRef]
40. Song, X.J.; Kuroha, T.; Ayano, M.; Furuta, T.; Nagai, K.; Komeda, N.; Segami, S.; Miura, K.; Ogawa, D.; Kamura, T.; et al. Rare allele of a previously unidentified histone H4 acetyltransferase enhances grain weight, yield, and plant biomass in rice. *Proc. Natl. Acad. Sci. USA* **2015**, *112*, 76–81. [CrossRef] [PubMed]
41. Wang, A.; Hou, Q.; Si, L.; Huang, X.; Luo, J.; Lu, D.; Zhu, J.; Shangguan, Y.; Miao, J.; Xie, Y.; et al. The PLATZ Transcription Factor GL6 Affects Grain Length and Number in Rice. *Plant Physiol.* **2019**, *180*, 2077–2090. [CrossRef]
42. Wang, Y.; Xiong, G.; Hu, J.; Jiang, L.; Yu, H.; Xu, J.; Fang, Y.; Zeng, L.; Xu, E.; Xu, J.; et al. Copy number variation at the GL7 locus contributes to grain size diversity in rice. *Nat. Genet.* **2015**, *47*, 944–948. [CrossRef]
43. Zhao, D.S.; Li, Q.F.; Zhang, C.Q.; Zhang, C.; Yang, Q.Q.; Pan, L.X.; Ren, X.Y.; Lu, J.; Gu, M.H.; Liu, Q.Q. GS9 acts as a transcriptional activator to regulate rice grain shape and appearance quality. *Nat. Commun.* **2018**, *9*, 1240. [CrossRef] [PubMed]
44. Song, X.J.; Huang, W.; Shi, M.; Zhu, M.Z.; Lin, H.X. A QTL for rice grain width and weight encodes a previously unknown RING-type E3 ubiquitin ligase. *Nat. Genet.* **2007**, *39*, 623–630. [CrossRef] [PubMed]
45. Ruan, B.; Shang, L.; Zhang, B.; Hu, J.; Wang, Y.; Lin, H.; Zhang, A.; Liu, C.; Peng, Y.; Zhu, L.; et al. Natural variation in the promoter of TGW2 determines grain width and weight in rice. *New Phytol.* **2020**, *227*, 629–640. [CrossRef] [PubMed]
46. Li, Y.; Fan, C.; Xing, Y.; Jiang, Y.; Luo, L.; Sun, L.; Shao, D.; Xu, C.; Li, X.; Xiao, J.; et al. Natural variation in GS5 plays an important role in regulating grain size and yield in rice. *Nat. Genet.* **2011**, *43*, 1266–1269. [CrossRef] [PubMed]
47. Shi, C.L.; Dong, N.Q.; Guo, T.; Ye, W.W.; Shan, J.X.; Lin, H.X. A quantitative trait locus GW6 controls rice grain size and yield through the gibberellin pathway. *Plant J.* **2020**, *103*, 1174–1188. [CrossRef] [PubMed]
48. Wang, S.; Wu, K.; Yuan, Q.; Liu, X.; Liu, Z.; Lin, X.; Zeng, R.; Zhu, H.; Dong, G.; Qian, Q.; et al. Control of grain size, shape and quality by OsSPL16 in rice. *Nat. Genet.* **2012**, *44*, 950–954. [CrossRef] [PubMed]
49. Jia, B.; Zhao, X.; Qin, Y.; Irfan, M.; Kim, T.H.; Wang, B.; Wang, S.; Keun Sohn, J. Quantitative trait loci mapping of panicle traits in rice. *Mol. Biol. Res. Commun.* **2019**, *8*, 9–15. [CrossRef] [PubMed]
50. Ahmadizadeh, M.; Babaeian-Jelodar, N.; Mohammadi-Nejad, G.; Bagheri, N.; Kumar Singh, R. High-density linkage mapping for agronomic and physiological traits of rice (*Oryza sativa* L.) under reproductive-stage salt stress. *J. Genet.* **2021**, *100*, 51. [CrossRef]
51. Li, G.; Cheng, Y.; Yin, M.; Yang, J.; Ying, J.; Zhu, C. Detection of QTLs for panicle-related traits using an indica x japonica recombinant inbred line population in rice. *PeerJ* **2021**, *9*, e12504. [CrossRef]
52. Huang, M.; Zou, Y.-B.; Jiang, P.; Xia, B.; Md, I.; Ao, H.-J. Relationship between Grain Yield and Yield Components in Super Hybrid Rice. *Agric. Sci. China* **2011**, *10*, 1537–1544. [CrossRef]
53. Marathi, B.; Guleria, S.; Mohapatra, T.; Parsad, R.; Mariappan, N.; Kurungara, V.K.; Atwal, S.S.; Prabhu, K.V.; Singh, N.K.; Singh, A.K. QTL analysis of novel genomic regions associated with yield and yield related traits in new plant type based recombinant inbred lines of rice (*Oryza sativa* L.). *BMC Plant Biol.* **2012**, *12*, 137. [CrossRef]
54. Ying, J.Z.; Shan, J.X.; Gao, J.P.; Zhu, M.Z.; Shi, M.; Lin, H.X. Identification of quantitative trait Loci for lipid metabolism in rice seeds. *Mol. Plant* **2012**, *5*, 865–875. [CrossRef]
55. Su, J.; Xu, K.; Li, Z.; Hu, Y.; Hu, Z.; Zheng, X.; Song, S.; Tang, Z.; Li, L. Genome-wide association study and Mendelian randomization analysis provide insights for improving rice yield potential. *Sci. Rep.* **2021**, *11*, 6894. [CrossRef]
56. Ying, J.Z.; Gao, J.P.; Shan, J.X.; Zhu, M.Z.; Shi, M.; Lin, H.X. Dissecting the genetic basis of extremely large grain shape in rice cultivar 'JZ1560'. *J. Genet. Genom.* **2012**, *39*, 325–333. [CrossRef] [PubMed]
57. Zhang, H.W.; Fan, Y.Y.; Zhu, Y.J.; Chen, J.Y.; Yu, S.B.; Zhuang, J.Y. Dissection of the qTGW1.1 region into two tightly-linked minor QTLs having stable effects for grain weight in rice. *BMC Genet.* **2016**, *17*, 98. [CrossRef] [PubMed]
58. Zheng, K.; Huang, N.; Bennett, J.; Khush, G.S. PCR-based marker-assisted selection in rice breeding. In *IRRI Discussion Paper Series No. 12*; International Rice Research Institute: Los Banos, Philippines, 1995.
59. Lander, E.S.; Green, P.; Abrahamson, J.; Barlow, A.; Daly, M.J.; Lincoln, S.E.; Newberg, L.A. MAPMAKER: An interactive computer package for constructing primary genetic linkage maps of experimental and natural populations. *Genomics* **1987**, *1*, 174–181. [CrossRef] [PubMed]
60. Wang, S.; Basten, C.J.; Zeng, Z.B. *Windows QTL Cartographer 2.5*; Department of Statistics, North Carolina State University: Raleigh, NC, USA, 2006.

plants

Article

Simultaneous Application of Several Exogenous dsRNAs for the Regulation of Anthocyanin Biosynthesis in *Arabidopsis thaliana*

Konstantin V. Kiselev, Andrey R. Suprun, Olga A. Aleynova, Zlata V. Ogneva and Alexandra S. Dubrovina *

Laboratory of Biotechnology, Federal Scientific Center of the East Asia Terrestrial Biodiversity, Far Eastern Branch of the Russian Academy of Sciences, 690022 Vladivostok, Russia; kiselev@biosoil.ru (K.V.K.); suprun.hi@gmail.com (A.R.S.); aleynova@biosoil.ru (O.A.A.); zlata.v.ogneva@gmail.com (Z.V.O.)

* Correspondence: dubrovina@biosoil.ru; Tel.: +8-423-2310410; Fax: +8-4232-310193

Abstract: Plant surface treatment with double-stranded RNAs (dsRNAs) has gained recognition as a promising method for inducing gene silencing and combating plant pathogens. However, the regulation of endogenous plant genes by external dsRNAs has not been sufficiently investigated. Also, the effect of the simultaneous application of multiple gene-specific dsRNAs has not been analyzed. The aim of this study was to exogenously target five genes in *Arabidopsis thaliana*, namely, three transcription factor genes (*AtCPC, AtMybL2, AtANAC032*), a calmodulin-binding protein gene (*AtCBP60g*), and an anthocyanidin reductase gene (*AtBAN*), which are known as negative regulators of anthocyanin accumulation. Exogenous dsRNAs encoding these genes were applied to the leaf surface of *A. thaliana* either individually or in mixtures. The mRNA levels of the five targets were analyzed using qRT-PCR, and anthocyanin content was evaluated through HPLC-MS. The results demonstrated significant downregulation of all five target genes by the exogenous dsRNAs, resulting in enhanced expression of chalcone synthase (*AtCHS*) gene and increased anthocyanin content. The simultaneous foliar application of the five dsRNAs proved to be more efficient in activating anthocyanin accumulation compared to the application of individual dsRNAs. These findings hold considerable importance in plant biotechnology and gene function studies.

Keywords: dsRNA plant surface treatment; gene expression; RNAi; anthocyanin biosynthesis; dsRNA foliar application

check for updates

Citation: Kiselev, K.V.; Suprun, A.R.; Aleynova, O.A.; Ogneva, Z.V.; Dubrovina, A.S. Simultaneous Application of Several Exogenous dsRNAs for the Regulation of Anthocyanin Biosynthesis in *Arabidopsis thaliana*. *Plants* **2024**, *13*, 541. https://doi.org/10.3390/plants13040541

Academic Editors: Zhiyong Li and Jian Zhang

Received: 29 December 2023
Revised: 10 February 2024
Accepted: 14 February 2024
Published: 16 February 2024

1. Introduction

The task of developing innovative approaches to modify various plant traits, while keeping their genome intact, is of great importance in modern plant biotechnology. While transgenic plants and transgene-free genome editing [1] have proven to be effective approaches to managing plant traits, there is still a significant lack of understanding of the potential consequences of genetic modifications and genome editing in plants. This has sparked intense debates surrounding the safety of genetically modified or edited organisms, leading to numerous countries imposing legal restrictions on the production and cultivation of such organisms [2,3].

Plant surface treatment with aqueous solutions of double-stranded RNAs (dsRNAs), which is known as spray-induced gene silencing or SIGS, is an emerging strategy for manipulating plant characteristics. This innovative technique is currently being extensively researched as a viable alternative to genetically modified plants [4–6]. dsRNAs are a key inducer of the RNA interference (RNAi)-based gene silencing processes in eukaryotic organisms. During RNAi-mediated gene silencing, specialized enzymes known as Dicer-like or DCL, convert the dsRNAs into short interfering RNAs or siRNAs [7,8]. Subsequently, these siRNAs are integrated into the RNA-induced silencing complex (RISC). The primary function of the RISC is to degrade any RNA molecules that possess a similarity to the dsRNAs. Consequently, the mRNAs of genes that share significant homology with the dsRNA-derived siRNAs are destroyed, resulting in decreased expression of these genes.

At present, the RNAi phenomenon has become a prevalent technique utilized in experimental biology to effectively inhibit gene expression in plants. This technique is instrumental in studying gene functions as well as modifying desirable plant characteristics to meet specific needs and requirements [9–11]. However, the application of this approach necessarily involves the stage of obtaining a transgenic plant transformed by a dsRNA-encoding vector construct or the application of dsRNA-encoding constructs based on attenuated plant viruses for subsequent plant infection without viral infection symptoms [9,12].

At present, multiple studies have been conducted reporting on plant pathogen protection using the SIGS approach with the application of exogenous dsRNAs designed to inhibit the expression of virulence-associated genes in the attacking pathogens [4,11,13,14]. It has been established that fungal pathogens can efficiently internalize exogenous dsRNAs applied to the plant surface. Once internalized, these dsRNAs undergo processing and subsequently trigger the silencing of the targeted pathogen genes [13,15,16]. Furthermore, several intriguing studies have demonstrated that external dsRNAs, which are designed to target plant genes, may lead to the down-regulation of the gene targets in the plant genome with the consequent biochemical or phenotypic changes, such as altered flower morphology, reductions in glucose and fructose content, drought stress tolerance, changed anthocyanin production, or fungal stress resistance [17–22]. There have also been studies demonstrating that the suppression of targets in plants was attained through nanoparticle-mediated or laser light-guided external treatments with dsRNAs [23–25]. The externally applied dsRNAs have been detected in the plant cells and the plant vascular system [15,19,23]. However, the number of reports on externally induced silencing of plant targets is scarce in the current literature, and further studies are required. The discovery of the ability to regulate the level of plant gene expression by applying dsRNA to the surface of the plant opens up new possibilities for functional studies on plant genes and for regulating plant traits for the desired time in agriculture.

Our recent work has shown that external treatment of the *Arabidopsis thaliana* leaf surface with aqueous solutions of dsRNAs and siRNAs that specifically target the chalcone synthase *AtCHS* and two transcription repressor *AtMybL2* and *AtANAC032* genes, which are implicated in anthocyanin biosynthesis and its regulation, led to a significant target down-regulation and corresponding changes in anthocyanin levels [19]. Anthocyanins, which are derived from the phenylpropanoid biosynthetic pathway (as shown in Figure 1), are a type of colored secondary metabolites. These compounds are well-known for their valuable properties that have significant applications in various fields such as medicine, the food industry, cosmetology, and ornamental horticulture [26,27]. It has been shown that exogenous *AtCHS*-dsRNA penetrates into the vascular system and individual plant cells of *A. thaliana* and spreads through the vascular system and in groups of parenchymal cells [19]. We also found that the decrease in the *AtCHS* expression levels was associated with the appearance of a fraction of *AtCHS*-specific small RNAs against this gene and was presumably induced by RNAi-mediated processes [28].

This investigation was aimed to increase the content of anthocyanins in *A. thaliana* by utilizing exogenous dsRNAs that specifically target five genes in *A. thaliana*, namely, three transcription factor genes (*AtCPC, AtMybL2, AtANAC032*), a calmodulin-binding protein gene (*AtCBP60g*), and an anthocyanidin reductase gene (*AtBAN*). The genes of single-repeat R3 MYB transcription factors AtCPC and AtMybL2 [29,30], a NAC transcription factor AtANAC032 [31], and a calmodulin-binding protein AtCBP60g [32] are known as negative regulators (Figure 1) of anthocyanin accumulation in *A. thaliana* [29–31], while the *AtBAN* (*AtANR*) gene encodes an anthocyanin reductase (Figure 1) converting anthocyanidins to 2,3-*cis*-flavan-3-ols in a competing pathway [33]. According to data in the literature, mybl2, cpc-1, cbp60g mutant plant lines and chimeric repressor line ANAC032-SRDX (phenotype similar to loss-of-function mutants) exhibited upregulated expression levels of

AtCHS and some other anthocyanin biosynthetic genes in *A. thaliana* [29–32]. This was also accompanied by an increased content of anthocyanins. At the same time, overexpression lines were, in turn, characterized by downregulated *AtCHS*. Therefore, downregulation of these "negative regulators" should lead to increased mRNA levels of the *AtCHS* gene and anthocyanin production. To our understanding, there is no information regarding the simultaneous application of several dsRNAs in a mixture for the downregulation of gene targets in plants. We also evaluated individual effects of the five dsRNAs on target gene expression, anthocyanin content, and *AtCHS* expression.

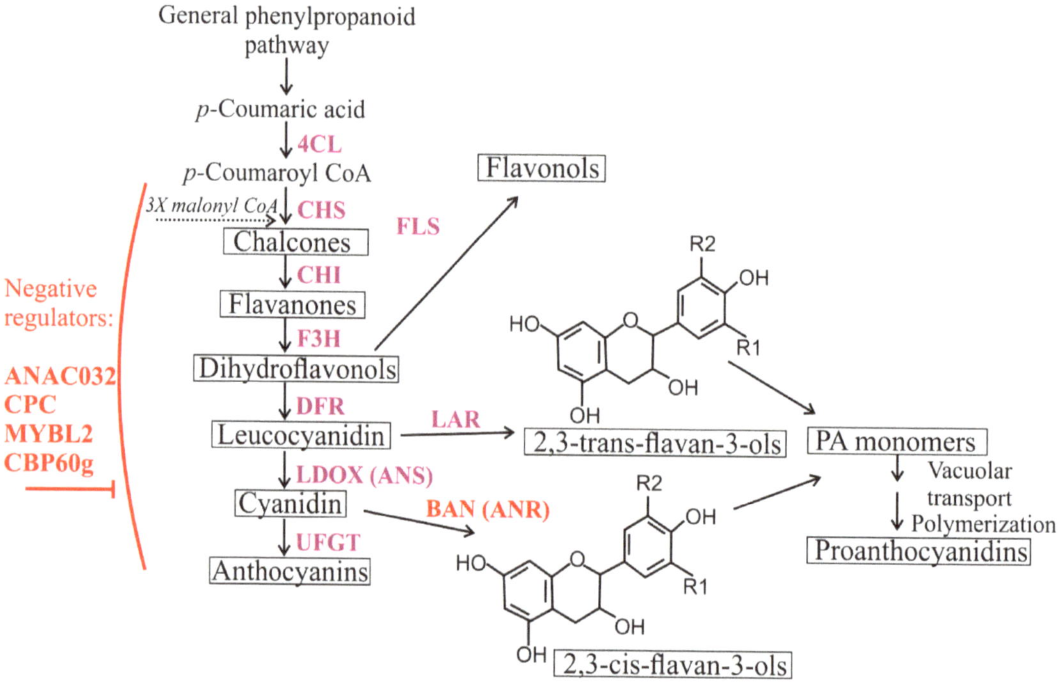

Figure 1. The pathway of anthocyanin biosynthesis. The 4-coumarate-CoA ligase (4CL) is the enzyme involved in general phenylpropanoid pathway. The enzymes participating in the anthocyanin biosynthesis pathway encompass chalcone synthase (CHS), chalcone isomerase (CHI), flavanone 3-hydroxylase (F3H), dihydroflavonol 4-reductase (DFR), anthocyanidin synthase (ANS or LDOX), anthocyanidin/flavonol 3-*O*-glucosyltransferase (UFGT), flavanonol synthase (FLS), leucoanthocyanidin reductase (LAR), and anthocyanidin reductase (ANR or BAN). Enzymes of each step are shown in purple. Negative regulators of the anthocyanin biosynthesis and the gene target from a competing pathway are shown in red.

2. Results

2.1. Effect of Exogenous dsRNAs Encoding AtANAC032, AtBAN, AtCBP60g, AtCPC, and AtMYBL2, Applied Separately, on the Accumulation of mRNAs of These Target Genes in A. thaliana

In order to assess the effectiveness of five external dsRNAs in inhibiting gene expression in *A. thaliana*, five genes of *AtANAC032*, *AtCBP60g*, *AtCPC*, and *AtMYBL2*, and *AtBAN* were selected. Five gene-specific dsRNAs of 762, 724, 218, 588, and 486-bp in length, respectively, were synthesized using PCR and an in vitro transcription protocol based on the *AtANAC032*, *AtCBP60g*, *AtCPC*, *AtMYBL2*, and *AtBAN* gene sequences of *A. thaliana* for the exogenous RNAi experiments (Figure 2). In order to confirm the specificity of the observed effects induced by dsRNAs, we synthesized dsRNA targeting an unrelated gene of a bacterial neomycin phosphotransferase II *NPTII* (Figure 2). The full-length cDNAs of *AtANAC032*,

AtCPC, and *AtMYBL2*, and large cDNA fragments of *AtCBP60g* and *AtBAN*, were amplified using primers that contained the T7 RNA polymerase promoter (Figure 2; Table S1).

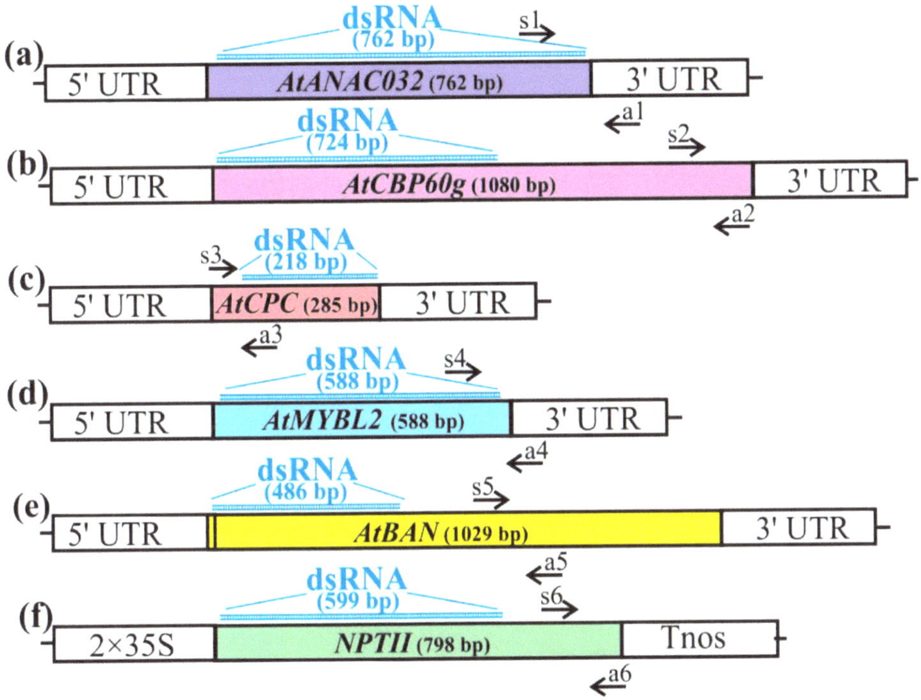

Figure 2. The dsRNA design and primer positions used to assess the impact of exogenous dsRNAs on the *AtANAC032*, *AtCBP60g*, *AtCPC*, *AtMYBL2*, and *AtBAN* gene expressions. The coding regions of each gene along with the dsRNA and primer positions are depicted as follows: (**a**) *AtANAC032* along with the dsRNA and primer positions; (**b**) *AtCBP60g* along with the sRNA and primer positions; (**c**) *AtCPC* along with the dsRNA and primer positions; (**d**) *AtMYBL2* along with the dsRNA and primer positions; (**e**) *AtBAN* along with the dsRNA and primer positions; (**f**) *NPTII* along with the dsRNA and primer positions. The black arrows show the primer pair positions (s1-a1, s2-a2, s3-a3, s4-a4, s5-a5, s6-a6) employed for the analysis of the mRNA levels of the *AtANAC032*, *AtCBP60g*, *AtCPC*, *AtMYBL2*, and *AtBAN* genes. UTR—untranslated region, 2 × 35S—the double 35S promoter of the cauliflower mosaic virus (CaMV), Tnos—nopaline synthase terminator.

PCR products were subjected to modification by incorporating T7 promoters at both ends, facilitating their utilization as templates during the in vitro transcription process. To treat external plant surfaces, a solution containing 35 µg of a dsRNA was prepared by diluting it in 100 µL of water (0.35 µg/µL). This solution was then gently spread onto the leaves of each individual *A. thaliana* plant using sterile soft brushes [19,34]. According to our previous analyses [19,34,35], our observations showed that considerable gene silencing of a transgene in *A. thaliana* can be achieved by treating four-week-old rosettes during the late hours of the day (21:00–21:30) under conditions of low soil moisture conditions. Thus, we decided to focus on these parameters during the current experiments.

We then investigated the effect of the individual exogenous *AtANAC032*, *AtCBP60g*, *AtCPC*, *AtMYBL2*, and *AtBAN* applied separately on the accumulation of mRNAs of the target genes compared to the control water and *NPTII*-dsRNAs treatments (Figure 3). Due to the insufficient anthocyanin levels and limited activity of anthocyanin biosynthesis genes observed under normal growth conditions, the treated *A. thaliana* rosettes were categorized into two groups and subjected to post-treatment incubation. The first group was exposed to control conditions (+22 °C, 16 h light), while the second group was exposed to anthocyanin-inducing conditions (+7 °C, 23 h light) for two and seven days. This differential treatment aimed to stimulate anthocyanin biosynthesis for further analysis.

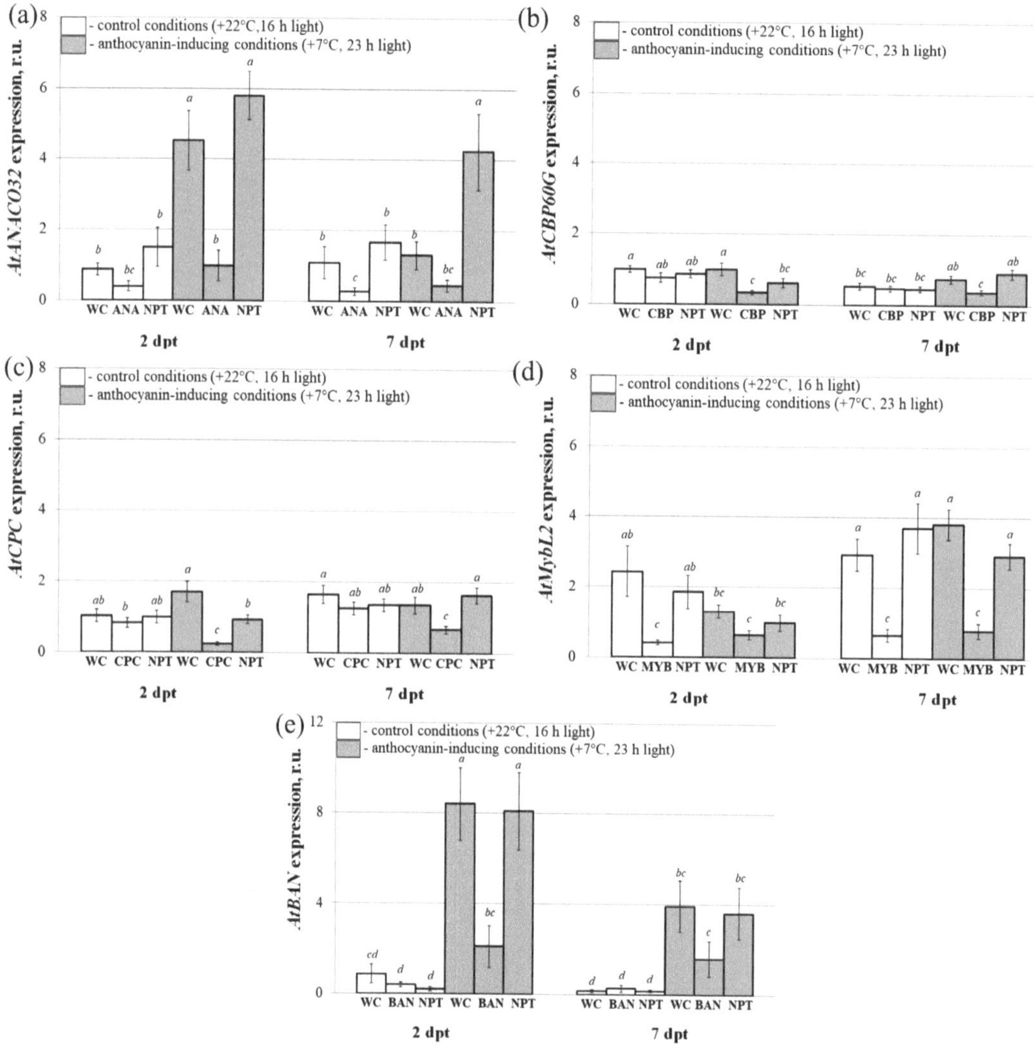

Figure 3. The relative fold change in the mRNA levels of *AtANAC032* (**a**), *AtCBP60g* (**b**), *AtCPC* (**c**), *AtMYBL2* (**d**), and *AtBAN* (**e**) after treatment with individual dsRNAs in *Arabidopsis thaliana* compared to untreated plants. The treatment groups included *A. thaliana* treated with sterile water (WC), *ANAC032*-dsRNA (ANA), *AtCBP60g*-dsRNA (CBP), *AtCPC*-dsRNAs (CPC), *AtMYBL2*-dsRNAs (MYB), *AtBAN*-dsRNAs (BAN), and *NPTII*-dsRNA (NPT). Total RNA was extracted at two and seven days following the application of dsRNA, with subsequent gene expression analysis performed using qRT-PCR. The treated *A. thaliana* rosettes were separated into two groups for incubation. The first group was exposed to control conditions (+22 °C, 16 h light), while the second group to anthocyanin-inducing conditions (+7 °C, 23 h light). The data are presented as the mean ± SE (three independent experiments). Means in each figure followed by the same letter were not different when Student's *t* test was used ($p < 0.05$).

qRT-PCR was utilized to examine the expression of the five target genes (Figure 3). According to the analysis, the expression levels of the *AtANAC032*, *AtCBP60g*, *AtCPC*, *AtMYBL2*, and *AtBAN* target genes were lower after foliar treatments with the five individual dsRNAs compared to the water or the non-specific *NPTII*-dsRNA treatments. Spreading the *AtANAC032*, *AtCBP60g*, *AtCPC*, *AtMYBL2*, and *AtBAN* dsRNAs under the conditions of reduced temperature and prolonged illumination lowered the expression of the target genes by 2–5.9-fold. It is important to note that the external application of the non-specific *NPTII*-dsRNA to plants did not have a substantial impact on the expression levels of *AtANAC032*, *AtCBP60g*, *AtCPC*, *AtMYBL2*, and *AtBAN* genes when compared to the control group treated with water. This finding confirms that the gene silencing effects induced by dsRNA were sequence-specific.

2.2. Simultaneous Application of Five dsRNAs Encoding the Genes AtANAC032, AtCBP60g, AtCPC, AtMYBL2, and AtBAN in Mixtures for the Target Gene Silencing in A. thaliana

To analyze the effect of the five dsRNAs applied in combination externally for the regulation of the *AtANAC032*, *AtCBP60g*, *AtCPC*, *AtMYBL2*, and *AtBAN* genes, we prepared mixtures of the five gene-specific dsRNAs containing 10, 20, or 30 µg of each dsRNA, i.e., 50 µg, 100 µg, or 150 µg of the dsRNAs mixed together (in 100 µL of water per individual plant). The leaf surface of four-week-old *A. thaliana* rosettes was treated externally with three different dsRNA mixtures. The mixtures of five dsRNAs targeting the *AtANAC032*, *AtCBP60g*, *AtCPC*, *AtMYBL2*, and *AtBAN* genes were applied to the leaf surface of four-week-old *A. thaliana* rosettes, both on the upper (adaxial) and lower (abaxial) sides, following the procedure described above for the individual application of the five dsRNAs.

We then investigated the effect of the exogeneous application of the multiple dsRNAs in mixtures on the expression of the five target genes in *A. thaliana* compared to the control group treated with water or *NPTII*-dsRNA two and seven days after treatment (Figure 4). For this purpose, the treated *A. thaliana* rosettes were separated into two groups for incubation. The first group was kept under control conditions, with a temperature of +22 °C and 16 h of light. The second group was exposed to conditions that induce anthocyanin production, with a temperature of +7 °C and 23 h of light.

Exogenous application of 100 and 150 µg dsRNA mixtures resulted in a marked decrease in the transcript level of *AtANAC032* (Figure 4a). This down-regulation effect was observed to be statistically significant at both two and seven days after treatment under the conditions of modulating anthocyanin biosynthesis. However, under control conditions, the down-regulation effect was only observed at two days post-treatment (Figure 4a). *AtCBP60G* and *AtBAN* expression exhibited a significant decrease after two and seven days of treatment with dsRNA under anthocyanin-inducing conditions for all types of mixtures (Figure 4b,e). We also observed significantly reduced *AtCBP60G* and *AtBAN* transcript levels for control conditions, and the effect was statistically significant for 50 and 150 µg for *AtCBP60G* and for 100 and 150 µg for *AtBAN*. *AtCPC* expression was considerably downregulated two days after treatment only under the anthocyanin-inducing conditions for all types of mixtures and for 50 µg seven days after treatment (Figure 4c). Similar to *AtANAC032*, mRNA levels of *AtMYBL2* were markedly downregulated two days after the application of 100 and 150 µg dsRNA mixtures under both control and anthocyanin-inducing conditions compared to the water-treated control, while the down-regulation effect of 50 µg dsRNA mixture was not statistically significant (Figure 4d). We also observed a similar effect of a decrease in the expression of this gene seven days after the application of all types of mixtures under control conditions and for 50 µg and 150 µg under anthocyanin-inducing conditions.

2.3. The Effect of Individual dsRNAs on the AtCHS mRNA Levels and Anthocyanin Content in A. thaliana

Using qRT-PCR, we also analyzed the effect of exogenous *AtANAC032*, *AtBAN*, *AtCBP60g*, *AtCPC*, and *AtMYBL2*-coding dsRNAs (separately) on the transcript levels of *AtCHS*, which encodes an important enzyme involved in anthocyanin biosynthesis

(Figure 1). The anthocyanin content was analyzed using the HPLC with a high-resolution mass spectrometry (HPLC-MS) analysis (Figures 5 and 6).

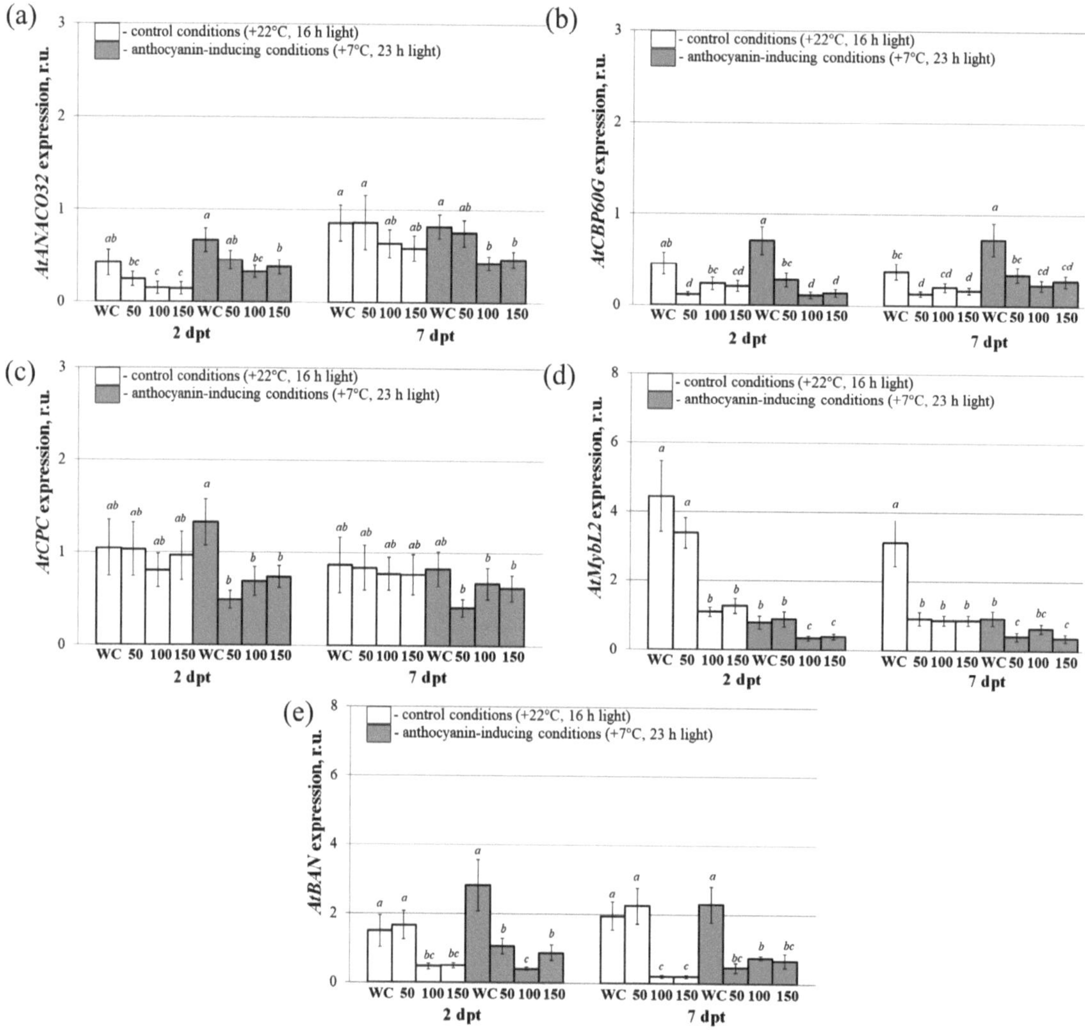

Figure 4. The relative fold change in the mRNA levels of *AtANAC032* (**a**), *AtCBP60g* (**b**), *AtCPC* (**c**), *AtMYBL2* (**d**), and *AtBAN* (**e**) after treatment with dsRNA mixtures in *Arabidopsis thaliana* compared to untreated plants. The dsRNA mixtures used were as follows: 50—mixture of the five gene-specific *dsRNAs* containing 10 μg of each dsRNA (*AtANAC032*, *AtCBP60g*, *AtCPC*, *AtMYBL2*, and *AtBAN*-specific dsRNAs); 100—mixture of the five gene-specific dsRNAs containing 20 μg of each dsRNA; 150—mixture of the five gene-specific dsRNAs containing 30 μg of each dsRNA; WC—plants treated with sterile water. Total RNA was extracted at two and seven days following the application of dsRNA, with subsequent gene expression analysis performed using qRT-PCR. The treated *A. thaliana* rosettes were separated into two groups for incubation. The first group was exposed to control conditions (+22 °C, 16 h light), while the second group to anthocyanin-inducing conditions (+7 °C, 23 h light). The data are presented as the mean ± SE (three independent experiments). Means in each figure followed by the same letter were not different when the Student's *t* test was used ($p < 0.05$).

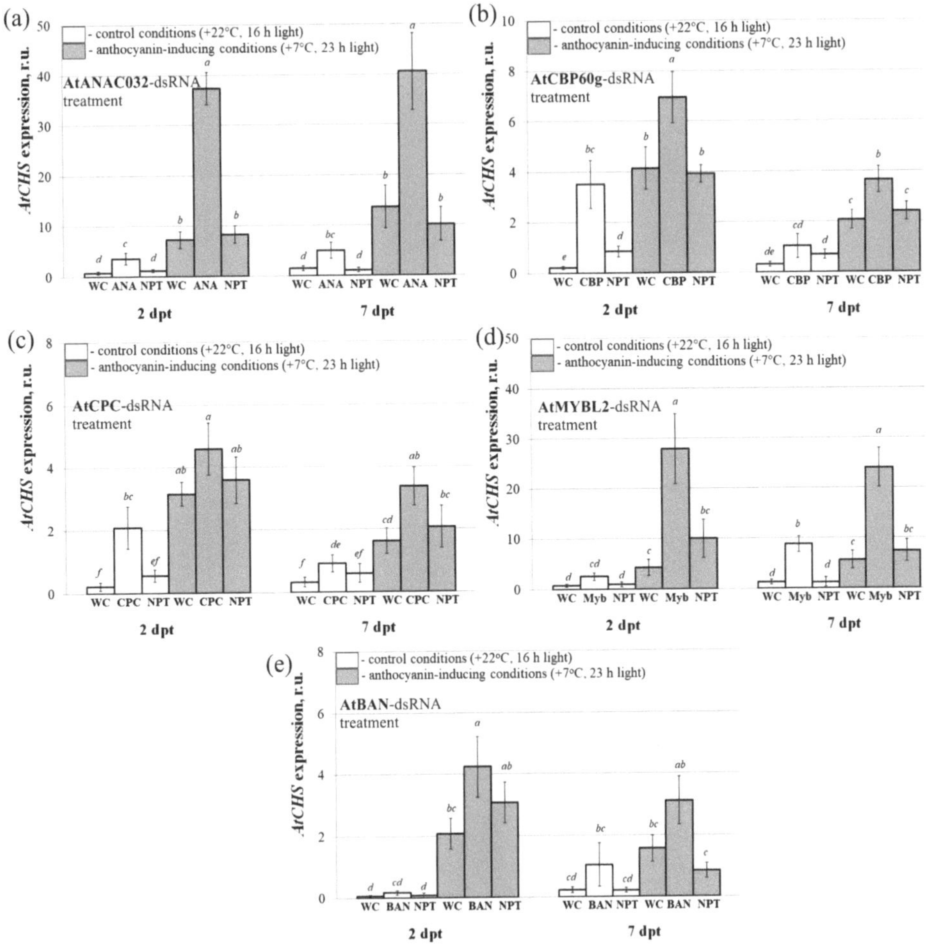

Figure 5. The relative fold change in *AtCHS* mRNA level after treatment of *Arabidopsis thaliana* with individual dsRNAs encoding the *AtANAC032* (**a**), *AtCBP60g* (**b**), *AtCPC* (**c**), *AtMYBL2* (**d**), *AtBAN* (**e**), and *NPTII* genes compared to untreated plants. The treatment groups included *A. thaliana* treated with sterile water (WC), *ANAC032*-dsRNA (ANA), *AtCBP60g*-dsRNA (CBP), *AtCPC*-dsRNAs (CPC), *AtMYBL2*-dsRNAs (MYB), *AtBAN*-dsRNAs (BAN), and *NPTII*-dsRNA (NPT). Total RNA was extracted at two and seven days following the application of dsRNA, with subsequent gene expression analysis performed using qRT-PCR. The treated *A. thaliana* rosettes were separated into two groups for incubation. The first group was exposed to control conditions (+22 °C, 16 h light), while the second group to anthocyanin-inducing conditions (+7 °C, 23 h light). The data are presented as the mean ± SE (three independent experiments). Means in each figure followed by the same letter were not different using Student's *t* test ($p < 0.05$).

The results demonstrated that under anthocyanin-inducing conditions, the cultivation of *A. thaliana* resulted in a notable increase in *AtCHS* transcript levels compared to the control conditions (Figure 5a–e). This indicated the successful activation of anthocyanin biosynthesis under the applied conditions. Specifically, we observed a 1.1–16.5-fold increase in *AtCHS* mRNA levels two days after treatment, and a 1.1–6.7-fold increase seven days after treatment following the application of the gene-specific dsRNAs.

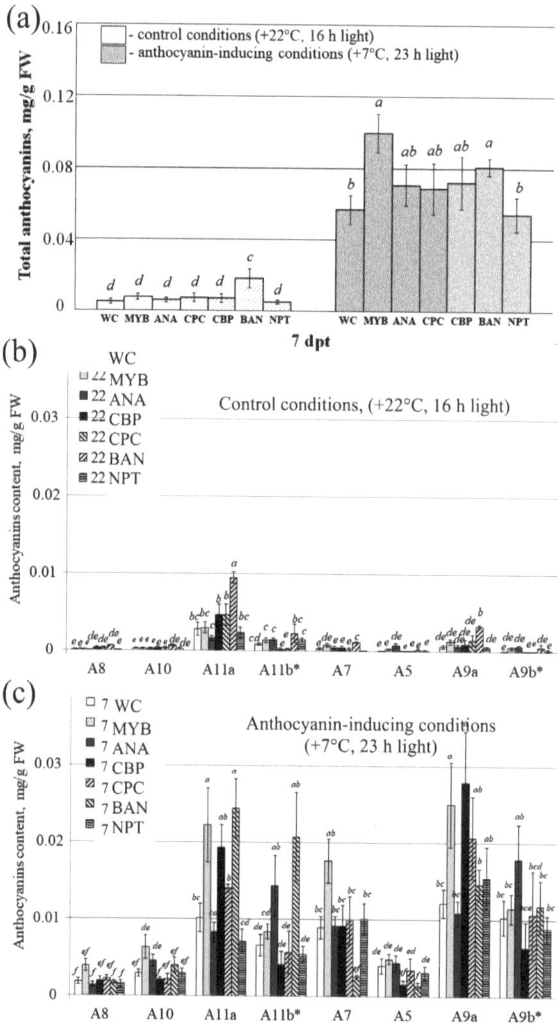

Figure 6. Anthocyanin content in the leaves of *Arabidopsis thaliana* seven days after treatment with the individual dsRNAs. (**a**) The total anthocyanin content; (**b**) the content of individual anthocyanins under control conditions; (**c**) the content of individual anthocyanins under the anthocyanin modulatory conditions. The treatment groups included *A. thaliana* treated with sterile water (WC), *ANAC032*-dsRNA (ANA), *AtCBP60g*-dsRNA (CBP), *AtCPC*-dsRNAs (CPC), *AtMYBL2*-dsRNAs (MYB), *AtBAN*-dsRNAs (BAN), and *NPTII*-dsRNA (NPT). The data are presented as the mean ± SE (three independent experiments). Means in each figure followed by the same letter were not different when using Student's *t* test ($p < 0.05$). A8—Cyanidin 3-*O*-[2″-*O*-(xylosyl) 6″-*O*-(*p*-O-(glucosyl) *p*-coumaroyl) glucoside] 5-*O*-[6‴-*O*-(malonyl) glucoside]; A10—Cyanidin 3-*O*-[2″-*O*-(2‴-*O*-(sinapoyl) xylosyl) 6″-*O*-(*p*-*O*-(glucosyl) *p*-coumaroyl) glucoside] 5-*O*-glucoside; A11a, A11b*—Cyanidin 3-*O*-[2″-*O*-(6‴-*O*-(sinapoyl) xylosyl) 6″-*O*-(*p*-*O*-(glucosyl)-*p*-coumaroyl) glucoside] 5-*O*-(6⁗-*O*-malonyl) glucoside; A7—Cyanidin 3-*O*-[2″-*O*-(2‴-*O*-(sinapoyl) xylosyl) 6″-*O*-(*p*-coumaroyl) glucoside] 5-*O*-glucoside; A5—Cyanidin 3-*O*-[2″-*O*-(xylosyl)-6″-*O*-(*p*-coumaroyl) glucoside] 5-*O*-malonylglucoside; A9a, A9b*—Cyanidin 3-*O*-[2″-*O*-(2‴-*O*-(sinapoyl) xylosyl) 6″-*O*-(*p*-*O*-coumaroyl) glucoside] 5-*O*-[6⁗-*O*-(malonyl) glucoside]. * Asterisk indicates a tautomer. The names of the anthocyanins are presented in accordance with previously published data [35].

The HPLC analysis of total anthocyanin profile revealed that plants treated with *AtMYBL2* and *AtBAN*-dsRNAs displayed a significantly higher total anthocyanin content compared to the control group treated with either water or *NPTII*-dsRNA (Figure 6a). The anthocyanin content in the dsRNA-treated plants reached 0.08–0.10 mg/g FW under anthocyanin-inducing conditions. While treatment with *AtANAC032*, *AtCBP60g*, and *AtCPC*-dsRNAs also resulted in increased total anthocyanin content (up to 0.07 mg/g FW), the increase was not statistically significant. The application of *AtMYBL2*-dsRNA resulted in the highest total anthocyanin level of 0.10 mg/g FW and was associated with a pronounced upregulation of the *AtCHS1* gene expression (Figures 5d and 6a). *AtBAN*-dsRNA modulated total anthocyanins both in the control and anthocyanin-modulated groups (Figure 6a).

Eight individual anthocyanin compounds were identified through the HPLC-MS analysis in the externally treated leaves of *A. thaliana* of all groups both under control conditions (Figures 6b and S1, Table S2) and conditions where anthocyanin levels were modulated (Figures 6c and S1, Table S2). Compared to the control conditions, we found that the plants grown under the high-light and cold-stress conditions exhibited higher levels of most individual anthocyanins (Figure 6b,c). Plant treatments with exogenous *AtMYBL2*, *AtANAC032*, *AtCBP*, and *AtBAN*-dsRNAs significantly enhanced the levels of several individual anthocyanins, with these changes being statistically significant for A11a and A9a (Figure 6b,c). It is important to note that the control non-specific dsRNA on *NPTII* did not affect the total level of anthocyanins (Figure 6a), the variety of anthocyanin compounds (Figure 6b,c), or *AtCHS* transcript levels (Figure 5). The findings provide further evidence of the targeted gene silencing effects mediated by gene-specific dsRNA.

2.4. The Effect of the Five dsRNAs Applied in Mixtures on the AtCHS mRNA and Anthocyanin Levels in A. thaliana

We then analyzed the influence of exogenous *AtANAC032*, *AtBAN*, *AtCBP60g*, *AtCPC*, and *AtMYBL2*-encoding dsRNAs applied simultaneously in mixtures on the expression of the *AtCHS* gene through qRT-PCR (Figure 7a) and anthocyanin content through HPLC-MS analysis (Figure 7b–d).

The expression of *AtCHS* and anthocyanin content were significantly increased in the control WC plants under the anthocyanin modulatory conditions, indicating that anthocyanin biosynthesis was successfully induced (Figure 7a,b). We found that the *AtCHS* transcript levels were increased 1.9–13.0 times two days post treatment and 3.8–5.7 times seven days post treatment after the application of the gene-specific dsRNA for both anthocyanin-inducing and control conditions (Figure 7a).

The HPLC analysis of the anthocyanin profile showed that *A. thaliana* treated with all three types of dsRNA mixtures (i.e., 50, 100, and 150 μg) contained significantly higher total amounts of anthocyanins, reaching 0.06–0.29 mg/g FW, than the plants treated with water under both control and anthocyanin-inducing conditions, except for 50 μg under control conditions (Figure 7b). Utilizing high doses of dsRNA under anthocyanin-inducing conditions resulted in the most notable accumulation of total anthocyanins at 0.29 mg/g FW, accompanied by the highest increase in expression of the *AtCHS* gene (Figure 7b).

HPLC-MS analysis demonstrated that leaves of *A. thaliana* treated with water and dsRNA contained identical eight anthocyanin compounds as observed under the control conditions (Figure 7c) and anthocyanin modulation conditions (Figure 7d). Plant treatment with the three dsRNA mixtures resulted in considerable elevation in the levels of most individual anthocyanins under anthocyanin-inducing condition, except for A8 (Figure 7d). Under control conditions, we observed less pronounced elevations in the content of anthocyanins with considerable changes in the content of A10, A11a, and A11b.

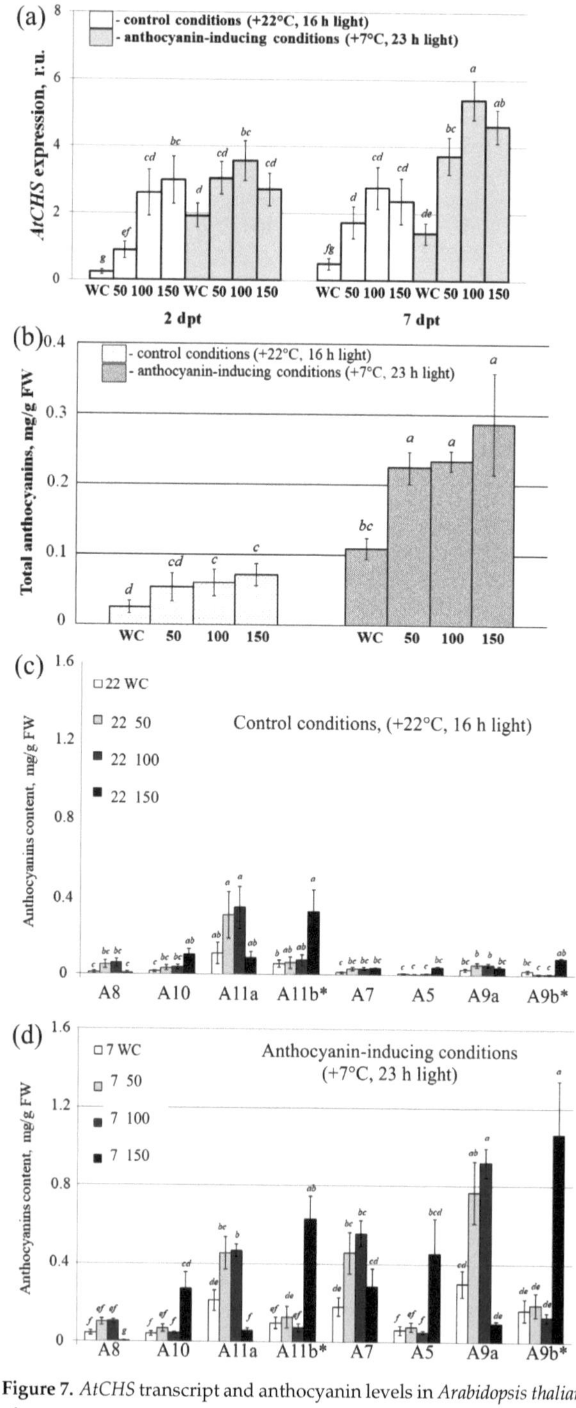

Figure 7. *AtCHS* transcript and anthocyanin levels in *Arabidopsis thaliana* leaves examined seven days after treatment with dsRNA mixtures of five dsRNAs of *AtANAC032*, *AtCBP60g*, *AtCPC*, *AtMYBL2*, and *AtBAN* genes. (**a**) Relative fold change in *AtCHS* mRNA level after treatment of *A. thaliana* with

dsRNAs mixtures; (**b**) the total anthocyanin content; (**c**) the content of individual anthocyanins under control conditions; (**d**) the content of individual anthocyanins under anthocyanin-inducing conditions. The dsRNA mixtures used were as follows: 50—mixture of the five gene-specific dsRNAs containing 10 µg of each dsRNA; 100—mixture of the five gene-specific dsRNAs containing 20 µg of each dsRNA; 150—mixture of the five gene-specific dsRNAs containing 30 µg of each dsRNA. Data are presented as the mean ± SE (three independent experiments). Means in each figure followed by the same letter were not different when using Student's *t* test ($p < 0.05$). A8—Cyanidin 3-*O*-[2″-*O*-(xylosyl) 6″-*O*-(*p*-O-(glucosyl) *p*-coumaroyl) glucoside] 5-*O*-[6‴-*O*-(malonyl) glucoside]; A10—Cyanidin 3-*O*-[2″-*O*-(2‴-*O*-(sinapoyl) xylosyl) 6″-*O*-(*p*-*O*-(glucosyl) *p*-coumaroyl) glucoside] 5-*O*-glucoside; A11a, A11b*—Cyanidin 3-*O*-[2″-*O*-(6‴-*O*-(sinapoyl) xylosyl) 6″-*O*-(*p*-*O*-(glucosyl)-*p*-coumaroyl) glucoside] 5-*O*-(6⁗-*O*-malonyl) glucoside; A7—Cyanidin 3-*O*-[2″-*O*-(2‴-*O*-(sinapoyl) xylosyl) 6″-*O*-(*p*-coumaroyl) glucoside] 5-*O*-glucoside; A5—Cyanidin 3-*O*-[2″-*O*-(xylosyl)-6″-*O*-(*p*-coumaroyl) glucoside] 5-*O*-malonylglucoside; A9a, A9b*—Cyanidin 3-*O*-[2″-*O*-(2‴-*O*-(sinapoyl) xylosyl) 6″-*O*-(*p*-*O*-coumaroyl) glucoside] 5-*O*-[6⁗-*O*-(malonyl) glucoside]. * Asterisk indicates a tautomer. The names of the anthocyanins are presented in accordance with previously published data [35].

3. Discussion

In modern society, there is a need to develop new effective approaches to protect and increase plant productivity using safe and environmentally friendly technologies. The increasing human population and the adverse effects of environmental stresses are driving the need for new molecular tools to improve and protect crops without altering the plant genome. Recently, a new approach, SIGS, has gained popularity for altering plant characteristics in the desired direction [4,5,36,37]. This approach includes treating plant surfaces with dsRNA solutions and induces RNAi-mediated silencing of a target gene in the plant genome or in the genome of the infecting pathogen, which is important for the regulated process or for the viability of the pathogen. There have been numerous studies where the utilization of exogenous dsRNA for virulence gene silencing in plant fungal pathogens has been documented [4,11,13], as well as studies on dsRNA antiviral effects [4,14,38]. However, there are only a few studies that have explored the effective silencing of plant genes through plant surface treatments with dsRNAs. Some studies and a patent have reported on the downregulation of plant genes through the external surface treatments with dsRNAs/siRNAs [17–22,39]. According to the studies, treating plants with dsRNAs encoding plant endogenous targets can downregulate mRNA levels of these genes. For instance, in tobacco and amaranth leaves, the transcript levels of the *EPSPS* gene were suppressed [39] and in orchid flower buds, the *Myb1* gene was silenced [17]. Similarly, treating grapevine with dsRNAs specific to the LBDIf7 [21] and GST40 [22] genes resulted in their downregulation. These studies have shown that exogenous plant treatments with target-specific dsRNAs can lead to desirable changes in plant phenotype or biochemistry, such as altered flower morphology, fungal resistance, and enhanced drought stress tolerance. In addition, the nanoparticle-mediated delivery of dsRNAs [24,26] and laser light-accompanied exogenous dsRNA treatments [25] have also been explored as a means to achieve plant gene silencing. However, exploring the efficient silencing of plant genes through surface treatments with dsRNAs is a relatively new area of research. Further research is needed to develop efficient approaches to inducing RNAi and specifically downregulating target genes in plants.

In this investigation, the implications of applying five different dsRNAs, both individually and simultaneously, to the leaves of *A. thaliana* were examined. The goal was to suppress the expression of five specific genes that play a role in blocking or competing with the biosynthesis of anthocyanins, thereby influencing anthocyanin accumulation in the leaves of *A. thaliana*. Although anthocyanins can be found in various plant species, they are often found in limited quantities or are nonexistent in numerous plants as a result of the restricted activity of the flavonoid biosynthetic pathway [40,41].

The process of anthocyanin biosynthesis in Arabidopsis and other plants is regulated by the coordinated action of various transcription factors. These transcription factors work together to form a protein complex called the MYB–bHLH–WD repeat (MBW) complex. This complex consists of basic helix–loop–helix (bHLH), R2R3-MYB, and WD40-repeat proteins [42]. However, there are also specific single repeat R3-MYB transcription factors that negatively regulate anthocyanin biosynthesis. These include TRIPTYCHON (TRY), MYBL2, CAPRICE (CPC), ENHANCER OF TRY AND CPC 1 (ETC1), and ETC2, which interfere with the formation of the MBW protein complex, thereby inhibiting anthocyanin accumulation [30,42]. Furthermore, other molecular players such as ubiquitin protein ligases or additional transcription factors also contribute to the regulation of anthocyanin biosynthesis in Arabidopsis and other plants. For instance, a recent discovery identified the NAC transcription factor ANAC032 as a suppressor of anthocyanin biosynthesis in *A. thaliana* [31]. In our study, we focused on four negative regulators of anthocyanin biosynthesis (*AtANAC032*, *AtCPC*, *AtMYBL2*, *AtCBP60g*) and a competing enzyme *AtBAN*, and we demonstrated that the applied individual dsRNAs downregulated their expression while simultaneously increasing the expression of *AtCHS*.

By applying dsRNAs encoding the five regulators of anthocyanin accumulation to the leaves of *A. thaliana* plants, we observed a decrease in the expression of the five targets, while there was a simultaneous elevation in the mRNA levels of anthocyanin synthesis genes. This resulted in a significant elevation in the anthocyanin levels, as confirmed through HPLC-MS analysis. This study demonstrated that *A. thaliana* leaves contained a total of 0.01 to 0.02 mg/g FW anthocyanins when the five dsRNAs were separately applied under control conditions. However, when dsRNAs were applied simultaneously, the anthocyanin levels reached between 0.05 and 0.07 mg/g FW under control conditions. The simultaneous foliar application of all five dsRNAs proved to be more effective in promoting anthocyanin accumulation compared to individual dsRNA treatments. Overall, under anthocyanin-modulating conditions, the application of the exogenous dsRNAs in our study successfully boosted the anthocyanin levels in *A. thaliana* leaves to 0.29 mg/g FW, representing a significant increase. Thus, the efficiency of the exogenously induced RNAi appears to be stronger under the anthocyanin-inducing conditions, presumably due to activated anthocyanin biosynthesis, where the overall negative regulation of anthocyanin accumulation is restrained. While the exogenous dsRNAs reduced the expression of the five genes of anthocyanin negative regulators, there were a larger number of blockers that continued to restrain the biosynthesis of anthocyanins under control conditions at 22 °C. However, under the conditions stimulating the biosynthesis of anthocyanins (+7 °C, 23 h light), the effect of other blockers decreased, and the RNAi effects were modulated.

Several research studies have presented evidence supporting the involvement of AtCPC [29], AtMYBL2 [30], and AtANAC032 [31] transcriptional factors in exerting inhibitory effects on the production of anthocyanins in *A. thaliana*. In addition, several reports indicate that the calmodulin-binding protein AtCBP60g has a role in plant defense, drought tolerance, and abscisic acid sensitivity [43,44]. Recently, AtCBP60g has been found to repress levels of anthocyanins induced by kinetin, sucrose, or drought [33], while the *AtBAN* gene acts as an anthocyanin reductase that converts anthocyanidins to 2,3-*cis*-flavan-3-ols through a competing pathway [34]. Our findings confirm that silencing of *AtANAC032*, *AtBAN*, *AtCBP60g*, *AtCPC*, and *AtMYBL2* genes, which are all involved in anthocyanin biosynthesis repression or competition, leads to the activation of anthocyanin biosynthesis in *A. thaliana*. Thus, this study showed that these factors act as negative regulators of anthocyanin biosynthesis or suppress the biosynthesis of anthocyanins due to activity in a close competing biosynthetic pathway. Importantly, the application of the nonspecific *NPTII*-dsRNAs to *A. thaliana* had no impact on the production of anthocyanins or the expression of the *AtCHS* gene in the plant. This illustrates that the observed dsRNA-induced silencing effect on anthocyanin biosynthesis-related genes was specific to the sequence and not due to the application of the dsRNA itself.

In conclusion, these results shed new light on the exogenous plant foliar treatments utilizing a combination of gene-specific dsRNAs in terms of future applications. This innovative approach can be considered as a rapid and effective tool for silencing plant genes, enabling a comprehensive investigation of plant gene function. Moreover, it holds great promise for inducing the yield of valuable secondary compounds in the field of plant biotechnology.

4. Methods and Materials

4.1. Plant Materials

To ensure sterility, the wild-type *A. thaliana* (cv. Columbia) seeds were subjected to vapor-phase sterilization according to a described procedure [45]. Subsequently, the sterized seeds were plated on solid $\frac{1}{2}$ Murashige and Skoog (MS) medium for two days at a temperature of 4 °C. Then, the plates were transferred to a growth chamber (Sanyo MLR-352, Panasonic, Osaka, Japan) set at a light intensity of approximately 120 µmol $m^{-2}s^{-1}$ for a daily light period of 16 h at 22 °C. After one week, the one-week-old *A. thaliana* seedlings were transplanted into pots measuring 7 cm × 7 cm and filled with 100 g of nutrient-rich soil (the soil was irrigated using filtered water applied at the bottom of each pot). Thereafter, the plants were cultivated in the growth chamber, covered with a plastic wrap, for additional three weeks without further irrigation at 22 °C. Subsequently, the dsRNA treatments were applied exogenously to the four-week-old *A. thaliana* plants. Following the dsRNA application, the plants were placed in a growth chamber (KS-200, Smolenskoye SKTB SPU, Smolensk, Russia) and incubated for an additional seven days under either control conditions (22 °C, 16 h daily light period) or anthocyanin-inducing conditions (7 °C, 23 h daily light period), without further irrigation. This was performed to trigger the expression of *AtCHS* and subsequent accumulation of anthocyanins.

4.2. Isolation, Cloning, and Sequencing of *AtANAC032, AtCPC, AtCBP60g, AtMYBL2,* and *AtBAN* Genes

Full-length coding cDNA sequences of *AtANAC032 (AT1G77450*, 762 bp), *AtCPC (AT2G46410*, 285 bp), AtCBP60g (AT5G26920, *1080 bp)*, AtMYBL2 (AT1G71030.1, 588 bp), and *AtBAN (AF092912*, 1029 bp) genes were amplified through RT-PCR using RNA samples extracted from the adult leaves of *A. thaliana*. The RT-PCRs were carried out in a Bis-M1105 Thermal Cycler (Bis-N, Novosibirsk, Russia) with the primers listed in Table S1. Subsequently, the RT-PCR products were subcloned into pJET1.2/blunt and subjected to sequencing following a previously described protocol [46].

4.3. Design and Synthesis of dsRNAs

In order to produce the dsRNAs, we utilized the T7 RiboMAX™ Express RNAi System (Promega, Madison, WI, USA) to amplify the cloned full-length cDNAs of *AtCPC*, *AtMYBL2*, and *AtANAC032* as well as the large cDNA fragments of *AtCBP60g* (724 bp out of 1080 bp) and *AtBAN* (486 bp out of 1029 bp) via PCR. Additionally, we amplified a large fragment of *NPTII* (GenBank AJ414108, 599 bp out of 798 bp) using pZP-RCS2-nptII plasmid [47]. To achieve this, the T7 promoter sequence was introduced into both the 5′ and 3′ ends of the amplified *AtCPC*, AtCBP60g, AtBAN, AtMYBL2, AtANAC032, and *NPTII* in a PCR for each gene using primers listed in Table S1. The PCRs were conducted in the Bis-M1105 Thermal Cycler following the instructions of the T7 RiboMAX™ Express RNAi System. Subsequently, the obtained PCR products were employed as templates for in vitro transcription and dsRNA synthesis, in accordance with the manufacturer's protocol. After in vitro transcription, the dsRNAs were treated with DNAse and RNAse and reprecipitated with ethanol according to the manufacturer's recommendations (Promega, Madison, WI, USA). We assessed the resulting dsRNAs using gel electrophoresis (Figure S2) and spectrophotometry to verify dsRNA purity, integrity, and amount.

4.4. Application of Exogenous dsRNAs

To apply the dsRNAs, including the *AtCPC*, AtCBP60g, AtBAN, AtMYBL2, AtANAC032, and *NPTII*-dsRNAs, to individual four-week-old rosettes of wild-type *A. thaliana*, we used sterilized individual soft brushes (made of natural pony hair) [34]. For a separate application of dsRNA, we diluted 35 µg of the dsRNA in 100 µL of nuclease-free water for each treated rosette and spread it on the foliar surface using the brushes. Both the adaxial (upper) and abaxial (lower) sides of all leaves of one rosette were treated for each type of dsRNA. In the case of mixture application, we diluted equal quantities of each dsRNA (10, 20, or 30 µg of each dsRNA or 50, 100, and 150 mg total dsRNA in mixtures, respectively) in 100 µL of nuclease-free water and spread them on the foliar surface of each treated rosette. This resulted in a final concentration of 0.1, 0.2, or 0.3 µg/µL of each dsRNA, equivalent to 0.5, 1.0, or 1.5 µg/µL of total dsRNAs. Both the adaxial and abaxial sides of all leaves of one rosette were treated for each type of mixture. In each independent experiment, one plant of *A. thaliana* was treated with 100 µL of the dsRNA of each type, while another plant received 100 µL of sterile filtered water. It should be noted that the dsRNAs were applied to four-week-old rosettes of *A. thaliana* late in the day (21:00–21:30) under low soil moisture conditions. This was based on our recent analysis, which determined these conditions as crucial parameters for successful target gene suppression in *A. thaliana* [34]. The soil water content before dsRNA treatment was 50–60%.

4.5. RNA Isolation and Reverse Transcription

To isolate total RNA, we collected a typical adult L3 leaf [48] from a single *A. thaliana* plant before any treatment. Additionally, we collected leaves two days and seven days after each specific treatment. The cetyltrimethylammonium bromide (CTAB)-based protocol [49] was employed to extract total RNA. Subsequently, complementary DNAs were synthesized as previously described [50].

4.6. Analysis of Gene Expression Using qRT-PCR

The PCR-amplified reverse transcription products were examined for the absence of DNA contamination using primers listed in Table S1. qRT-PCR was performed using SYBR Green I Real-time PCR dye and a real-time PCR kit (Evrogen, Moscow, Russia), following the methodology described by [51]. Two internal controls, GAPDH and UBQ, which were previously validated as suitable reference genes for qRT-PCR analysis in Arabidopsis [52], were included. The expression levels were determined using the $2^{-\Delta\Delta CT}$ method [53]. The obtained data for target gene expression at seven and two days post treatment were divided to the expression levels of the target gene before treatment, indicating fold change in target gene expression relative to the respective data before treatment. The gene identification numbers and the primers used in this study are listed in Table S1.

4.7. HPLC-MS Analysis of Anthocyanins

The frozen *A. thaliana* rosettes that underwent treatment were homogenized using a mortar and pestle. Shredded tissue was weighed and then subjected to extraction in 1 mL of 1% (*v*/*v*) hydrochloric acid and methanol for 24 h at 4 °C. Afterwards, the mixture was centrifuged at 13,200 rpm for 15 min. To prepare samples for HPLC–MS analysis, a 0.45-um nylon filter was used for sample filtration. All anthocyanins present were identified using an Agilent Technologies 1260 Infinity analytical HPLC system (Santa Clara, CA, USA) coupled with Bruker HCT ultra PTM Discovery System (Bruker Daltonik GmbH, Bremen, Germany) that featured an electrospray ionization (ESI) source, according to a protocol described previously [20]. A Shimadzu HPLC LC-20AD XR analytical system (Kyoto, Japan) equipped with diode array detection (HPLC–DAD) was employed for quantification of all anthocyanins following previously described procedures [20]. The anthocyanin contents were determined using an external standard method with the four-point regression calibration curves constructed using available standards. As a control, the commercial standard cyanidin chloride, obtained from Sigma-Aldrich (St. Louis, MO, USA), was used.

4.8. Statistical Analyses

The data are presented as mean ± standard error (SE) and were tested through a paired Student's *t*-test. The level of $p < 0.05$ was chosen as the point of minimal statistical significance in all analyses. Each type of analysis was repeated in a minimum of three independent experiments.

Supplementary Materials: The following supporting information can be downloaded at: https://www.mdpi.com/article/10.3390/plants13040541/s1, Figure S1: HPLC chromatogram of anthocyanins in *Arabidopsis thaliana* leaves examined seven days after treatment with dsRNA mixture of five dsRNAs of *AtANAC032*, *AtCBP60g*, *AtCPC*, *AtMYBL2*, and *AtBAN* gene under anthocyanin-induction conditions (A. 150 mg total dsRNA, +7 °C, 23 h light) or with water (B. H_2O); Figure S2: Electrophoretic separation of used dsRNAs in 2% agarose gel. *ANAC032-dsRNA* (a), *AtCBP60g*-dsRNA (b), *AtCPC*-dsRNA (c), *AtMYBL2*-dsRNA (d), *AtBAN*-dsRNA (e), and *NPTII*-dsRNA (f); Table S1: Primers used in RT-PCR and qRT-PCRs; Table S2: List of anthocyanins identified in the methanol extracts of *Arabidopsis thaliana* rosettes.

Author Contributions: A.S.D. and K.V.K. were responsible for the research design, experiments, RNA isolation, cDNA preparation, dsRNA synthesis, data analysis and interpretation, and paper preparation. A.R.S. performed HPLC analysis, qRT-PCRs, data analysis, plant management, and treatments. O.A.A. and Z.V.O. performed qRT-PCRs and data analyses. All authors have read and agreed to the published version of the manuscript.

Funding: This work was supported by a grant from the Russian Science Foundation (grant number 23-26-00253, https://rscf.ru/project/23-26-00253/, accessed on 16 February 2024).

Data Availability Statement: The data presented in this study are available within the article and Supplementary Materials.

Conflicts of Interest: The authors declare no conflicts of interest.

References

1. Gu, X.; Liu, L.; Zhang, H. Transgene-free genome editing in plants. *Front. Genome Ed.* **2021**, *3*, 805317. [CrossRef]
2. Azadi, H.; Ghanian, M.; Ghoochani, O.M.; Rafiaani, P.; Taning, C.N.T.; Hajivand, R.Y.; Dogot, T. Genetically modified crops: Towards agricultural growth, agricultural development, or agricultural sustainability? *Food Rev. Int.* **2015**, *31*, 195–221. [CrossRef]
3. Turnbull, C.; Lillemo, M.; Hvoslef-Eide, T.A.K. Global regulation of genetically modified crops amid the gene edited crop boom—A review. *Front. Plant Sci.* **2021**, *12*, 630396. [CrossRef]
4. Dubrovina, A.S.; Kiselev, K.V. Exogenous RNAs for gene regulation and plant resistance. *Int. J. Mol. Sci.* **2019**, *20*, 2282. [CrossRef]
5. Hoang, B.T.L.; Fletcher, S.J.; Brosnan, C.A.; Ghodke, A.B.; Manzie, N.; Mitter, N. RNAi as a foliar spray: Efficiency and challenges to field applications. *Int. J. Mol. Sci.* **2022**, *23*, 6639. [CrossRef]
6. Koeppe, S.; Kawchuk, L.; Kalischuk, M. RNA interference past and future applications in plants. *Int. J. Mol. Sci.* **2023**, *24*, 9755. [CrossRef]
7. Wilson, R.C.; Doudna, J.A. Molecular mechanisms of RNA interference. *Annu. Rev. Biophys.* **2013**, *42*, 217–239. [CrossRef]
8. Borges, F.; Martienssen, R.A. The expanding world of small RNAs in plants. *Nat. Rev. Mol. Cell Biol.* **2015**, *507*, 727–741. [CrossRef]
9. Kamthan, A.; Chaudhuri, A.; Kamthan, M.; Datta, A. Small RNAs in plants: Recent development and application for crop improvement. *Front. Plant Sci.* **2015**, *6*, 208. [CrossRef]
10. Morozov, S.Y.; Solovyev, A.G.; Kalinina, N.O.; Taliansky, M.E. Double-stranded RNAs in plant protection against pathogenic organisms and viruses in agriculture. *Acta Nat.* **2019**, *11*, 13–21. [CrossRef]
11. Gebremichael, D.E.; Haile, Z.M.; Negrini, F.; Sabbadini, S.; Capriotti, L.; Mezzetti, B.; Baraldi, E. RNA interference strategies for future management of plant pathogenic fungi: Prospects and challenges. *Plants* **2021**, *10*, 650. [CrossRef]
12. Tiwari, M.; Sharma, D.; Trivedi, P.K. Artificial microRNA mediated gene silencing in plants: Progress and perspectives. *Plant Mol. Biol.* **2014**, *86*, 1–18. [CrossRef]
13. Padilla-Roji, I.; Ruiz-Jiménez, L.; Bakhat, N.; Vielba-Fernández, A.; Pérez-García, A.; Fernández-Ortuño, D. RNAi Technology: A new path for the research and management of obligate biotrophic phytopathogenic fungi. *Int. J. Mol. Sci.* **2023**, *24*, 9082. [CrossRef]
14. Mitter, N.; Worrall, E.A.; Robinson, K.E.; Xu, Z.P.; Carroll, B.J. Induction of virus resistance by exogenous application of double-stranded RNA. *Curr. Opin. Virol.* **2017**, *26*, 49–55. [CrossRef]

15. Koch, A.; Biedenkopf, D.; Furch, A.; Weber, L.; Rossbach, O.; Abdellatef, E.; Linicus, L.; Johannsmeier, J.; Jelonek, L.; Goesmann, A.; et al. An RNAi-based control of *Fusarium graminearum* infections through spraying of long dsRNAs involves a plant passage and is controlled by the fungal silencing machinery. *PLoS Pathog.* **2016**, *12*, e1005901. [CrossRef]
16. Wang, M.; Weiberg, A.; Lin, F.M.; Thomma, B.P.H.J.; Huang, H.D.; Jin, H.L. Bidirectional cross-kingdom RNAi and fungal uptake of external RNAs confer plant protection. *Nat. Plants* **2016**, *2*, 16151. [CrossRef]
17. Lau, S.E.; Schwarzacher, T.; Othman, R.Y.; Harikrishna, J.A. dsRNA silencing of an R2R3-MYB transcription factor affects flower cell shape in a Dendrobium hybrid. *BMC Plant Biol.* **2015**, *15*, 194. [CrossRef]
18. Warnock, N.D.; Wilson, L.; Canet-Perez, J.V.; Fleming, T.; Fleming, C.C.; Maule, A.G.; Dalzell, J.J. Exogenous RNA interference exposes contrasting roles for sugar exudation in host-finding by plant pathogens. *Int. J. Parasitol.* **2016**, *46*, 473–477. [CrossRef]
19. Kiselev, K.V.; Suprun, A.R.; Aleynova, O.A.; Ogneva, Z.V.; Kalachev, A.V.; Dubrovina, A.S. External dsRNA downregulates anthocyanin biosynthesis-related genes and affects anthocyanin accumulation in *Arabidopsis thaliana*. *Int. J. Mol. Sci.* **2021**, *22*, 6749. [CrossRef]
20. Suprun, A.R.; Kiselev, K.V.; Dubrovina, A.S. Exogenously induced silencing of four MYB transcription repressor genes and activation of anthocyanin accumulation in *Solanum lycopersicum*. *Int. J. Mol. Sci.* **2023**, *24*, 9344. [CrossRef]
21. Marcianò, D.; Ricciardi, V.; Fassolo, E.M.; Passera, A.; BIANCO, P.A.; Failla, O.; Casati, P.; Maddalena, G.; De Lorenzis, G.; Toffolatti, S.L. RNAi of a putative grapevine susceptibility gene as a possible downy mildew control strategy. *Front. Plant Sci.* **2021**, *12*, 667319. [CrossRef]
22. Nerva, L.; Guaschino, M.; Pagliarani, C.; De Rosso, M.; Lovisolo, C.; Chitarra, W. Spray-induced gene silencing targeting a glutathione S-transferase gene improves resilience to drought in grapevine. *Plant Cell Environ.* **2022**, *45*, 347–361. [CrossRef]
23. Jiang, L.; Ding, L.; He, B.; Shen, J.; Xu, Z.; Yin, M.; Zhang, X. Systemic gene silencing in plants triggered by fluorescent nanoparticle-delivered double-stranded RNA. *Nanoscale* **2014**, *6*, 9965–9969. [CrossRef]
24. Killiny, N.; Gonzalez-Blanco, P.; Gowda, S.; Martini, X.; Etxeberria, E. Plant functional genomics in a few days: Laser-assisted delivery of double-stranded RNA to higher plants. *Plants* **2021**, *10*, 93. [CrossRef]
25. Molesini, B.; Pennisi, F.; Cressoni, C.; Vitulo, N.; Dusi, V.; Speghini, A.; Pandolfini, T. Nanovector-mediated exogenous delivery of dsRNA induces silencing of target genes in very young tomato flower buds. *Nanoscale Adv.* **2022**, *4*, 4542–4553. [CrossRef]
26. Kong, J.M.; Chia, L.S.; Goh, N.K.; Chia, T.F.; Brouillard, R. Analysis and biological activities of anthocyanins. *Phytochemistry* **2003**, *64*, 923–933. [CrossRef]
27. Khoo, H.E.; Azlan, A.; Tang, S.T.; Lim, S.M. Anthocyanidins and anthocyanins: Colored pigments as food, pharmaceutical ingredients, and the potential health benefits. *Food Nutr. Res.* **2017**, *61*, 1361779. [CrossRef]
28. Nityagovsky, N.N.; Kiselev, K.V.; Suprun, A.R.; Dubrovina, A.S. Exogenous dsRNA induces RNA interference of a chalcone synthase gene in *Arabidopsis thaliana*. *Int. J. Mol. Sci.* **2022**, *23*, 5325. [CrossRef]
29. Zhu, H.F.; Fitzsimmons, K.; Khandelwal, A.; Kranz, R.G. CPC, a single-repeat R3 MYB, is a negative regulator of anthocyanin biosynthesis in Arabidopsis. *Mol. Plant.* **2009**, *2*, 790–802. [CrossRef]
30. Matsui, K.; Umemura, Y.; Ohme-Takagi, M. AtMYBL2, a protein with a single MYB domain, acts as a negative regulator of anthocyanin biosynthesis in Arabidopsis. *Plant J.* **2008**, *55*, 954–967. [CrossRef]
31. Mahmood, K.; Xu, Z.; El-Kereamy, A.; Casaretto, J.A.; Rothstein, S.J. The Arabidopsis transcription factor ANAC032 represses anthocyanin biosynthesis in response to high sucrose and oxidative and abiotic stresses. *Front. Plant Sci.* **2016**, *7*, 1548. [CrossRef]
32. Zou, B.; Wan, D.; Li, R.; Han, X.; Li, G.; Wang, R. Calmodulin-binding protein CBP60g functions as a negative regulator in Arabidopsis anthocyanin accumulation. *PLoS ONE* **2017**, *12*, e0173129. [CrossRef]
33. Xie, D.Y.; Sharma, S.B.; Paiva, N.L.; Ferreira, D.; Dixon, R.A. Role of anthocyanidin reductase, encoded by BANYULS in plant flavonoid biosynthesis. *Science* **2003**, *299*, 396–399. [CrossRef]
34. Kiselev, K.V.; Suprun, A.R.; Aleynova, O.A.; Ogneva, Z.V.; Dubrovina, A.S. Physiological conditions and dsRNA application approaches for exogenously induced RNA interference in *Arabidopsis thaliana*. *Plants* **2021**, *10*, 264. [CrossRef]
35. Tohge, T.; Nishiyama, Y.; Hirai, M.Y.; Yano, M.; Nakajima, J.; Awazuhara, M.; Inoue, E.; Takahashi, H.; Goodenowe, D.B.; Kitayama, M.; et al. Functional genomics by integrated analysis of metabolome and transcriptome of Arabidopsis plants over-expressing an MYB transcription factor. *Plant J.* **2005**, *42*, 218–235. [CrossRef]
36. Kiselev, K.V.; Suprun, A.R.; Aleynova, O.A.; Ogneva, Z.V.; Kostetsky, E.Y.; Dubrovina, A.S. The specificity of transgene suppression in plants by exogenous dsRNA. *Plants* **2022**, *11*, 715. [CrossRef]
37. Wang, M.; Jin, H. Spray-induced gene silencing: A powerful innovative strategy for crop protection. *Trends Microbiol.* **2017**, *25*, 4–6. [CrossRef]
38. Akbar, S.; Wei, Y.; Zhang, M.-Q. RNA Interference: Promising Approach to Combat Plant Viruses. *Int. J. Mol. Sci.* **2022**, *23*, 5312. [CrossRef]
39. Sammons, R.; Ivashuta, S.; Liu, H.; Wang, D.; Feng, P.; Kouranov, A.; Andersen, S. Polynucleotide Molecules for Gene Regulation in Plants. U.S. Patent 20110296556, 8 March 2011.
40. Swanson, B.G. Tannins and polyphenols. In *Encyclopedia of Food Sciences and Nutrition*, 2nd ed.; Caballero, B., Ed.; Academic Press: Cambridge, MA, USA, 2003; pp. 5729–5733.
41. Sakuta, M.; Tanaka, A.; Iwase, K.; Miyasaka, M.; Ichiki, S.; Hatai, M.; Inoue, Y.T.; Yamagami, A.; Nakano, T.; Yoshida, K.; et al. Anthocyanin synthesis potential in betalain-producing Caryophyllales plants. *J. Plant Res.* **2021**, *134*, 1335–1349. [CrossRef]
42. Li, S. Transcriptional control of flavonoid biosynthesis. *Plant Signal. Behav.* **2014**, *9*, e27522. [CrossRef]

43. Wang, L.; Tsuda, K.; Sato, M.; Cohen, J.D.; Katagiri, F.; Glazebrook, J. Arabidopsis CaM binding protein CBP60g contributes to MAMP-induced SA accumulation and is involved in disease resistance against *Pseudomonas syringae*. *PLoS Pathog.* **2009**, *5*, e1000301. [CrossRef]
44. Wan, D.; Li, R.; Zou, B.; Zhang, X.; Cong, J.; Wang, R.; Xia, Y.; Li, G. Calmodulin-binding protein CBP60g is a positive regulator of both disease resistance and drought tolerance in Arabidopsis. *Plant Cell Rep.* **2012**, *31*, 1269–1281. [CrossRef]
45. Dubrovina, A.S.; Aleynova, O.A.; Suprun, A.R.; Ogneva, Z.V.; Kiselev, K.V. Transgene suppression in plants by foliar application of in vitro-synthesized small interfering RNAs. *Appl. Microbiol. Biotechnol.* **2020**, *104*, 2125–2135. [CrossRef]
46. Dubrovina, A.S.; Aleynova, O.A.; Ogneva, Z.V.; Suprun, A.R.; Ananev, A.A.; Kiselev, K.V. The effect of abiotic stress conditions on expression of calmodulin (*CaM*) and calmodulin-like (*CML*) genes in wild-growing grapevine *Vitis amurensis*. *Plants* **2019**, *8*, 602. [CrossRef]
47. Tzfira, T.; Tian, G.W.; Lacroix, B.; Vyas, S.; Li, J.; Leitner-Dagan, Y.; Krichevsky, A.; Taylor, T.; Vainstein, A.; Citovsky, V. pSAT vectors: A modular series of plasmids for autofluorescent protein tagging and expression of multiple genes in plants. *Plant Mol. Biol.* **2005**, *57*, 503–516. [CrossRef]
48. Kiselev, K.V.; Aleynova, O.A.; Ogneva, Z.V.; Suprun, A.R.; Dubrovina, A.S. 35S promoter-driven transgenes are variably expressed in different organs of *Arabidopsis thaliana* and in response to abiotic stress. *Mol. Biol. Rep.* **2021**, *48*, 2235–2241. [CrossRef]
49. Kiselev, K.V.; Dubrovina, A.S.; Shumakova, O.A.; Karetin, Y.A.; Manyakhin, A.Y. Structure and expression profiling of a novel calcium-dependent protein kinase gene, CDPK3a, in leaves, stems, grapes, and cell cultures of wild-growing grapevine *Vitis amurensis* Rupr. *Plant Cell Rep.* **2013**, *32*, 431–442. [CrossRef]
50. Aleynova, O.A.; Kiselev, K.V.; Ogneva, Z.V.; Dubrovina, A.S. The grapevine calmodulin-like protein gene CML21 is regulated by alternative splicing and involved in abiotic stress response. *Int. J. Mol. Sci.* **2020**, *21*, 7939. [CrossRef]
51. Dubrovina, A.S.; Aleynova, O.A.; Kalachev, A.V.; Suprun, A.R.; Ogneva, Z.V.; Kiselev, K.V. Induction of transgene suppression in plants via external application of synthetic dsRNA. *Int. J. Mol. Sci.* **2019**, *20*, 1585. [CrossRef]
52. Czechowski, T.; Stitt, M.; Altmann, T.; Udvardi, M.K.; Scheible, W.R. Genome-wide identification and testing of superior reference genes for transcript normalization in Arabidopsis. *Plant Physiol.* **2005**, *139*, 5–17. [CrossRef]
53. Livak, K.J.; Schmittgen, T.D. Analysis of relative gene expression data using real-time quantitative PCR and the 2(-Delta Delta C(T)) method. *Methods* **2001**, *25*, 402–408. [CrossRef] [PubMed]

plants

Article

Genome-Wide Association Study for Agronomic Traits in Gamma-Ray-Derived Mutant Kenaf (*Hibiscus cannabinus* L.)

Woon Ji Kim [1], Baul Yang [1], Ye-jin Lee [1], Jae Hoon Kim [1], Sang Hoon Kim [1], Joon-Woo Ahn [1], Si-Yong Kang [2], Seong-Hoon Kim [3] and Jaihyunk Ryu [1,*]

1 Advanced Radiation Technology Institute, Korea Atomic Energy Research Institute, Jeongeup 56212, Republic of Korea; wjkim0101@kaeri.re.kr (W.J.K.); byang@kaeri.re.kr (B.Y.); yjinlee@kaeri.re.kr (Y.-j.L.); jaehun@kaeri.re.kr (J.H.K.); shkim80@kaeri.re.kr (S.H.K.); joon@kaeri.re.kr (J.-W.A.)
2 Department of Horticulture, College of Industrial Sciences, Kongju National University, Yesan 32439, Republic of Korea; sykang@kongju.ac.kr
3 National Agrobiodiversity Center, National Institute of Agricultural Sciences, Rural Development Administration, Jeonju 5487, Republic of Korea; shkim0819@korea.kr
* Correspondence: jhryu@kaeri.re.kr; Tel.: +82-63-570-3311

Abstract: Kenaf (*Hibiscus cannabinus* L.), in the Malvaceae family, is an important crop for not only fiber production, but also various other industrial materials. We performed phylogenetic analysis and a genome-wide association study (GWAS) of seven agronomic traits: days to flowering, plant height, fresh weight, dry weight, flower color, stem color, and leaf shape, using 96 kenaf genotypes, including gamma-irradiation-derived mutant lines. Genotypes were determined by genotyping-by-sequencing (GBS) and a total of 49,241 single-nucleotide polymorphisms (SNPs) were used in the analysis. Days to flowering, plant height, fresh weight, and dry weight were positively correlated with each other, and stem color was also correlated with fresh weight and dry weight. The phylogenetic analysis divided the 96 lines into nine related groups within two independent groups, and the GWAS analysis detected a total of 49 SNPs for days to flowering, plant height, fresh weight, dry weight, flower color, stem color, and leaf shape with $-\log_{10}(P) \geq 4$, of which 22 were located in genic regions. The detected SNPs were located in genes with homology ranging from 45% to 96% to plants of the Malvaceae and Betulaceae, and these genes were found to be involved in plant growth and development via various pathways. Our identification of SNP markers related to agronomic traits is expected to help improve the quality of selective breeding programs for kenaf.

Keywords: kenaf; genome-wide association study; genotyping-by-sequencing; single-nucleotide polymorphism

Citation: Kim, W.J.; Yang, B.; Lee, Y.-j.; Kim, J.H.; Kim, S.H.; Ahn, J.-W.; Kang, S.-Y.; Kim, S.-H.; Ryu, J. Genome-Wide Association Study for Agronomic Traits in Gamma-Ray-Derived Mutant Kenaf (*Hibiscus cannabinus* L.). *Plants* **2024**, *13*, 249. https://doi.org/10.3390/plants13020249

Academic Editors: Mario Ciaffi, Jian Zhang and Zhiyong Li

Received: 16 November 2023
Revised: 14 December 2023
Accepted: 10 January 2024
Published: 16 January 2024

1. Introduction

A member of the Malvaceae family, kenaf (*Hibiscus cannabinus* L.), is a diploid (2n = 2X = 36) annual herbaceous plant native to Africa [1]. Originating in North Africa, kenaf is now cultivated in many countries, including India, Russia, China, and the United States, and is known to thrive in temperate and tropical environments with abundant sunlight and precipitation [1,2]. Kenaf has a variety of valuable industrial uses, including as a source of edible seed oil, stem fiber, plastic raw materials, and pharmaceuticals. In the past, the main purpose of kenaf production was to produce fiber for the manufacture of carpets, canvas bags, and rope. During World War II, the use of kenaf fiber for rope production came to the forefront, leading to research into its cultivation, production, and processing [3,4]. In Africa, kenaf leaves and stems were used to treat Guinea-worm disease and anemia, while in Ayurvedic medicine, an ancient Indian Hindu tradition, kenaf leaves were used to treat bile, blood, coughing, and diabetes [5,6]. Kenaf is known to have a crude protein content of 6% to 23% in the whole plant, and in particular, the crude protein content in the leaves reaches 14% to 34%, making it suitable as a feed source for livestock [2].

Recently, kenaf fiber has been attracting attention as a material for biodegradable, eco-friendly composites that can replace glass fiber composites [7,8]. Additionally, kenaf is an important medicinal crop as it contains high amounts of phytochemical compounds including polyphenols and anthocyanins. Thus, kenaf has been used in Indian traditional medicine (Ayurvedic medicine) as an aphrodisiac, purgative, and digestive aid including antioxidant, hepatoprotective, and anticancer activities [5].

The yield of kenaf stalks varies from 11 to 18 tons/ha, with cropping season, temperature, and soil moisture acting as major factors [2]. Kenaf plants are mainly known for their light yellow and cream-colored flowers with green stems [2]. Kenaf varieties can be categorized into three maturation groups, determined by their photosensitivity: ultra-early maturing, early-to-medium maturing, and late maturing [3]. Kenaf maturity is closely related to yield: to increase fiber yield, it is advantageous to delay flowering to increase growth rate [9]. However, if flowering is delayed too much, it is disadvantageous for seed maturation, so the photoperiod of the cultivation area according to latitude must be taken into consideration [4]. Mutation breeding involves the use of a mutagen to develop plants, exhibiting novel mutated characteristics that do not disturb elite cultivar traits [10]. Novel kenaf cultivars generated by radiation mutagenesis and showing improved seed and biomass yield characteristics have been developed using radiation breeding techniques. The cultivars of Hibiscus species have been generated via hybridization, but the floriculture and/or pharmaceutical industry relies on a limited number of mutated traits established based on specific flower quality parameters and consumer palatability [11–13].

Genetic change, or mutation, is a natural process that creates new genetic variants. Since the frequency of natural mutations is quite low, it is difficult to discover useful genetic variations in a short period of time. Therefore, generating genetic variations by treating crops with physical or chemical mutagens has been used successfully in mutation breeding [14,15]. Among physical mutagens, gamma rays have been used to induce mutations in various crops, such as rice, soybean, and kenaf, and to breed new cultivars [10,14,16]. Anthocyanins play an important role in plant pigmentation, which is strongly associated with the coloration in kenaf. Our research group has also developed novel kenaf cultivars via gamma irradiation on seeds of the introduced cultivars and registered them in the Korea Seed and Variety Service. In our previous work, we explored the potential of a novel flower or stem color mutation cultivar, 'Jeokbong' and 'Bora', as a functional food [14].

Detection of genomic sequences related to particular traits has the potential not only to study the function of genes, but also to accelerate the speed of future breeding [17–19]. Advances in next-generation sequencing (NGS) have made the rearrangement of plant genomes efficient and economical, and have also made it possible to detect the loci, types, and rates of mutations on a genome-wide scale [20,21]. GBS is a sequence-based genotyping method that is characterized by the use of restriction enzymes to sequence multiple samples simultaneously on an NGS platform using a reduced label of the target genome and a DNA barcode adapter [22,23]. GWAS is a method for detecting associations between phenotypes and genes in a population. While linkage mapping methods use bi-parental populations, GWAS is an approach that exploits diverse natural populations [24–26]. This approach has been applied to a variety of crops, including rice [27], soybean [28], maize [29], and sorghum [20], and it is being used in breeding programs. To date, no GWAS analyses of kenaf have been reported.

Although the potential industrial value of kenaf is quite high, genetic diversity and genetic studies of the crop are scarce. Therefore, in this study, 96 kenaf genetic resources, including gamma ray mutant lines, were sequenced using GBS, and GWAS analyses were performed for seven agronomic traits: days to flowering, dry weight, fresh weight, plant height, flower color, stem color, and leaf shape.

2. Results

2.1. Phenotypic Variation and Correlation Analysis

The 96 kenaf lines were determined for four quantitative traits: days to flowering, plant height, fresh weight, and dry weight, and three qualitative traits: flower color, stem color, and leaf shape (Table 1 and Figure 1). The days to flowering ranged from 72 to 125 d (mean 92 d) and plant height was 223–444 cm (mean 315 cm). The fresh and dry weights were 472–2400 g and 110–672 g, respectively (mean 1209 and 312 g). The coefficients of variation for days to flowering and plant height were 16.1% and 14.6%, respectively, which were significantly lower than for fresh and dry weight (43.4% and 47.9%). The skewness of all four traits was close to zero, and the height of the distribution was lower than the normal distribution. Ivory was the most common flower color (55 lines), followed by white (25) and purple (1). The most common stem color was green (89 lines), followed by dark purple (6) and brown (1). For leaf shape, there were 53 palmate types and 43 entire types. The correlations between the seven agricultural traits are shown in Table 2. The four agronomic traits, days to flowering, plant height, fresh weight, and dry weight, each had a positive correlation ($p \leq 0.01$). In contrast, stem color showed a negative correlation with fresh and dry weight ($p \leq 0.05$).

Table 1. Descriptive statistics for agronomic traits in 96 kenaf lines.

Trait	Mean	Min	Max	SD	CV (%)	Skew	Kur
Days to flowering (d)	92	72	125	14.77	16.1	0.463	−0.410
Fresh weight (g)	1209	472	2400	525.45	43.4	−0.046	−1.211
Dry weight (g)	311	110	672	149.12	47.9	0.117	−1.043
Plant height (cm)	315	223	444	46.14	14.6	0.229	−0.309

Min, minimum; Max, maximum; SD, standard deviation; CV, coefficient of variation; Skew, skewness; Kur, Kurtosis.

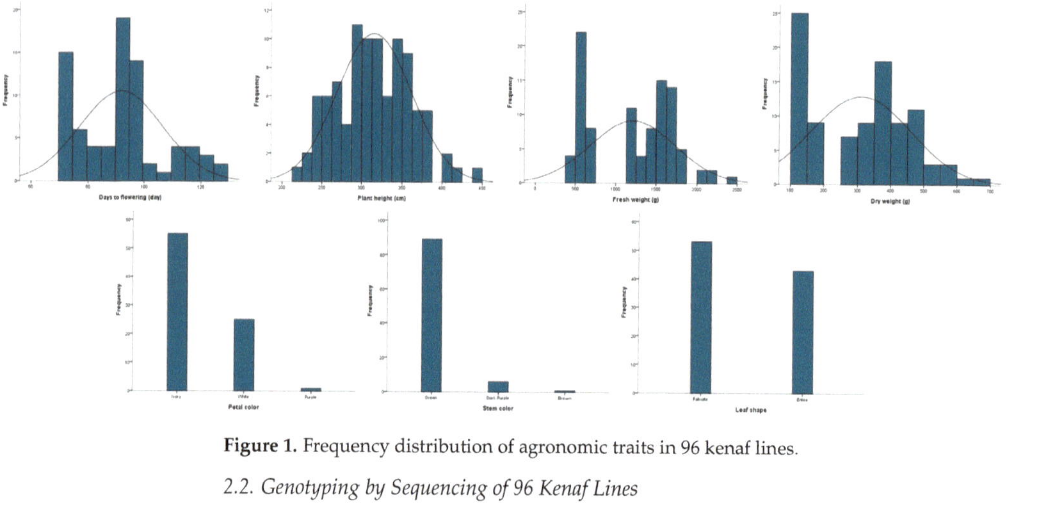

Figure 1. Frequency distribution of agronomic traits in 96 kenaf lines.

2.2. Genotyping by Sequencing of 96 Kenaf Lines

The GBS library was constructed from 96 kenaf lines, including gamma-ray-derived mutations, and sequenced using the Illumina Hiseq 2000 platform. A summary of the GBS results is presented in Table 3. Using two biological replicates, a total of 702 million reads comprising 106,096,097,764 nucleotides (106 Gb) were generated, with 7.3 million reads per genotype on average (Table 3). After trimming low quality sequences, 664,405,534 clean reads remained, with 6.6 million reads per genotype on average. The total length of the mapped region was 3,263,929,064 bp, with an average of 33,999,261 bp per sample, which covered approximately 3.17% of the whole genome; the sequences were mapped to the reference genome sequence. Among the 96 lines, the average depth of read mapping ranged from 10.84× to 21.49×.

Table 2. Correlation coefficients for agronomic traits in 96 kenaf lines.

	Days to Flowering	Flower Color	Stem Color	Leaf Shape	Fresh Weight	Dry Weight
Flower color	−0.159					
Stem color	−0.128	−0.123				
Leaf shape	0.05	−0.067	−0.039			
Fresh weight	0.744 **	−0.054	−0.243 *	0.012		
Dry weight	0.725 **	−0.084	−0.229 *	−0.022	0.992 **	
Plant height	0.876 **	0.054	−0.192	−0.002	0.909 **	0.905 **

* significant at $p \leq 0.05$, ** significant at $p \leq 0.01$.

Table 3. Summary of GBS sequence data and alignment to the reference genome sequence.

	Total	Mean/Genotype
	Raw data	
Reads	702,623,164	7,318,991
Bases (bp)	106,096,097,764	1,105,167,685
After trimming		
Reads	635,581,208	6,620,638
Bases (bp)	63,782,931,221	664,405,534
Mapped to reference genome		
Reads	631,887,178	6,582,158
Bases (bp)	3,263,929,064	33,999,261
Reference genome coverage (%)		3.17

2.3. Construction of Phylogenetic Tree and Genome-Wide Association Study for Agronomic Traits

Phylogenetic analysis was performed on 96 kenaf lines, including gamma-ray-derived mutants, based on the UPGMA method, and a dendrogram was generated with 49,241 filtered SNPs using the neighbor-joining method (Figure 2). In the cluster analysis, the 96 kenaf genotypes were divided into nine related groups within two independent groups, with the exception of seven lines (2012_WFM_2_3, C_11_P, C12, jangbaek_72, A19_dae, C15, and Z_1) that could not be grouped.

By plotting r^2 against the distance (in kb) between a pair of SNPs, it was a very modest decrease in LD decay. When we estimated LD decay for the entire genome, the maximum r^2 was halved at around 1433 kb, and as a result, the size of the LD block was considered quite large (Figure S1).

The GWAS was conducted using a multi-locus mixed model for seven agronomic traits: days to flowering, dry weight, fresh weight, plant height, flower color, stem color, and leaf shape. Using a threshold of $-\log_{10}(P) \geq 4$, a total of 49 SNPs were detected across 13 chromosomes for seven agronomic traits (Table 4 and Figure 3). Of these, 12 SNPs associated with days to flowering were detected on chromosomes 1, 4, 5, 7, 10, 14, 16, and 18; 11 SNPs associated with plant height were detected on chromosomes 1, 5, 15, 16, and 18; three SNPs associated with fresh weight and dry weight were detected on chromosomes 1, 2, and 13; seven SNPs associated with flower color were detected on chromosomes 1, 2, 4, 7, and 18; five SNPs associated with stem color were detected on chromosomes 1, 11, and 17; and eight SNPs associated with leaf shape were detected on chromosomes 1, 2, 9, 11, 13, and 16. Among the 49 SNPs detected, 27 were located in intergenic regions and 22 in genic regions.

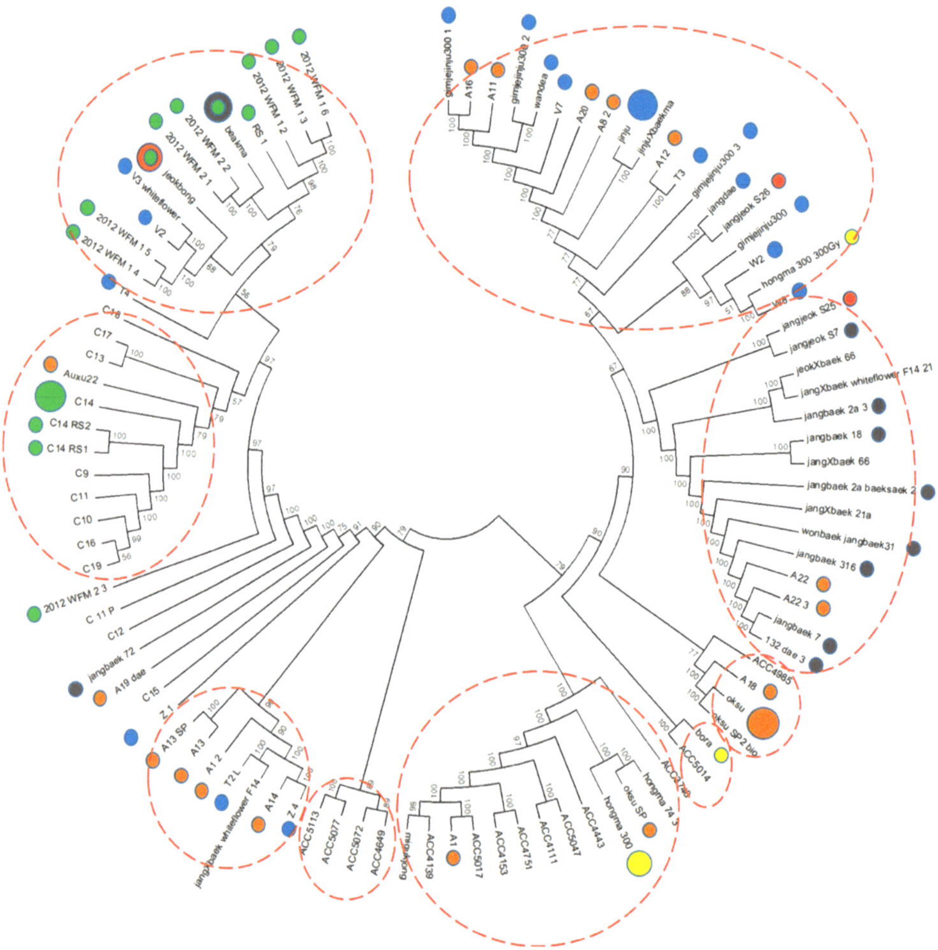

Figure 2. Phylogenetic tree constructed using 49,241 filtered SNP loci from 96 kenaf genotypes. Colored circles represent mutant lines; the larger colored circles denote the respective original genotypes.

Table 4. SNPs associated with seven agronomic traits detected in GWAS analysis.

Trait	Chr	Position	$-\log_{10}(P)$	Reference	Allele	MAF	Genic/Intergenic
Days to flowering	1	1,024,914	18.66	A	G/A	0.058	Genic
	1	42,517,384	19.30	G	A/G	0.209	Intergenic
	4	6,472,359	6.86	T	C/T	0.435	Genic
	5	6,215,497	18.86	G	A/G	0.167	Genic
	7	47,923,332	6.31	T	T/C	0.065	Genic
	10	41,332,261	4.20	T	A/T	0.310	Intergenic
	14	9,697,002	5.15	T	G/T	0.137	Genic
	14	31,710,078	14.47	C	A/C	0.069	Intergenic
	16	13,124,040	4.12	T	C/T	0.441	Intergenic
	16	23,595,487	4.51	C	T/C	0.383	Intergenic
	16	38,590,756	9.43	A	G/A	0.368	Intergenic
	18	1,013,873	7.86	A	G/A	0.291	Intergenic

Table 4. *Cont.*

Trait	Chr	Position	$-\log_{10}(P)$	Reference	Allele	MAF	Genic/Intergenic
Plant height	1	21,588,612	4.02	G	C/G	0.149	Intergenic
	1	48,009,505	4.40	C	T/C	0.375	Genic
	5	23,514,194	4.59	T	T/G	0.060	Intergenic
	15	48,486,135	4.25	T	A/T	0.284	Genic
	15	50,639,175	4.14	G	A/G	0.295	Genic
	15	51,634,174	4.13	G	A/G	0.327	Intergenic
	16	7,227,452	4.17	C	T/C	0.113	Genic
	16	7,227,677	4.31	A	T/A	0.122	Genic
	16	7,227,739	4.37	C	C/T	0.117	Genic
	18	42,282,127	4.70	A	G/A	0.298	Genic
	18	42,294,729	4.15	C	T/C	0.266	Intergenic
Fresh weight	1	30,956,363	4.96	A	G/A	0.218	Intergenic
	2	34,266,740	4.70	C	T/C	0.071	Intergenic
	13	36,194,770	4.57	T	G/T	0.880	Intergenic
Dry weight	1	30,956,363	4.96	A	G/A	0.218	Intergenic
	2	34,266,740	4.70	C	T/C	0.071	Intergenic
	13	36,194,770	4.57	T	G/T	0.880	Intergenic
Flower color	1	32,661,009	4.34	G	A/G	0.169	Intergenic
	2	4,123,503	4.01	T	G/T	0.426	Intergenic
	4	26,157,173	4.15	C	C/T	0.163	Intergenic
	7	54,080,566	4.02	T	C/T	0.500	Genic
	7	54,146,827	4.49	C	A/C	0.489	Genic
	7	56,094,620	5.66	A	G/A	0.240	Genic
	18	1,155,161	4.07	G	A/G	0.486	Intergenic
Stem color	1	59,001,068	10.81	G	T/G	0.069	Genic
	11	13,785,162	4.13	G	A/G	0.085	Genic
	11	13,785,316	4.47	A	A/G	0.065	Genic
	11	14,130,866	4.34	G	T/G	0.089	Intergenic
	17	44,726,748	4.23	T	A/T	0.121	Intergenic
Leaf shape	1	76,940,299	11.01	A	T/A	0.059	Genic
	2	70,676,156	6.67	A	G/A	0.161	Genic
	9	22,789,371	6.17	G	T/G	0.109	Intergenic
	11	4,977,859	4.89	C	T/C	0.307	Genic
	11	10,223,305	4.42	T	T/A	0.095	Intergenic
	13	28,565,139	5.38	C	A/C	0.104	Intergenic
	13	33,269,519	8.96	T	C/T	0.175	Intergenic
	16	6,361,215	4.59	A	G/A	0.317	Genic

Chr, chromosome; MAF, minor allele frequency.

2.4. Gene Annotation

The gene annotations of SNPs located in genic regions detected in the GWAS analysis are shown in Table 5. The genic regions had a 45.53% to 96.34% identity with genes from the Malvaceae family (*Herrania umbratica*, *Gossypium mustelinum*, *Hibiscus syriacus*, *Durio zibethinus*, *Gossypium hirsutum*, and *Gossypium australe*) and the Betulaceae family (*Carpinus fangiana*). Except for fresh weight and dry weight, where no SNPs were detected in the genic region, the majority of SNPs were located in the exon of the respective genic regions: three out of five SNPs were detected for days to flowering, four out of five SNPs for plant height, two out of three SNPs for flower color, one out of two SNPs for stem color, and four SNPs for leaf shape. The highest level of SNPs located in an exon of a gene associated with days to flowering was found on chromosome 5 with $-\log_{10}(P)$ = 18.86 and a 90.50% homology to the gene encoding an uncharacterized protein LOC120130459 from *Hibiscus syriacus*. The next highest $-\log_{10}(P)$ values were found for SNPs associated with other traits: plant height showed a 64.05% homology with the gene encoding the abrin-b-like of *Durio zibethinus* on chromosome 18, flower color showed a 66.89% homology with the gene encoding the hypothetical protein FH972_000750 of *Carpinus fangiana* on chromosome 7, stem color showed an 88.18% homology with the gene encoding the pentatricopeptide repeat-containing protein of *Hibiscus syriacus* on chromosome 11, and leaf shape showed an 82.24% homology with the gene encoding the protein DETOXIFICATION 3-like of *Hibiscus syriacus* on chromosome 1.

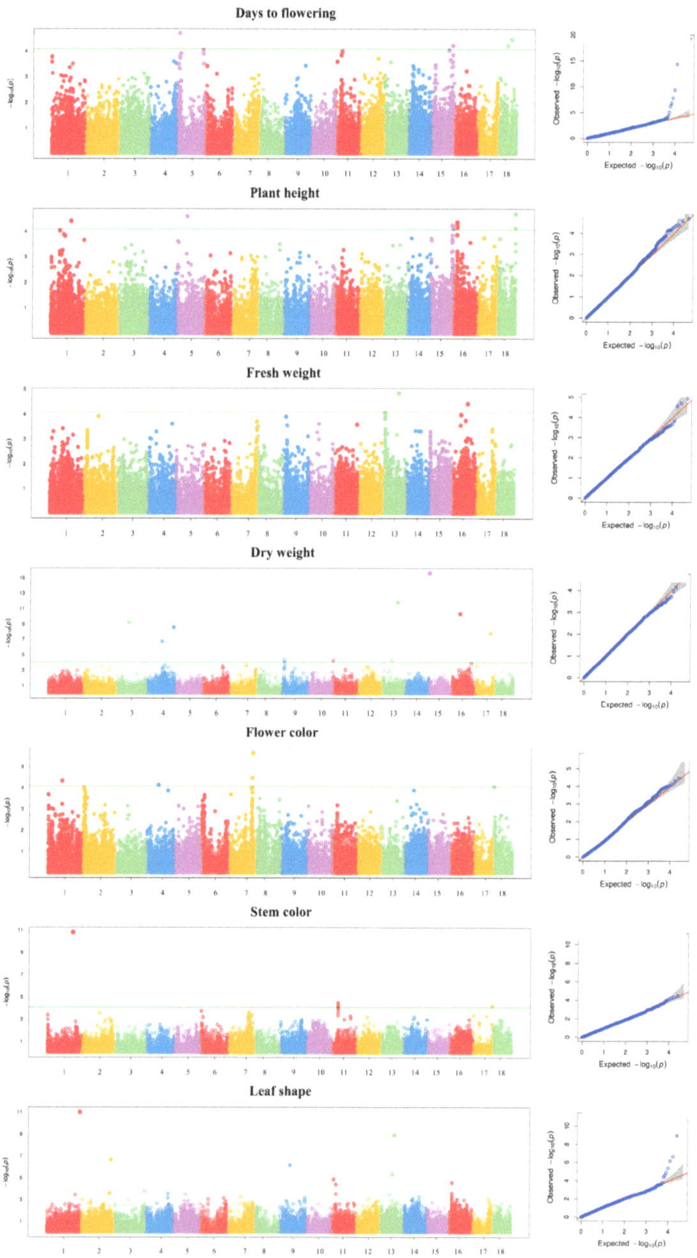

Figure 3. Manhattan plots and quantile–quantile plots for seven agronomic traits in 96 kenaf genotypes.

Table 5. SNPs and candidate genes associated with seven agronomic traits in 96 kenaf genotypes.

Trait	Chr	Position	Gene ID	Feature	Identity (%)	Location (bp)	Gene Description	Species
Days to flowering	1	1,024,914	GWHTACDB000043.1	Intron	81.36	1,021,982–1,029,723	serine/threonine-protein kinase STY46	Herrania umbratica
	4	6,472,359	GWHTACDB042287.1	Exon, CDS	45.53	6,472,165–6,473,223	hypothetical protein E1A91_D11G253500v1	Gossypium mustelinum
	5	6,215,497	GWHTACDB046286.1	Exon, CDS	90.50	6,207,495–6,233,844	uncharacterized protein LOC120130459	Hibiscus syriacus
	7	47,923,332	GWHTACDB055842.1	Intron	90.04	47,918,661–47,925,788	serine/threonine-protein kinase CTR1-like	Hibiscus syriacus
	14	9,697,002	GWHTACDB019052.1	Exon, CDS	90.14	9,690,988–9,697,964	putative LRR receptor-like serine/threonine-protein kinase	Hibiscus syriacus
Plant height	1	48,009,505	GWHTACDB002361.1	Intron	92.05	47,985,685–48,016,384	DNA polymerase I-like isoform X2	Hibiscus syriacus
	15	48,486,135	GWHTACDB023623.1	Exon, CDS	87.60	48,483,244–48,490,781	E3 ubiquitin-protein ligase BRE1-like 1	Hibiscus syriacus
	15	50,639,175	GWHTACDB023849.1	Exon, CDS	87.81	50,636,950–50,639,279	IQ domain-containing protein IQM2-like	Hibiscus syriacus
	16	7,227,452	GWHTACDB024773.1	Exon, UTR	92.59	7,227,353–7,229,462	uncharacterized protein LOC120203503	Hibiscus syriacus
	18	42,282,127	GWHTACDB032363.1	Exon, CDS	64.05	42,280,976–42,289,390	abrin-b-like	Durio zibethinus
Flower color	7	54,080,566	GWHTACDB056744.1	Exon, CDS	94.14	54,076,502–54,080,857	coatomer subunit epsilon-1	Durio zibethinus
	7	54,146,827	GWHTACDB056753.1	Exon, CDS	66.89	54,135,157–54,148,228	hypothetical protein FH972_000750	Carpinus fangiana
	7	56,094,620	GWHTACDB057049.1	Intron	94.14	56,091,648–56,099,189	PREDICTED: stomatal closure-related actin-binding protein 1-like	Gossypium hirsutum
Stem color	1	59,001,068	GWHTACDB003133.1	Intron	96.34	59,000,335–59,006,569	serine/threonine-protein kinase tricornered-like	Hibiscus syriacus
	11	13,785,162	GWHTACDB010323.1	Exon, UTR	88.18	13,784,863–13,787,951	Pentatricopeptide repeat-containing protein	Hibiscus syriacus
Leaf shape	1	76,940,299	GWHTACDB005392.1	Exon, CDS	82.24	76,939,400–76,941,687	protein DETOXIFICATION 3-like	Hibiscus syriacus
	2	70,676,156	GWHTACDB036463.1	Exon, CDS	52.79	70,673,908–70,679,842	uncharacterized protein LOC111277501	Durio zibethinus
	11	4,977,859	GWHTACDB009089.1	Exon, CDS	91.19	4,975,419–4,978,563	glutathione S-transferase DHAR3, chloroplastic-like isoform X1	Hibiscus syriacus
	16	6,361,215	GWHTACDB024720.1	Exon, UTR	94.85	6,358,025–6,361,424	protein FAR1-RELATED SEQUENCE 11	Gossypium australe

Chr, chromosome; CDS, coding sequence; UTR, untranslated region.

3. Discussion

Kenaf has wide utility value as it can be used as a source of fiber, edible and pharmaceutical ingredients, as well as for plastic materials, but the crop's genetics remain underresearched. The main objective of kenaf breeding has been to develop new cultivars that have high yields, are resistant to pests and diseases, are drought tolerant, or can be locally adapted to different environmental and growing conditions [5]. Breeding methods mainly used in kenaf include introduction breeding, exploiting natural variation, hybridizations, and mutation breeding. Recently, cultivars with new characteristics have become required because of changes in various cultivation environments and industrial demands [5,11]. Therefore, remote crossing and modular and mutation breeding are gradually being more widely used. However, the identification of genetic information and relationships using NGS technologies has been limited in kenaf. It is necessary to better understand the genetic basis of important traits in kenaf to improve production and to lay the foundation for molecular breeding efforts [11–13]. Our current study was the first to attempt GBS analysis targeting kenaf genetic resources and breeding lines for various agronomic traits and/or useful traits for the pharmaceutical industry.

In this study, we performed GWAS to investigate candidate genes for agronomic traits using 96 kenaf genotypes, including gamma-irradiation-derived mutant lines. Mutation induced using gamma radiation has been widely used to create genetic diversity via a variety of gene mutations [27]. In GWAS analyses, progeny lines can be highly correlated between samples and are liable to overestimate the genetic association because of false positives [30–32]. In addition, because progeny lines are derived from one or a few parent plants, they do not represent a wide range of genetic diversity, which makes it difficult to capture the many variations in the plant genome and the unusual variants that can occur via any pathway. Therefore, it is common to apply GWAS analyses to natural populations. In our results, among the lines used in the GWAS analysis, we found the phylogeny of the radiation mutant lines was sometimes close to the original line, such as the C14 lines, while in other cases, such as the Auxu and Jinju lines, they diverged into several branches (Table S1 and Figure 2). These results suggest that gamma-irradiation-derived mutant lines can be phylogenetically separated by GWAS, which uses genetic diversity wider than the original lines.

On the basis of their geographic origin, the phylogenetic analysis revealed an unclear pattern of division among the genotypes. Given that the kenaf genotypes evaluated in this study originated from various breeding methods and mixed pedigrees, it is likely that the phylogenetic analysis using many origins and races was not able to differentiate among all the genotypes. Kenaf originated in South Africa and was then introduced into India, China, Russia, and the Americas in the eighteenth century. Nowadays, kenaf is commercially cultivated in more than 20 countries, but the dissemination of genetic information about the crop worldwide is limited. Genotypes collected from Asia and central and North America were found to have close genetic relationships [33].

Correlations between agronomic traits can be used to predict yields or control harvest timing, but they are also important for breeding programs. In various crops, including kenaf, flowering time and biomass traits, such as fresh weight, dry weight, and plant height, are known to be positively correlated [34], and our results in Table 2 show the same pattern. Korean Kenaf cultivars are divided into three maturation groups depending on the flowering date: early maturing, mid–late maturing, and late maturing. Early-maturing groups mature in 70–80 d after sowing, which enables seed harvests, but at the cost of lower biomass. Late-maturing groups grow vegetatively for 130–140 d and yield significantly higher biomass, but late maturation reduces seed quality [14,15]. Therefore, a breeding goal in new Korean kenaf cultivars is to increase both biomass and seed yields per unit area [35,36]. Mutation breeding has the merits of creating new mutant characteristics and adding only few traits without disturbing the other characteristics of a cultivar [37]. The mid–late cultivars 'Jangdae' and 'Wandae', which afford both high biomass and high seed yield, have been registered with the Korea Seed and Variety Service.

Notably, there is a weak correlation between stem color and biomass. The population used in the analysis had three stem colors: green, dark purple, and brown, with stem colors other than green having a slightly lower biomass content. Although the non-green stem color was associated with lower biomass, such lines are valuable as breeding material because they produce functional substances such as anthocyanins [14]. In general, plant pigments are functional substances with various activities, such as antioxidant, antidiabetic, and anti-inflammatory compounds. Therefore, changes in stem color and flower color are accompanied by changes in the contents of functional compounds that might be valuable in various food and pharmaceutical industries. The Jeokbong, RS1, and RS2 genotypes has distinctive morphological characteristics such as dark purple color. The high levels of phenolic compounds, such as anthocyanin, observed in the previous study for the purple genotypes and their antioxidant activity are approximately 4~5 times higher than other cultivars caused by anthocyanin [14]. In the kenaf plant, the genes of phenylpropanoid biosynthetic enzymes, including *HcPAL*, *Hc4CL*, *HcC4H*, and *HcCHS*, have been isolated in previous studies [38,39]. Recent progress, including NGS technology, has facilitated various transcriptome analyses in kenaf. Previously, we successfully undertook transcriptome analysis of leaf coloration in 'Jeokbong', which exhibited dramatic changes in eight flavonoid structural genes. Identifying genes related to the biosynthesis of anthocyanins and kaempferitrin in kenaf leaves, 29 differentially expressed genes were assigned to eight structural genes, namely *4CL, CHS, CHI, F3H, DFR, ANS, FLS,* and *3GT* [12,14]. In addition, we performed comparative transcriptome analysis using RNA sequencing and identified putative genes (CHS, F3′H, FLS, DFR, MAT, UFGTs, TT12, GST, and RNS) involved in flower coloration at different flower developmental stages in three kenaf mutants, namely white flower, ivory flower, and purple flower [13]. When the color of petals was changed to purple or white by radiation mutation breeding, the total amount of flavonoids and phenolic compounds was more [12,13,36]. It was assumed that the accumulation of flavonoids and phenolic acids was accelerated. In the reports on the function of inside the plants, flavonoid glycosides were involved in protection against radiation stress and in producing a purple pigment [5,12,13]. We selected the SNPs in the flowers of new kenaf mutants, including white petal, purple petal, and moral types (ivory petal), as well as candidate gene data to better understand the mechanism of flower coloration via radiation.

GWAS analysis confirmed that the gene regions in which the SNPs detected for each trait were located were highly homologous to known gene regions in the Malvaceae and Betulaceae families, but no clearly identified genes for the traits were found. The LRR receptor-like serine/threonine-protein kinase (LRR-RLKs) detected as associated with flowering might be involved in several aspects of plant development, such as cell growth, differentiation, organ formation, and stress responses. Its roles might be similar to those of LRR-RLK BRASSINOSTEROID INSENSITIVE 1 (BRI1), a receptor for brassinosteroids, a class of plant hormones that regulate cell elongation and division [40–42]. The E3 ubiquitin-protein ligase BRE1-like 1 (HUB1), detected as associated with plant height, is known to regulate floral development in *Arabidopsis* by monoubiquitinating H2B [43]. The pentatricopeptide repeat-containing protein (PPR), detected as associated with stem color, is involved in the regulation of genes related to chlorophyll biosynthesis, photosynthesis, and flowering. It is also known to be involved in defense mechanisms against abiotic stresses, such as drought and salinity, and pathogens, such as viruses and bacteria [44]. The Korea Atomic Energy Research Institute discovered a variant with dark purple stems generated using gamma radiation as a mutation source and registered a cultivar 'Jeokbong' that is rich in anthocyanins (delphinidin-3-O-sambubioside). This is the same as the anthocyanin reported in roselle *(Hibiscus sabdariffa* L.). The antioxidant (2,2-diphenyl-1-picrylhydrazyl radical scavenging activities) and angiotensin-converting enzyme inhibitory activity of 'Jeokbong' was approximately four times higher than that of three other cultivars [14]. However, 'Jeokbong' and other purple stem color mutants (RS1, C14_RS1, jangjeok_S7, jangjeok_S26, and jangjeok_S25) showed poor salt tolerance. These results suggest it should be possible to develop an SNP marker for selection of stem anthocyanin and salt tolerance in kenaf.

The protein FAR1-RELATED SEQUENCE 11 (FRS11), detected as associated with leaf shape, is a member of the FRS family of transcription factors that play important roles in plant growth and development, light signaling, plant hormone responses, and stress resistance. Using a mutant with knocked out FRS11, it was confirmed that FRS11 regulates plant growth and development in response to light via its response to far-infrared light. It delayed flowering in potatoes [45] and is involved in leaf senescence in *Arabidopsis* [46]. It is clear that most of the genes identified in our study might play important roles in growth and development. However, it is likely that they influence each trait via a complex mechanism rather than a direct causal link, so further research is needed. Our findings will provide useful information for a better understanding of kenaf genetics and demonstrate the utility of mutations, either directly in breeding programs, or indirectly as a research tool.

4. Materials and Methods

4.1. Plant Materials and DNA Isolation

Ninety-six genotypes were studied (Table S1). These included 55 mutant genotypes derived from gamma irradiation (300 Gy) of seed. Thirty genotypes were obtained from the Genebank of Rural Development Administration (RDA) in Korea and the Bangladesh Jute Mills Corporation (BJC) (Dhaka, Bangladesh). These kenaf lines were collected from South Korea (1), United States of America (1), Iran (1), India (2), Russia (2), China (7), Italy (2), and Bangladesh (14). Eight white flower mutant genotypes were generated from selection breeding (natural variants). Four genotypes were developed from hybridization between 'Jangdae' and 'Baekma' (JangXbaek21, jangXbaek21a), 'Jeokbong' and 'Baekma' (jeokXbaek66), and 'Jinju' and 'Baekma' (jinjuXbaekma). The agronomic traits investigated were days to flowering, plant height, fresh weight, dry weight, flower color, stem color, and leaf shape. Genomic DNA was isolated from 20 mg of lyophilized leaf tissue using a DNeasy Plant Mini Kit (QIAGEN, Valencia, CA, USA) according to the manufacturer's protocol.

4.2. Sequence Pre-Processing and Alignment to Reference Genome

For demultiplexing, we used barcode sequences, adapter sequence trimming with cutadapt (version 1.8.3) [47], and sequence quality trimming with DynamicTrim and LengthSort programs from the SolexaQA (v.1.13) package [48]. DynamicTrim trims bad quality bases at both ends of short reads on the basis of their phred score and refines them into good-quality cleaned reads, while LengthSort removes reads with too many bases trimmed by DynamicTrim. DynamicTrim uses a phred score $\geq$ 20 and the LengthSort process uses a short read length $\geq$ 25 bp. The cleaned reads that passed the pre-processing were mapped to the reference genome using the BWA (0.7.17-r1188) program [49].

4.3. Raw SNP Detection and Generation of SNP Matrix

BAM-formatted files generated by mapping clean reads to the reference genome were used to perform SNP validation and raw SNP (In/Del) detection using the SAMtools (0.1.16) program [50] and SEEDERS own scripts [51], as well as to extract matching sequences. Default values were used except for the following options: minimum mapping quality for SNPs ($-$Q) = 30, minimum mapping quality for gaps ($-$q) = 15, minimum read depth ($-$d) = 3, maximum read depth ($-$D) = 398, minimum indel score for filtering nearby SNPs ($-$G) = 30, SNPs within INT bp around gaps to filter ($-$w) = 15, and window size for filtering high density SNPs ($-$W) = 15. We created a unified SNP matrix between samples to perform SNP comparison analysis between analytes. In this method, the raw SNP positions obtained by comparing each sample with the reference genome are used as candidate SNP positions to build a unified list, and the empty regions (non-SNP loci) are filled with the consensus sequence of the sample to create a matrix. The final SNP matrix is then created by filtering out mis-called SNP (In/Del) loci via SNP comparisons between samples. Among these loci, SNPs (In/Del) are classified according to the following type classification criteria: Homozygous, SNP read depth $\geq$ 90%; Heterozygous, 40% $\leq$ SNP read depth $\leq$ 60%; etc.: 20% $\leq$ read rate < 40% and 60% < read rate < 90%.

4.4. Linkage Disequilibrium Estimation

LD decay analysis with PopLDdecay program (https://github.com/BGI-shenzhen/PopLDdecay accessed on 5 December 2023) using a VCF file of 49,241 SNP loci from 96 kenaf samples. PopLDdecay Option was set as follows: Min minor allele frequency filter: 0.05, Max ratio of het allele filter: 0.6, and Max ratio of miss allele filter: 0.3. Using the calculated LD values, the LD decay plot was plotted using the HW method (the Hill and Weir method) [52] in R package.

4.5. Genome-Wide Association Study and Phylogenetic Analysis

To perform association analysis, we used an SNP filter process with the following conditions: SNP loci of biallelic, minor allele frequency > 5%, and missing data < 30%. The representative SNP loci and trait information selected via the filter process were used for association analysis using GAPIT [53]. The GWAS analysis was performed using a multiple mixed linear model. A total of 49,241 positions were used in the final dataset and the phylogenetic analysis was performed in MEGA6 [54]. Bootstrap consensus trees inferred from 1000 replicates were used to represent the evolutionary history of the analyzed taxa using the UPGMA method.

5. Conclusions

In this study, we performed phylogenetic and GWAS analyses using GBS data to investigate candidate genes for agronomic traits in 96 kenaf genotypes, including mutant lines derived from gamma irradiation. Gamma-irradiation-derived mutant lines were phylogenetically separated using wider genetic diversity compared with the original varieties. As a result of our GWAS analysis of agronomic traits, kenaf genes showing homology to known genes related to growth were discovered. The results of our study identified mutant genotypes and 19 candidate genes significantly associated with five traits of kenaf. This is the first study describing the SNPs generated using GBS and an association analysis to identify genes related to the large variability in morphological characteristics and yield-related traits among kenaf mutants derived from gamma-ray irradiation. The mutants obtained in this study will be useful genetic resources for the development of novel kenaf cultivars with improved ornamental value and chemical compound profiles. Our results may help breeders select kenaf mutants with suitable SNPs as optimal genotypes for the bioplastics and pharmaceutical industry. These findings highlight the value of mutant lines in genetic analyses and should prove useful for kenaf breeding programs.

Supplementary Materials: The following supporting information can be downloaded at: https://www.mdpi.com/article/10.3390/plants13020249/s1, Table S1: Country of origin for the 96 kenaf assessed in this study. Figure S1: Distribution of the marker density and decay of the linkage disequilibrium (LD) over distance presented as an accumulative distribution.

Author Contributions: Writing—original draft preparation, W.J.K.; methodology, B.Y., Y.-j.L. and S.-H.K.; writing—review and editing, J.H.K.; resources, S.H.K., J.-W.A. and S.-Y.K.; supervision, J.R. All authors have read and agreed to the published version of the manuscript.

Funding: This work was supported by the research program of KAERI, Republic of Korea (Project No. 523410-24) and the National Research Foundation of Korea (NRF) grant funded by the Korea government (RS-2022-00156231).

Data Availability Statement: The original contribution presented in the study are publicly available.

Conflicts of Interest: The authors declare that the research was conducted in the absence of any commercial or financial relationships that could be construed as potential conflicts of interest.

References

1. Afzal, M.Z.; Ibrahim, A.K.; Xu, Y.; Niyitanga, S.; Li, Y.; Li, D.; Yang, X.; Zhang, L. Kenaf (*Hibiscus cannabinus* L.) breeding. *J. Nat. Fibers* **2022**, *19*, 4063–4081. [CrossRef]
2. Webber, C.L., III; Bhardwaj, H.L.; Bledsoe, V.K. Kenaf production: Fiber, feed, and seed. *Trends New Crops New Uses* **2002**, *13*, 327–339.
3. Dempsey, J.M. *Fiber Crops*; Univ. Presses of Florida: Gainesville, FL, USA, 1975.
4. Scott, A. *Kenaf Seed Production: 1981–82*; Biennial Report for 1980; Rio Farms, Inc.: Monte Alto, TX, USA, 1981; pp. 60–63.
5. Alexopoulou, E.; Papatheohari, Y.; Christou, M.; Monti, A. Origin, description, importance, and cultivation area of kenaf. In *Kenaf: A Multi-Purpose Crop for Several Industrial Applications: New Insights from the Biokenaf Project*; Springer: Berlin/Heidelberg, Germany, 2013; pp. 1–15.
6. Khare, C.P. *Indian Medicinal Plants: An Illustrated Dictionary*; Springer Science & Business Media: Berlin, Germany, 2008.
7. Taufiq, M.; Mansor, M.R.; Mustafa, Z. Characterisation of wood plastic composite manufactured from kenaf fibre reinforced recycled-unused plastic blend. *Compos. Struct.* **2018**, *189*, 510–515. [CrossRef]
8. Ramesh, P.; Durga Prasad, B.; Narayana, K. Characterization of kenaf fiber and its composites: A review. *J. Reinf. Plast. Compos.* **2018**, *37*, 731–737. [CrossRef]
9. Dryer, J. In Kenaf seed cultivars. In Proceedings of the First Conference on Kenaf for Pulp, Gainesville, FL, USA, 31 October–1 November 1967; pp. 44–46.

10. Ha, B.-K.; Lee, K.J.; Velusamy, V.; Kim, J.-B.; Kim, S.H.; Ahn, J.-W.; Kang, S.-Y.; Kim, D.S. Improvement of soybean through radiation-induced mutation breeding techniques in Korea. *Plant Genet. Resour.* **2014**, *12* (Suppl. S1), S54–S57. [CrossRef]
11. Li, D.; Huang, S. The Breeding of Kenaf. In *Kenaf: A Multi-Purpose Crop for Several Industrial Applications: New Insights from the Biokenaf Project*; Springer: Berlin/Heidelberg, Germany, 2013; pp. 45–58.
12. Lyu, J.I.; Choi, H.-I.; Ryu, J.; Kwon, S.-J.; Jo, Y.D.; Hong, M.J.; Kim, J.-B.; Ahn, J.-W.; Kang, S.-Y. Transcriptome analysis and identification of genes related to biosynthesis of anthocyanins and kaempferitrin in kenaf (*Hibiscus cannabinus* L.). *J. Plant Biol.* **2020**, *63*, 51–62. [CrossRef]
13. Lyu, J.I.; Ryu, J.; Kim, D.-G.; Kim, J.M.; Ahn, J.-W.; Kwon, S.-J.; Kim, S.H.; Kang, S.-Y. Comparative Transcriptome Analysis Identified Potential Genes and Transcription Factors for Flower Coloration in Kenaf (*Hibiscus cannabinus* L.). *Agronomy* **2023**, *13*, 715. [CrossRef]
14. Ryu, J.; Kwon, S.-J.; Ahn, J.; Kim, S.H.; Lee, S.Y.; Kim, J.-B.; Jo, Y.; Ha, B.-K.; Kang, S.-Y. Development of a stem-color mutant kenaf (*Hibiscus cannabinus* L.) cultivar, 'Jeokbong', and analysis of its functional compounds. *Hortic. Sci. Technol.* **2018**, *36*, 77–81. [CrossRef]
15. Kim, D.-G.; Ryu, J.; Lee, M.-K.; Kim, J.M.; Ahn, J.-W.; Kim, J.-B.; Kang, S.-Y.; Bae, C.-H.; Kwon, S.-J. Nutritional properties of various tissues from new kenaf cultivars. *J. Crop Sci. Biotechnol.* **2018**, *21*, 229–239. [CrossRef]
16. Kang, S.; Shin, I.; Kim, D.; Lee, G.; Kim, J.; Lee, D.; Lee, S.; Lee, D. A new green-kerneled glutinous rice mutant variety, "Nogwonchalbyeo" developed by gamma ray irradiation. *Korean J. Breed. Sci.* **2008**, *40*, 303–307.
17. Lu, L.; Zhai, X.; Li, X.; Wang, S.; Zhang, L.; Wang, L.; Jin, X.; Liang, L.; Deng, Z.; Li, Z. Met1-specific motifs conserved in OTUB subfamily of green plants enable rice OTUB1 to hydrolyse Met1 ubiquitin chains. *Nat. Commun.* **2022**, *13*, 4672. [CrossRef]
18. Jiang, M.; Chen, S.; Lu, X.; Guo, H.; Chen, S.; Yin, X.; Li, H.; Dai, G.; Liu, L. Integrating Genomics and Metabolomics for the Targeted Discovery of New Cyclopeptides with Antifungal Activity from a Marine-Derived Fungus Beauveria felina. *J. Agric. Food Chem.* **2023**, *71*, 9782–9795. [CrossRef] [PubMed]
19. Li, D.; Zaman, W.; Lu, J.; Niu, Q.; Zhang, X.; Ayaz, A.; Saqib, S.; Yang, B.; Zhang, J.; Zhao, H. Natural lupeol level variation among castor accessions and the upregulation of lupeol synthesis in response to light. *Ind. Crops Prod.* **2023**, *192*, 116090. [CrossRef]
20. Lee, Y.-J.; Yang, B.; Kim, W.J.; Kim, J.; Kwon, S.-J.; Kim, J.H.; Ahn, J.-W.; Kim, S.H.; Rha, E.-S.; Ha, B.-K. Genome-Wide Association Study (GWAS) of the Agronomic Traits and Phenolic Content in Sorghum (*Sorghum bicolor* L.) Genotypes. *Agronomy* **2023**, *13*, 1449. [CrossRef]
21. Behjati, S.; Tarpey, P.S. What is next generation sequencing? *Arch. Dis. Child. -Educ. Pract.* **2013**, *98*, 236–238. [CrossRef] [PubMed]
22. Poland, J.A.; Rife, T.W. Genotyping-by-sequencing for plant breeding and genetics. *Plant Genome* **2012**, *5*, 92–101. [CrossRef]
23. Deschamps, S.; Llaca, V.; May, G.D. Genotyping-by-sequencing in plants. *Biology* **2012**, *1*, 460–483. [CrossRef]
24. Josephs, E.B.; Stinchcombe, J.R.; Wright, S.I. What can genome-wide association studies tell us about the evolutionary forces maintaining genetic variation for quantitative traits? *New Phytol.* **2017**, *214*, 21–33. [CrossRef]
25. Ayaz, A.; Huang, H.; Zheng, M.; Zaman, W.; Li, D.; Saqib, S.; Zhao, H.; Lü, S. Molecular cloning and functional analysis of GmLACS2-3 reveals its involvement in cutin and suberin biosynthesis along with abiotic stress tolerance. *Int. J. Mol. Sci.* **2021**, *22*, 9175. [CrossRef]
26. Ayaz, A.; Saqib, S.; Huang, H.; Zaman, W.; Lü, S.; Zhao, H. Genome-wide comparative analysis of long-chain acyl-CoA synthetases (LACSs) gene family: A focus on identification, evolution and expression profiling related to lipid synthesis. *Plant Physiol. Biochem.* **2021**, *161*, 1–11. [CrossRef]
27. Hwang, S.-G.; Lee, S.C.; Lee, J.; Lee, J.W.; Kim, J.-H.; Choi, S.Y.; Kim, J.-B.; Choi, H.-I.; Jang, C.S. Resequencing of a core rice mutant population induced by gamma-ray irradiation and its application in a genome-wide association study. *J. Plant Biol.* **2020**, *63*, 463–472. [CrossRef]
28. Kim, W.J.; Kang, B.H.; Kang, S.; Shin, S.; Chowdhury, S.; Jeong, S.-C.; Choi, M.-S.; Park, S.-K.; Moon, J.-K.; Ryu, J. A Genome-Wide Association Study of Protein, Oil, and Amino Acid Content in Wild Soybean (*Glycine soja*). *Plants* **2023**, *12*, 1665. [CrossRef]
29. Tian, F.; Bradbury, P.J.; Brown, P.J.; Hung, H.; Sun, Q.; Flint-Garcia, S.; Rocheford, T.R.; McMullen, M.D.; Holland, J.B.; Buckler, E.S. Genome-wide association study of leaf architecture in the maize nested association mapping population. *Nat. Genet.* **2011**, *43*, 159–162. [CrossRef]
30. Clayton, D.G.; Walker, N.M.; Smyth, D.J.; Pask, R.; Cooper, J.D.; Maier, L.M.; Smink, L.J.; Lam, A.C.; Ovington, N.R.; Stevens, H.E. Population structure, differential bias and genomic control in a large-scale, case-control association study. *Nat. Genet.* **2005**, *37*, 1243–1246. [CrossRef]
31. Korte, A.; Vilhjálmsson, B.J.; Segura, V.; Platt, A.; Long, Q.; Nordborg, M. A mixed-model approach for genome-wide association studies of correlated traits in structured populations. *Nat. Genet.* **2012**, *44*, 1066–1071. [CrossRef]
32. Brachi, B.; Morris, G.P.; Borevitz, J.O. Genome-wide association studies in plants: The missing heritability is in the field. *Genome Biol.* **2011**, *12*, 232. [CrossRef]
33. Chen, Y.; Lin, L.; Wu, J.; Qi, J.; Zhou, R. Genetic effect analysis of some field and quality traits of kenaf hybrid and parents. *Plant Fibers Prod.* **2004**, *26*, 261–266.
34. Cheng, Z.; Lu, B.-R.; Sameshima, K.; Fu, D.-X.; Chen, J.-K. Identification and genetic relationships of kenaf (*Hibiscus cannabinus* L.) germplasm revealed by AFLP analysis. *Genet. Resour. Crop Evol.* **2004**, *51*, 393–401. [CrossRef]
35. Kang, S.; Kwon, S.; Jeong, S.; Kim, J.; Kim, S.; Ryu, J. An improved kenaf cultivar 'Jangdae' with seed harvesting in Korea. *Korean J. Breed. Sci.* **2016**, *48*, 349–354. [CrossRef]

36. Ryu, J.; Kwon, S.-J.; Kim, D.-G.; Lee, M.-K.; Kim, J.M.; Jo, Y.D.; Kim, S.H.; Jeong, S.W.; Kang, K.-Y.; Kim, S.W. Morphological characteristics, chemical and genetic diversity of kenaf (*Hibiscus cannabinus* L.) genotypes. *J. Plant Biotechnol.* **2017**, *44*, 416–430. [CrossRef]
37. Spencer-Lopes, M.; Forster, B.P.; Jankuloski, L. *Manual on Mutation Breeding*; Food and Agriculture Organization of the United Nations (FAO): Rome, Italy, 2018.
38. Lyu, J.I.; Ramekar, R.; Kim, D.-G.; Kim, J.M.; Lee, M.-K.; Hung, N.N.; Kim, J.-B.; Ahn, J.-W.; Kang, S.-Y.; Choi, I.-Y. Characterization of gene isoforms related to cellulose and lignin biosynthesis in Kenaf (*Hibiscus cannabinus* L.) mutant. *Plants* **2020**, *9*, 631. [CrossRef]
39. Ryu, J.; Kwon, S.-J.; Sung, S.Y.; Kim, W.-J.; Kim, D.S.; Ahn, J.-W.; Kim, J.-B.; Kim, S.H.; Ha, B.-K.; Kang, S.-Y. Molecular cloning, characterization, and expression analysis of lignin biosynthesis genes from kenaf (*Hibiscus cannabinus* L.). *Genes Genom.* **2016**, *38*, 59–67. [CrossRef]
40. Lease, K.A.; Lau, N.Y.; Schuster, R.A.; Torii, K.U.; Walker, J.C. Receptor serine/threonine protein kinases in signalling: Analysis of the erecta receptor-like kinase of Arabidopsis thaliana. *New Phytol.* **2001**, *151*, 133–143. [CrossRef]
41. Lin, F.; Li, S.; Wang, K.; Tian, H.; Gao, J.; Zhao, Q.; Du, C. A leucine-rich repeat receptor-like kinase, OsSTLK, modulates salt tolerance in rice. *Plant Sci.* **2020**, *296*, 110465. [CrossRef]
42. Liu, P.-L.; Du, L.; Huang, Y.; Gao, S.-M.; Yu, M. Origin and diversification of leucine-rich repeat receptor-like protein kinase (LRR-RLK) genes in plants. *BMC Evol. Biol.* **2017**, *17*, 47. [CrossRef]
43. Cao, Y.; Dai, Y.; Cui, S.; Ma, L. Histone H2B monoubiquitination in the chromatin of FLOWERING LOCUS C regulates flowering time in Arabidopsis. *Plant Cell* **2008**, *20*, 2586–2602. [CrossRef]
44. Barkan, A.; Small, I. Pentatricopeptide repeat proteins in plants. *Annu. Rev. Plant Biol.* **2014**, *65*, 415–442. [CrossRef]
45. Chen, Q.; Song, Y.; Liu, K.; Su, C.; Yu, R.; Li, Y.; Yang, Y.; Zhou, B.; Wang, J.; Hu, G. Genome-Wide Identification and Functional Characterization of FAR1-RELATED SEQUENCE (FRS) Family Members in Potato (*Solanum tuberosum*). *Plants* **2023**, *12*, 2575. [CrossRef]
46. Ma, L.; Li, G. FAR1-related sequence (FRS) and FRS-related factor (FRF) family proteins in Arabidopsis growth and development. *Front. Plant Sci.* **2018**, *9*, 692. [CrossRef]
47. Martin, M. Cutadapt removes adapter sequences from high-throughput sequencing reads. *EMBnet. J.* **2011**, *17*, 10–12. [CrossRef]
48. Cox, M.P.; Peterson, D.A.; Biggs, P.J. SolexaQA: At-a-glance quality assessment of Illumina second-generation sequencing data. *BMC Bioinform.* **2010**, *11*, 485. [CrossRef] [PubMed]
49. Li, H. Aligning sequence reads, clone sequences and assembly contigs with BWA-MEM. *arXiv* **2013**, arXiv:1303.3997.
50. Li, H.; Handsaker, B.; Wysoker, A.; Fennell, T.; Ruan, J.; Homer, N.; Marth, G.; Abecasis, G.; Durbin, R.; Subgroup, G.P.D.P. The sequence alignment/map format and SAMtools. *Bioinformatics* **2009**, *25*, 2078–2079. [CrossRef] [PubMed]
51. Kim, J.-E.; Oh, S.-K.; Lee, J.-H.; Lee, B.-M.; Jo, S.-H. Genome-wide SNP calling using next generation sequencing data in tomato. *Mol. Cells* **2014**, *37*, 36. [CrossRef] [PubMed]
52. Hill, W.; Weir, B. Variances and covariances of squared linkage disequilibria in finite populations. *Theor. Popul. Biol.* **1988**, *33*, 54–78. [CrossRef]
53. Lipka, A.E.; Tian, F.; Wang, Q.; Peiffer, J.; Li, M.; Bradbury, P.J.; Gore, M.A.; Buckler, E.S.; Zhang, Z. GAPIT: Genome association and prediction integrated tool. *Bioinformatics* **2012**, *28*, 2397–2399. [CrossRef]
54. Tamura, K.; Stecher, G.; Peterson, D.; Filipski, A.; Kumar, S. MEGA6: Molecular evolutionary genetics analysis version 6.0. *Mol. Biol. Evol.* **2013**, *30*, 2725–2729. [CrossRef]

plants

Article

Variability of Glucosinolates in Pak Choy (*Brassica rapa* subsp. *chinensis*) Germplasm

Seong-Hoon Kim [1,*], Kingsley Ochar [1,2], Aejin Hwang [1], Yoon-Jung Lee [1] and Hae Ju Kang [3]

[1] National Agrobiodiversity Center, National Institute of Agricultural Sciences, Rural Development Administration, Jeonju 5487, Republic of Korea; ocharocharking@yahoo.com (K.O.); hyj6138@korea.kr (A.H.); yoon112@korea.kr (Y.-J.L.)
[2] Council for Scientific and Industrial Research, Plant Genetic Resources Research Institute, Bunso P.O. Box 7, Ghana
[3] Department of Agrofood Resources, National Institute of Agricultural Sciences, Rural Development Administration, Wanju 55365, Republic of Korea; kanghaeju5@korea.kr
* Correspondence: shkim0819@korea.kr

Abstract: Glucosinolates are sulfur-containing phytochemicals generally abundant in cruciferous vegetables such as pak choy. Glucosinolates participate in a range of biological activities essential for promoting a healthy human body. In this study, we aimed to elucidate glucosinolate variability present in pak choy germplasm that are under conservation at the Rural Development Administration Genebank, Jeonju, Republic of Korea. The Acquity Ultra-Performance Liquid Chromatography (UHPLC) analytical system was used in profiling the glucosinolate content in leaf samples of various accessions. We identified a total of 17 glucosinolates in the germplasm. Based on principal component analysis performed, three separate groups of the accessions were obtained. Group 1 contained the cultivar cheongsacholong which recorded high content of glucobrassicin (an indole), glucoerucin (aliphatic), gluconasturtiin (aromatic) and glucoberteroin (aliphatic). Group 2 consisted of six accessions, BRA77/72, Lu ling gaogengbai, 9041, Wuyueman, RP-75 and DH-10, predominatly high in aliphatic compounds including glucoiberin, glucocheirolin, and sinigrin. Group 3 comprised the majority of the accessions which were characterized by high content of glucoraphanin, epiprogoitrin, progoitrin, and glucotropaeolin. These results revealed the presence of variability among the pak choy germplasm based on their glucosinolate content, providing an excellent opprtunity for future breeding for improved glucosinolate content in the crop.

Keywords: Brassica; compounds; germplasm; glucosinolates; pak choy; phytochemicals

Citation: Kim, S.-H.; Ochar, K.; Hwang, A.; Lee, Y.-J.; Kang, H.J. Variability of Glucosinolates in Pak Choy (*Brassica rapa* subsp. *chinensis*) Germplasm. *Plants* **2024**, *13*, 9. https://doi.org/10.3390/plants13010009

Academic Editor: Ulrike Lohwasser

Received: 22 November 2023
Revised: 16 December 2023
Accepted: 18 December 2023
Published: 19 December 2023

1. Introduction

Plant bioactive compounds or phytochemicals continue to attract extensive multidisciplinary research attention due to their known essential roles in the human body [1,2]. In particular their antioxidant prospects and consequently their involvements in many human disease prevention therapies make these compounds worthy of research [3,4]. A broad diversity of phytochemicals is reported in fruit and vegetable species which underscores the importance of consuming diets that are rich in essential bioactive compounds [5,6]. Pak choy (*Brassica rapa* subsp. *chinensis*), a member of the family Brassicaceae, is widely recognized as an important leafy vegetable not only for its remarkable culinary adaptability and nutritional richness but also its association with health benefits in humans [7,8]. Pak choy has characteristic non-heading leaves, which makes the crop distinct from its close relatives, mainly Chinese cabbage and yellow sarson [9,10]. As a result of the presence of a wide range of variability in the crop such as its morphological features, there are diverse accessions of pak choy found in many major Genebanks across the globe [9]. As a cruciferous vegetable, pak choy contains glucosinolates (GSL), a category of sulfur-containing secondary plant metabolites that have come under intense scrutiny for their bioactive properties and potential health-enhancing effects [11,12]. GSLs are famous based on their linkage with a broad range of health benefits such as their antioxidant, antimicrobial, and anti-inflammatory activities, and perhaps most notably their anticancer properties [12–14]. By examining GSLs within pak choy germplasm, researchers can gain a thorough understanding of their composition and concentrations across different genotypes and populations [15,16]. Such a comprehensive chemical profiling facilitates the identification of germplasm with high glucosinolate content. Germplasm containing GSLs at a desirable level of composition can be leveraged to develop functional foods or dietary strategies aimed at harnessing specific health benefits [17]. This further presents a treasurable resource for research aimed at investigating the composition and variations of GSLs present in different accessions of the crop for the purpose of crop improvement.

Technological advancements, particularly in the domain of analytical techniques like Ultra-Performance Liquid Chromatography-Tandem Mass Spectrometry (UPLC) integrated with mass spectrometry (MS), have considerably increased the precision and accuracy level of GSLs analyses [18,19]. Such cutting-edge techniques ease the identification, separation, and quantification process of the numerous GSLs present in pak choy tissues. The enhanced ability to accurately measure and quantify GSLs leads to comprehensive profiling, fostering a profound understanding of the GSL content in pak choy germplasm [20,21]. We are of the view that investigating GSL content in pak choy germplasm stored in Genebanks coincides with the growing global interest in enhancing nutritionally rich diets and promoting the consumption of functional foods [22,23]. By extracting the composition and assessing the presence of variability of GSLs in pak choy, researchers can help devise strategies that maximize the potential health benefits associated with these bioactive compounds. Therefore, in this study, we profiled GSLs content in pak choy germplasm selected from the Rural Development Administration genebank (RDA-Genebank), the Republic of Korea using the Acquity Ultra-Performance Liquid Chromatography (Milford, CT, USA) [24]. We analyzed the compositional variation in the compounds and identified accessions that exhibited desirably high GSLs contents, thus providing substantial insights for breeders. Overall, our results confirmed that the information on GSLs profiled in this study is useful for metabolic differentiation within the pak choy germplasm. Detailed knowledge about the diversity of GSLs in pak choy will shed light on the phytochemical composition of this vegetable; also, the study underscores the potential of the compound as a functional food with significant health implications, emphasizing the importance of incorporating this vegetable into a balanced diet for disease prevention and overall well-being.

2. Materials and Methods

2.1. Chemicals and Glucosinolate Standards

We used analytical-grade chemicals sourced from ThermoFisher Scientific Korea Ltd. (Seoul, Republic of Korea) and Sigma-Aldrich (St. Louis, MO, USA) for the extraction and analyses of the GSLs. Seventeen commercially available GSL standards used for the analyses were procured either from Phytoplan (Neuenheimer, Heidelberg, Germany) or Phytolab (Martin Baue, KG, Vestenbergsgreuth, Germany). The entire GSL standards had a purity $\geq$98%. Table 1 shows details of GSL standards used in this study.

2.2. Plant Materials, Sample Preparation and Compound Extraction

The National Agrobiodiversity Center of the Rural Development Administration (RDA), Jeonju, Republic of Korea has over 90 pak choy accessions under conservation in its National Genebank (RDA-Genebank). In this study, we selected 65 different pak choy accessions, classified as either cultivar or landrace, originating from different geographical locations (Table S1). First, seeds of the various accessions were multiplied under greenhouse conditions from February to June, 2019–2021. High-quality seeds from the various pak choy accessions were then selected for planting in the experimental field of the National Institute of Agricultural Sciences, Jeonju, and Republic of Korea form September to November 2021. For each accession, leaves of 18 plants, all of uniform phenotype, were collected, pooled together and kept in polyvinyl bags and stored at −80 °C until they were used for GSL extraction. The mixed leaf samples per each of three replications were analyzed. The extraction of the GSLs was performed based on the procedure as used previously in the work of Kim et al. [24]. Briefly, a 0.1 g sample containing 5 mL methanol (80%) that had been stored in a temperature of 25 °C for 30 min was thoroughly shaken to mix the contents at 120 rpm for 30 min at room temperature. The mixture was centrifuged (14,000 rpm at 4 °C for 10 min) and the supernatants transferred to fresh vials for GSL analyses using UPLC-MS/MS.

Table 1. List of glucosinolate standards procured and used in the study.

Class	Glucosinolate	Abbreviation	Molecular Formula	Molecular Weight (g/mol)	Source
Aliphatic GSL	Glucoiberin	GIB	C11H21NO10S3	423.5	Phytolab
	Sinigrin	SIN	C10H16KNO9S2	397.5	Phytoplan
	Glucocheirolin	GCH	C11H20KNO11S3	477.6	Phytoplan
	Glucoerucin	GER	C12H23NO9S3	421.5	Phytoplan
	Glucoraphanin	GRA	C12H23NO10S3	437.5	Phytoplan
	Gluconapin	GNA	C11H10NO9S2	373.4	Phytoplan
	Progoitrin	PRO	C11H19NO10S2	389.4	Phytolab
	Epiprogoitrin	EPI	C11H19NO10S2	389.4	Phytolab
	Glucoraphasatin	GRH	C12H21NO10S3	435.5	Phytoplan
	Glucoraphanin	GRE	C12H23NO10S3	437.5	Phytolab
	Glucoberteroin	GBE	C13H25NO9S3	435.5	Phytoplan
	Glucobrassicanapin	GBN	C12H21NO9S2	387.4	Phytolab
Aromatic GSL	Glucotropaeolin	GTL	C14H19NO9S2	409.4	Phytoplan
	Gluconasturtiin	GNS	C15H21NO9S2	423.5	Phytoplan
	Glucobarbarin	GBB	C15H21NO10S2	439.5	Phytoplan
	Sinalbin	SNB	C14H19NO10S2	425.4	Phytolab
Indolic GSL	Glucobrassicin	GBC	C16H20N2O9S2	448.5	Phytoplan

2.3. *Identification and Quantification of GSLs Using UPLC-MS/MS*

The Acquity Ultra-Performance Liquid Chromatography (UPLC) System (Waters, Milford, CT, USA) provides an advanced level of chromatographic performance for the separation and detection of compounds [25]. This system reduces the likelihood of undetected analytes, consequently ensuring increased efficiency and confidence of results [20,21]. In this experimental study, analyses of the GSL content of each sample was performed using the Acquity UPLC System coupled to a Xevo™ TQ-S system (Waters, MS Technologies, Wilmslow, UK). Here, 5 µL of the GLS sample was analyzed with the Acquity UPLC BEH C18 (1.7 µm, 2.1 × 100 mm) column. For the elution, 0.1% trifluoroacetic acid in water was used as Eluent A, with Eluent B mobile phase of 0.1% trifluoroacetic acid in methanol. The flow rate was maintained at 0.5 mL/min, column temperature at 35 °C and injection volume of 5 µL. The elution conditions were set as 100% of A from 0.0 to 1.0 min, 100% of A from 1.0 to 7.0 min, 100–80% of A from 7.0 to 10 min, 80–0% of A from 10 to 11 min, 0–100% of A from 11 to 15 min, and 100% of A thereafter. For the detection of the GSLs, negative ion electrospray ionization (ESI-) and multiple reaction monitoring (MRM) modes were employed. The MS/MS parameters were set using capillary and con voltages set at 3 kV and 54 V, respectively, for ionization. The ionization source was set at a 150 °C temperature while the dissolution temperature was set at 350, 150 °C. For cone and dissolution gas, temperature was set at 150 and 650 Lh-, respectively. The identification of the GSLs was carried out through direct comparison involving the retention times and MS and MS/MS fragmentation spectra with the commercially procured standards. We measured the linear, intraday, and interday precision in order to validate the precision and accuracy of the method used. For standards preparation, 10 mg of each GSL was dissolved in methanol to obtain stock solutions (1 mg mL^{-1}). To calculate GSL concentrations, calibration curves were plotted based on the corresponding standards and the results expressed as μmol GSLs kg^{-1} sample dry weight (DW). The LOD (limit of detection) and LOQ (limit of quantification) values were taken as three and ten times, respectively, the standard error of the intercept of the regression equation of the linear calibration curve divided by the slope (Table 2). Fresh batches of test solutions were always prepared before sample analysis.

Table 2. Results of the UPLC spectroscopy analysis showing the seventeen glucosinolates, retention time (RT), calibration curves, and multiple reaction monitoring (MRM) conditions for quantitation of glucosinolates by negative ion MRM.

Class	Name	Abbreviation	RT (min)	MRM Transition	CID (ev)	Dwell Time (sec)	Calibration Curve Parameters
Aliphatic	Progoitrin	PRO	5.94	387.77 > 194.85	25	0.029	Y = 8.2526X + 28.1501 (r^2 = 0.961)
	Sinigrin	SIN	6.56	357.75 > 161.84	25	0.029	Y = 12.7878X − 11.1181 (r^2 = 0.999)
	Gluconapin	GNA	7.78	371.74 > 258.74	25	0.029	Y = 8.36216X + 29.5397 (r^2 = 0.994)
	Glucoiberin	GIB	7.98	421.62 > 357.73	25	0.029	Y = 33.6632X + 446.334 (r^2 = 0.997)
	Epiprogoitrin	EPI	8.06	387.7 > 258.74	25	0.029	Y = 7.4939X − 6.76519 (r^2 = 0.999)
	Glucocheirolin	GCH	8.38	437.71 > 258.74	25	0.029	Y =20.7762X + 39.3608 (r^2 = 0.986)
	Glucoraphanin	GRA	8.39	435.59 > 177.78	25	0.029	Y = 25.0808X +60.584 (r^2 = 0.983)
	Glucoraphenin	GRE	8.53	433.66 > 258.81	25	0.029	Y = 15.2565X + 3.62242 (r^2 = 0.988)
	Glucobrassicanapin	GBN	8.60	385.71 > 258.87	25	0.029	Y = 7.2514X + 47.2841 (r^2 = 0.992)
	Glucoerucin	GER	8.73	419.69 > 258.74	25	0.029	Y = 6.77393X + 73.6679 (r^2 = 0.984)
	Glucoberteroin	GBE	9.18	433.72 > 275.06	25	0.029	Y = 6.09397X + 63.1212 (r^2 = 0.997)
	Glucoraphasatin	GRH	9.62	417.63 > 258.81	25	0.029	Y = 15.5149X − 5.95281 (r^2 = 0.997)
Aromatic	Glucobarbarin	GBB	8.64	437.71 > 274.75	25	0.029	Y = 9.29915X− 0.454779 (r^2 = 0.999)
	Glucotropaeolin	GTL	8.88	407.72 > 258.87	25	0.029	Y = 18.2122X − 3.93949 (r^2 = 0.999)
	Sinalbin	SNB	9.10	423.62 > 258.74	25	0.029	Y = 49.7228X − 33.0636 (r^2 = 0.999)
	Gluconasturtiin	GNS	9.34	421.69 > 274.87	25	0.029	Y = 4.36109X − 90.233 (r^2 = 0.961)
Indolyl	Glucobrassicin	GBC	9.31	446.69 > 204.94	25	0.029	Y = 6.39827X + 2.6232 (r^2 = 0.997)

2.4. Statistical Analysis

All the analyses were performed using three independent samples as biological replicates. The resulting data were subjected to analysis of variance (ANOVA) using the XLSTAT software v2019 (Addinsoft, Paris, France). The quantification data obtained were used for principal component analysis. The Pearson's correlation coefficient method was used to visualize the association of GSL compounds in the data. To obtain optimal clustering, we employed the K-means method to present a dendrogram. We also used the Orthogonal Partial Least Squares Discriminant Analysis (OPLS-DA), first to study the distribution and second to identify the key variables of GSL compounds responsible for cluster differentiation.

3. Results and Discussion

3.1. Variability of GSL Metabolite Composition in Pak Choy Germplasm

Glucosinolate compounds are famous for their potential health benefits. The breakdown of GSLs provides many useful products such as isothiocyanates, indoles, and sulforaphane which are associated with several biological activities in the human body [26,27]. The composition of GSLs in pak choy naturally differs across different geographical regions as well as on the basis of their type, such as cultivar or landrace. Details of the raw data of GSL content of the accessions used in the current study would be made available and accessed from the RDA-Genebank at http://genebank.rda.go.kr/. In order to satisfy our curiosity on the presence of variability of GSL content in pak choy, we employed the Acquity Ultra-Performance Liquid Chromatography-Tandem Mass Spectrometry (UPLC-MS/MS) analysis system to determine the GSL composition of diverse pak choy accessions that are under conservation in the national gene bank of the Rural Development Administration (RDA-Genebank), Jeonju, Republic of Korea. Our analyses detected a total of 17 glucosinolate-derived metabolites in the germplasm. Details of the Acquity UPLC spectroscopy analysis results of the 17 GSLs are shown in Table 3. In a previous study, Ju-Hee and co-workers [28] identified eight GSLs based on quantification of leaf samples from five commercial varieties and 45 accessions of diverse Brassica species including Kimchi cabbage, turnip and leaf mustard. Earlier, using 13 pak choy cultivars, Wiesner and colleagues [26] detected 11 GSLs that were predominantly composed of members of the aliphatic class. We examined the composition of the GSLs in the 65 pak choy accessions by calculating their concentration expressed in median values (μmol·kg^{-1} DW). Generally, significant variability was observed across the different classes of the GSLs (aliphatic, aromatic and indolic) (Table 3). The most predominant GSLs detected were in the aliphatic class. In line with a previous study by Wiesner et al. [26] in which pronounced variation in total aliphatic GSLs (18.7–61.7 μmol g^{-1} dw) was observed, we found significant variation among the aliphatic GSLs in the present study. A similar finding was presented in another study in which aliphatic compounds represented the predominant GSLs detected [28].

Table 3. Profile of individual glucosinolates in pak choy (μmol·kg^{-1} DW).

Class	Glucosinolates	Range	Median
Aliphatic GSL	Glucoiberin	0~35.069	0.375
	Sinigrin	0.162~7878.972	4.722
	Glucocheirolin	0.078~239.664	5.256
	Glucoerucin	0~2564.479	49.366
	Glucoraphanin	0.162~1558.413	172.591
	Gluconapin	117.379~19,009.896	6713.083
	Progoitrin	2.303~4116.955	1132.364
	Epiprogoitrin	1.629~3333.335	843.059
	Glucoraphasatin	0.025~6.134	0.231
	Glucoraphenin	0.0168~228.202	0.981
	Glucoberteroin	0~3491.342	148.188
	Glucobrassicanapin	0.263~8744.337	3139.729
Aromatic GSL	Glucotropaeolin	0.311~30.651	6.451
	Gluconasturtiin	74.282~2148.237	678.72
	Glucobarbarin	0.937~10.505	3.171
	Sinalbin	0~3.704	0.086
Indolic GSL	Glucobrassicin	77.984~1294.483	351.011

In this study, among the aliphatic GSLs, Gluconapin (GNA) recorded the highest concentration (median: 6713.083 μmol·kg^{-1} DW), while Glucoraphasatin represented the lowest concentration (median: 0.231 μmol·kg^{-1} DW). For the aromatic GSLs, Gluconasturtiin and Sinalbin recorded the highest and least concentrations with median values of 678.72 μmol·kg^{-1} DW and 0.086 μmol·kg^{-1} DW, respectively. An indolic GSL, Glucobrassicin had a concentration of 351.011 μmol·kg^{-1} DW higher than 50% and 75% of the aromatic and aliphatic compound, respectively, contrasting with that reported by Wiesner et al. [26], where indolic GSL occurred at low levels (0.6 to 2.35 μmol·g^{-1} dw). This contradiction perhaps resulted from differences in pak choy genetic materials as well as the extraction protocol used [28,29].

Previous studies on GSLs from various *B. rapa* vegetables revealed that choi sum has high GNA content compared to other *Brassica* vegetables [12,30]. However, the least detected GSL was GBC according to He and co-workers [30], contrasting with the Sinalbin (SNB) compound that is observed in our current study [31]. Clear variations in GSL content and individual GSL profiles exists among *Brassica* vegetables [30,32]. For example, leaf mustard (*B. juncea*), turnips (*B. rapa* var. rapa), collards (*B. oleracea* var. *viridis*), and kale (*B. oleracea* var. *sabellica*) exhibit higher total GSL levels compared to pak choy and choy sum (*B. chinensis* var. *parachinensis*) [30]. While aliphatic GSLs are generally abundant in *Brassica* species [33], choi sum and Chinese cabbage mainly contain GNA and GBN [24]. Pak choy, on the other hand, is very rich in GNA and PRO among aliphatic GSLs (Table 3). Furthermore, Gluconasturtiin displayed the highest median value (678.72 μmol·kg^{-1} DW) among aromatic GSLs. Glucotropaeolin and Glucobarbarin also demonstrated significant median values of 6.451 and 3.171 μmol·kg^{-1} DW, respectively. Glucobrassicin, the only compound detected as an indolic GSL in this study exhibited appreciably high median value (351.011 μmol·kg^{-1} DW).

Our results highlight promising genetic resources, offering breeders advanced cultivars to develop nutritionally superior vegetables in pak choy breeding programs. In the present finding, IT23558 had the highest GNA content with a median value of 19,009.9 μmol·kg^{-1} DW. This value surpassed the average GNA content in choy sum (*B. chinensis* var. *parachinensis*) by a notable margin of 2997.62 μmol·kg^{-1}. This germplasm, collected in 2001 from the RDA-Genebank to enhance genetic diversity, is available for purchase in the Republic of Korea. However, due to absence of detailed passport information, the origin or country of the breeding company remains unclear. Along with its high GNA content, this cultivar also contains progoitrin (PRO) levels of 2483 μmol·kg^{-1} DW, approximately twice the median value. PRO is recognized for its anti-inflammatory properties. As demonstrated by Jang and colleagues [31], it has increased antibacterial activity against Aeromonas hydrophilic [18]. Yet, it is crucial to note that PRO levels exceeding 3000 μmol·kg^{-1} DW can introduce a bitter taste, potentially affecting palatability and hindering growth in certain animals [34]. This might require selective use, depending on breeding objectives. The germplasm with the second highest GSL content, IT275755 (9039), contained 18,268.9 μmol·kg^{-1} DW. Cultivated in 2013 by the Asia Seed Company (Seoul, Korea), this germplasm was subsequently deposited at the RDA-Genebank. Morphologically, it is characterized as an overwintering type and developed as a red-headed cabbage cultivar. Its foliage is green, with a mid-season bolting pattern, displaying a dwarf growth habit. Significantly, its PRO content is exceptionally low at 5.993 μmol·kg^{-1}, signifying reduced bitterness, a distinguishing trait of this cultivar.

3.2. *Multivariate Analysis*

3.2.1. Correlation Analysis

Genetic correlation represents the relationship between two variables in a data set and provides a clue to the understanding of the shared biological network or the magnitude of association between variables [4,35]. Understanding the correlation or association between traits is important in deciding whether or not selection for a specific trait will influence the other [36,37]. Correlation analysis can be applied conveniently in determining the relationships among different phytochemical compounds [38,39]. Pearson's correlation coefficient (r) is used in measuring the linear association of two variables and has been used for establishing the association between different bioactive compounds [4,40]. In this study, the Pearson's correlation coefficient (r) of GSLs was estimated using the XLSTAT analysis software v.2019 (Addinsoft, Paris, France) (Figure 1). A very strong positive association characterized a number of aliphatic compounds including SIN and GIB (r = 0.988; $p \leq 0.0001$), GCH and GIB (r = 0.986; $p \leq 0.0001$), GCH and SIN (r = 0.968; $p \leq 0.0001$), GBE and GER (r = 0.845; $p \leq 0.0001$) and EPI and PRO (r = 0.956; $p \leq 0.0001$). A positive association between two GSL indicates that the two variables or compounds are located in biologically tightly linked pathways, and as a result of their shred biochemical nature, these compounds likely share the same cluster [4]. Thus, in crop breeding, indirect selection or improvement of one GSL compound could simultaneously contribute to enhancing the content of the other [41]. Similarly, a positive correlation was observed between GRH and GER (r = 0.614; $p \leq 0.0001$), SNB and GER (r = 0.641; $p \leq 0.0001$), and GRE and GNA (r = 0.624; $p \leq 0.0001$). Other noticeable positive associations were observed between GBS and GER (R = 0.485; $p \leq 0.0001$), SNB and GBE (r = 0.490; $p \leq 0.0001$), SNB and GRH (r = 0.468; $p \leq 0.0001$), GBC and GBE (r = 0.499; $p \leq 0.0001$), GTL and GNA (r = 0.476; $p \leq 0.0001$), GTL and GRE (r = 0.527; $p \leq 0.0001$), GNS and GTL (r = 0.468; $p \leq 0.0001$), GBC and GTL (r = 0.475; $p \leq 0.0001$), GCB and GNS (r = 0.542; $p \leq 0.0001$). The above results reveal that Brassica genotypes containing higher amount of one or more of the GSLs could be used in crop breeding [42] to increase the composition of other compounds since the compounds are likely to be biosynthetically linked. In correlation analyses, two variables display a negative correlation coefficient suggesting that indirect selection or improvement of one of the variables does not potentially contribute to enhanced expression of the other. In this study, generally, GSL compounds that are negatively correlated had weak correlation coefficient (Figure 1). The highest negative correlation coefficient was observed between GBN and GIB (r = −0.250; $p \leq 0.05$), GBN and GCH (r = −0.279; $p \leq 0.05$). Thus, genetic improvement of one of GBNs with the aim to enhance the composition of GIB or GCH is likely less useful.

Variable	GIB	SIN	GCH	GER	GBE	GNA	PRO	EPI	GRH	GRA	GRE	GBN	GTL	GNS	GBB	SNB	GBC
GIB																	
SIN	0.988***																
GCH	0.986***	0.968***															
GER	0.015	-0.059	-0.002														
GBE	-0.013	-0.073	-0.053	0.845***													
GNA	-0.117	-0.178	-0.025	-0.020	-0.206												
PRO	-0.190	-0.191	-0.136	-0.056	-0.041	0.045											
EPI	-0.166	-0.176	-0.114	-0.100	-0.085	0.038	0.956***										
GRH	-0.052	-0.054	-0.060	0.614***	0.316**	0.125	0.058	-0.007									
GRA	-0.025	-0.023	-0.025	0.026	0.021	-0.008	-0.145	-0.149	-0.025								
GRE	-0.014	-0.118	0.127	0.169	0.030	0.624***	0.256*	0.260	-0.038	0.015							
GBN	-0.250*	-0.195	-0.279*	-0.127	-0.066	0.217	0.094	0.026	0.130	0.019	-0.272*						
GTL	-0.007	-0.045	0.068	0.123	-0.007	0.476***	0.211	0.123	0.186	0.048	0.527***	0.148					
GNS	-0.001	-0.035	0.041	0.264*	0.275*	0.308**	0.328**	0.250*	0.159	0.004	0.324**	0.378**	0.468***				
GBB	0.280*	0.196	0.240*	0.521***	0.592***	-0.083	0.049	0.039	0.091	-0.062	0.092	-0.131	-0.031	0.277*			
SNB	-0.010	-0.043	-0.022	0.641***	0.490***	0.171	-0.049	-0.079	0.468***	-0.025	0.094	0.073	0.174	0.164	0.145		
GBC	-0.115	-0.156	-0.083	0.485***	0.499***	0.123	0.216	0.134	0.321**	0.029	0.321**	0.248*	0.475***	0.542***	0.357**	0.284*	

Figure 1. Pearson's correlation analysis of glucosinolates detected in pak choy. ***, **, * indicate correlation is significant at $p \leq 0.0001$, $p \leq 0.01$, and $p \leq 0.05$, respectively. Glucoiberin (GIB), Sinigrin (SIN), Glucocheirolin (GCH), Glucoerucin (GER), Glucoberteroin (GBE), Gluconapin (GNA), Progoitrin (PRO), Epiprogoitrin (EPI), Glucoraphasatin (GRH), Glucoraphanin (GRA), Glucoraphenin (GRE), Glucobrassicanapin (GBN), Glucotropaeolin (GTL), Gluconasturtiin (GNS), Glucobarbarin (GBB), Sinalbin (SNB)and Glucobrassicin (GBC).

3.2.2. Variability of GSLs in Pak Choy Based on PCA

Principal component analysis (PCA) is a typical chemometric tool commonly used in multivariate analysis for extracting and interpreting experimental results [4,43]. In order to investigate diversity in the GSL composition within the analytes, the quantification data of the 17 GSL compounds detected in the study were subjected to principal component analysis. PCA was used to determine the most relevant components with the largest variation. The data dimension was reduced to three principal components by employing the eigenvector values greater than or equal to one. The first three principal components accounted for 56.567% of the cumulative total variation (Table 4). PC1 accounted for 22.604% of the total variation while PC2 and PC3 accounted for 19.373 and 14.591% of the total variance, respectively. The two highest ranking components PC1 and PC2) accounted for 41.976% of the total variance with eigenvector values 3.843 and 3.293, respectively, lower than those obtained in some previous studies. For instance, Wiesner et al. [26] recorded 86% of total variation derived from the first three PCs, with the highest ranking PCs, PC1 and PC2, accounting for 49% and 22%, respectively. Generally, the PCs had many positive loading relative to the negative loadings (Table 4). The highest positive loading in PC1 corresponded with GBC (0.395), GER (0.365), GNS (0.331), and GBE (0.317), representing all the three classes of compounds. PC2 showed strong positive loading with three aliphatic GSL compounds, GIB (0.478), GCH (0.461) and SIN (0.455), which mainly accounted for the variability in GSLs profiled in this experiment. PCA revealed three distinct groups of all the pak choy accessions (Figures 2A and 3A,B) and the GSL compounds (Figure 2B), suggesting the presence of variability in GSLs among the genetic materials used in the study. A similar finding was previously reported in the works of Wiesner and colleagues who found aliphatic glucosinolates as the predominant GSL class in pak choi [26]. Nonetheless, the unexplained 44% of variability of the glucosinolates may be due to additional underlying biochemical pathways or metabolites not captured by the three principal components. This unaccounted variation could also be a result of the influence of environmental factors or complex interactions among multiple metabolites not emphasized by the primary components. Therefore, further investigations, potentially through targeted analyses or additional statistical methods, are necessary to elucidate the specific contributors to this remaining variance.

Table 4. Principal component analysis of GSL content in pak choy.

	Principal Component (Eigenvectors)		
GSL	PC1	PC2	PC3
GIB	−0.156	**0.478**	0.225
SIN	−0.192	**0.455**	0.210
GCH	−0.135	**0.461**	0.292
GER	**0.365**	0.251	−0.268
GBE	**0.317**	0.228	**−0.304**
GNA	0.180	−0.099	0.290
PRO	0.187	−0.187	**0.324**
EPI	0.145	−0.187	**0.327**
GRH	0.271	0.098	−0.167
GRA	−0.006	0.003	−0.064
GRE	0.230	0.001	**0.366**
GBN	0.119	−0.181	−0.007
GTL	0.268	−0.008	**0.320**
GNS	**0.331**	0.034	0.229
GBB	0.197	0.296	−0.053
SNB	0.285	0.150	−0.188
GBC	**0.395**	0.063	0.033
Eigen value	3.843	3.293	2.480
Proportion (%)	22.604	19.373	14.591
Cumulative (%)	22.604	41.976	56.567

The bold values represent the highest loadings in principal component analysis, which essentially accounted for the presence of variability in GSL profiles.

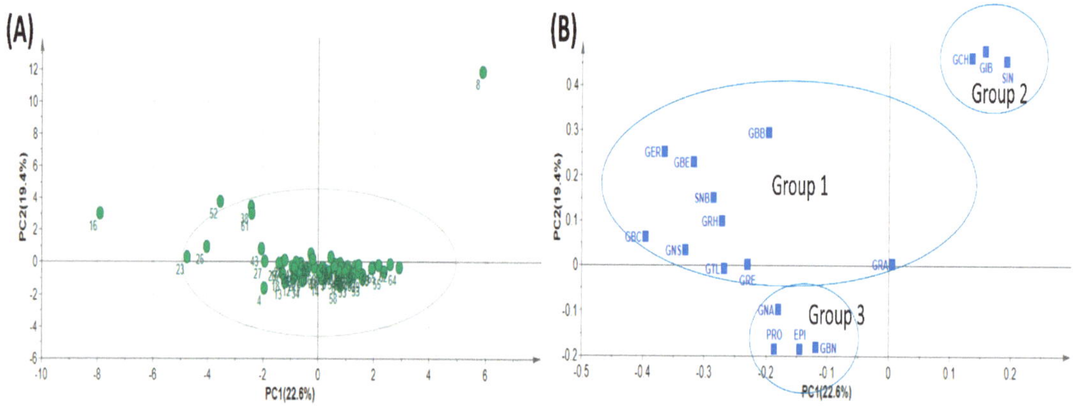

Figure 2. Principal Component Analysis showing Score plot of the 65 pak choy accessions (**A**) and Loading plot of the 17 pak choy accessions (**B**). Glucoiberin (GIB), Sinigrin (SIN), Glucocheirolin (GCH), Glucoerucin (GER), Glucoberteroin (GBE), Gluconapin (GNA), Progoitrin (PRO), Epiprogoitrin (EPI), Glucoraphasatin (GRH), Glucoraphanin (GRA), Glucoraphenin (GRE), Glucobrassicanapin (GBN), Glucotropaeolin (GTL), Gluconasturtiin (GNS), Glucobarbarin (GBB), Sinalbin (SNB) and Glucobrassicin (GBC).

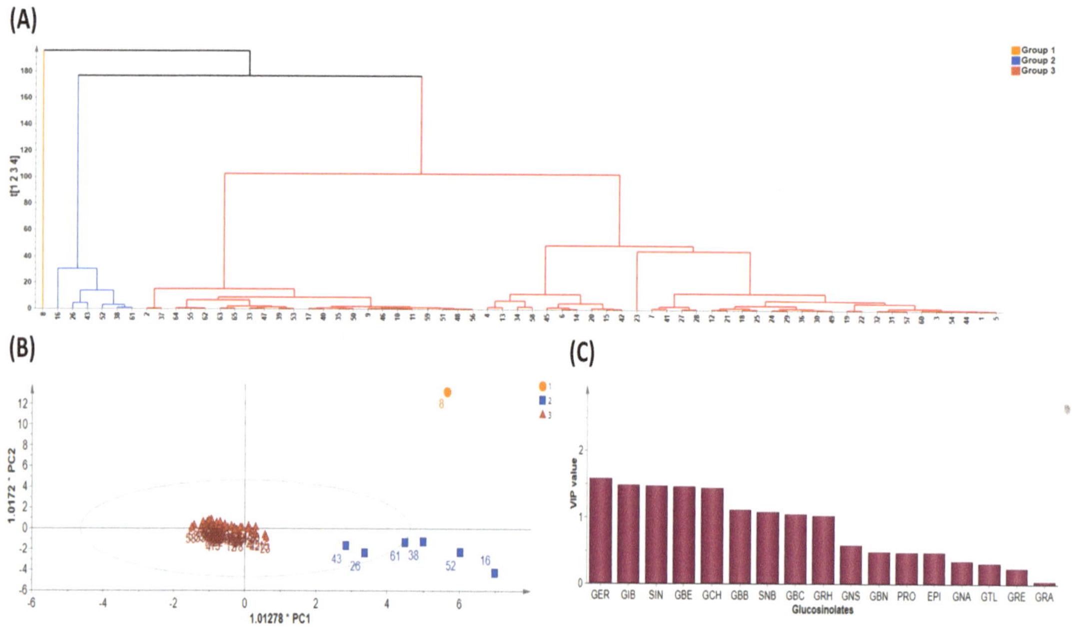

Figure 3. Diversity Analysis showing (**A**) cluster dendrogram of the 65 pak choy accessions, (**B**) Orthogonal Partial Least Squares Discriminant Analysis, and (**C**) Variable Importance in Projection value. Glucoerucin (GER), Glucoiberin (GIB), Sinigrin (SIN), Glucoberteroin (GBE), Glucocheirolin (GCH), Glucobarbarin (GBB), Sinalbin (SNB), Glucobrassicin (GBC), Glucoraphasatin (GRH), Gluconasturtiin (GNS), Glucobrassicanapin (GBN), Progoitrin (PRO), Epiprogoitrin (EPI), Gluconapin (GNA), Glucotropaeolin (GTL), Glucoraphenin (GRE) and Glucoraphanin (GRA).

Since PCA is limited by accuracy of clustering, we employed the K-means method to obtain optimal clustering [44] and this was presented as a dendrogram of three clusters (Figure 3A). By its dimension reduction property, the Orthogonal Partial Least Squares Discriminant Analysis (OPLS-DA) is more suitable for sample distinction relative to PCA [45]. So, in order to gain additional insight into the diversity of pak choy germplasm, the OPLS-DA was used to identify the key accessions that contributed to the cluster differentiation based on Variable Importance in Projection (VIP) values [46,47]. Here, three separate groups of pak choy were obtained as indicated with a yellow circle, a blue rectangle and a red triangle (Figure 3A,B). Group 1 contained the Korean cultivar cheongsacholong which recorded high content of GBC (indolic), GER (aliphatic), GNS (aromatic) and GBE (aliphatic) (Figures 2B and 3A,B). Conversely, this accession recorded lower levels of GIB, SIN, GCH, and GRA, all of which are aliphatic compounds. Group 2 consisted of six accessions, BRA77/72, Lu ling gaogengbai, 9041, Wuyueman, RP-75, DH-10, which were predominatly high in aliphatic compounds including GIB, GCH, and SIN. Also, these accessions recorded lower levels of aliphatic compounds GNA, PRO, EPI, GBN, and GTL. Group 3 comprised the majority of the accessions (Figures 2A and 3A,B). This group of accesstions showed a characteristic high content of aliphatic GSLs encompassing GRE, EPI, PRO, and GTL, but low content of GER, GBE, GRH, GRA, GBN, GBB and SNB. These results reveal the presence of distinct variation among the pak choy germplasm based on their GSL content, providing an excellent opprtunity for future breeding for improved GSL content in pak choy using reommended accesions [48]. The presence of diversity among pak choy accessions was reported in other studies [26,49]. Figure 3C indicates individual GSLs and their contibutions to the three clusters based on VIP values. According to Park et al. [50], in a given data set, variables with VIP values greater than one are the most significant contributors to the observed variablity. Based on the VIP values, the most influencial class of compounds responsible for variation in the germplasm were mainly attributed to nine metabolites including six aliphatic (GER, GIB, SIN, GBE, GCH and GRH), two aromatic (GBB and SNB) and one indolic (GBC) compounds (Figure 3C).

Overall, our results confirmed that the GSL compounds profiled in this study provide useful information for metabolic differentiation within the pak choy germplasm and provided useful information to facilitate breeding in pak choy.

4. Conclusions

In this study, we identify 17 GSLs among 65 pak choy accessions, indicating the presence of variability of the compound within pak choy germplasm. The identification and characterization of various GSLs, along with their potential health benefits, highlight the importance of pak choy as a valuable functional food with significant implications for human health. In the future, there is the need for further research on GSLs profiles in pak choy in relation to other cruciferous vegetables. Additionally, further investigations are required in elucidating the precise mechanisms through which these GSLs exert their health-promoting effects, particularly their roles in cancer prevention, anti-inflammatory activity, and detoxification pathways within the body. Studies are also needed to bridge the gap between laboratory analyses and practical applications, seeking methods to optimize GSL content in pak choy through agricultural practices or breeding strategies. This can enhance the nutritional value of the crop, thus increasing the crop's potential health benefits for consumers. Still, additional studies are warranted to understand the impact of cultivation conditions, and processing methods on the levels and types of these bioactive compounds in pak choy. Overall, the exploration of GSLs diversity in pak choy germplasm provides a promising avenue for harnessing the health benefits of these phytochemicals. The present study provides a basis for future research directions aimed at maximizing the potential of pak choy as a functional food for promoting human health and well-being.

Supplementary Materials: The following supporting information can be downloaded at: https://www.mdpi.com/article/10.3390/plants13010009/s1, Table S1: List of pak choy accessions used for GSL profiling.

Author Contributions: Methodology, Writing—original draft, Funding acquisition, S.-H.K.; Writing, editing and proofreading: K.O., H.J.K. A.H. and Y.-J.L. All authors have read and agreed to the published version of the manuscript.

Funding: This research was carried out with the support of the "Research Program for Agricultural Science and Technology Development (Project NO. PJ01425501/RS-2019-RD007776)", National Institute of Agricultural Sciences, Rural Development Administration, Republic of Korea.

Data Availability Statement: The data used to support the findings of this study can be made available by the corresponding author upon request.

Conflicts of Interest: Authors declare no conflict of interest and agreed to the published version of the manuscript.

Abbreviations

GSL: glucosinolate; GIB, glucoiberin; SIN, sinigrin; GCH, glucocheirolin; GER, glucoerucin; GRA, glucoraphanin; GNA, gluconapin; PRO, progoitrin; EPI, epiprogoitrin; GRH, glucoraphasatin; GRE, glucoraphenin; GBE, glucoberteroin; GBN, glucobrassicanapin; GTL, glucotropaeolin; GNS, gluconasturtiin; GBB, glucobarbarin; SNB, sinalbin; GBC, glucobrassicin.

References

1. Carpene, C.; Gomez-Zorita, S.; Deleruyelle, S.; Carpene, M.-A. Novel strategies for preventing diabetes and obesity complications with natural polyphenols. *Curr. Med. Chem.* **2015**, *22*, 150–164. [CrossRef] [PubMed]
2. Hedden, P.; Harrewijn, P.; Van Oosten, A.M.; Piron, P.G.M. Natural terpenoids as messengers: A multidisciplinary study of their production, biological functions, and practical applications. *Ann. Bot.* **2002**, *90*, 299–300. [CrossRef]
3. Samtiya, M.; Aluko, R.E.; Dhewa, T.; Moreno-Rojas, J.M. Potential health benefits of plant food-derived bioactive components: An overview. *Foods* **2021**, *10*, 839. [CrossRef] [PubMed]
4. Lee, W.; Yeo, Y.; Oh, S.; Cho, K.-S.; Park, Y.-E.; Park, S.K.; Lee, S.M.; Cho, H.S.; Park, S.-Y. Compositional analyses of diverse phytochemicals and polar metabolites from different-colored potato (*Solanum tubersum* L.) tubers. *Food Sci. Biotechnol.* **2017**, *26*, 1379–1389. [PubMed]
5. Sakauchi, F.; Mori, M.; Washio, M.; Watanabe, Y.; Ozasa, K.; Hayashi, K.; Miki, T.; Nakao, M.; Mikami, K.; Ito, Y. Dietary habits and risk of urothelial cancer incidence in the JACC Study. *J. Epidemiol.* **2005**, *15*, S190–S195.
6. Neuhouser, M.L.; Patterson, R.E.; Thornquist, M.D.; Omenn, G.S.; King, I.B.; Goodman, G.E. Fruits and vegetables are associated with lower lung cancer risk only in the placebo arm of the β-carotene and retinol efficacy trial (CARET). *Cancer Epidemiol. Biomark. Prev.* **2003**, *12*, 350–358.
7. Jahangir, M.; Kim, H.K.; Choi, Y.H.; Verpoorte, R. Health-affecting compounds in Brassicaceae. *Compr. Rev. Food Sci. Food Saf.* **2009**, *8*, 31–43. [CrossRef]
8. Chen, J.; Zhang, J.; Xiang, Y.; Xiang, L.; Liu, Y.; He, X.; Zhou, X.; Liu, X.; Huang, Z. Extracts of Tsai Tai (*Brassica chinensis*): Enhanced antioxidant activity and anti-aging effects both in vitro and in *Caenorhabditis elegans*. *Food Funct.* **2016**, *7*, 943–952. [CrossRef]
9. Li, P.; Su, T.; Zhao, X.; Wang, W.; Zhang, D.; Yu, Y.; Bayer, P.E.; Edwards, D.; Yu, S.; Zhang, F. Assembly of the non-heading pak choi genome and comparison with the genomes of heading Chinese cabbage and the oilseed yellow sarson. *Plant Biotechnol. J.* **2021**, *19*, 966–976. [CrossRef]
10. Sun, X. Morphological and Genetic Characterization of the Leafy Head of Chinese Cabbage *(Brassica rapa)*. Ph.D. Thesis, Wageningen University and Research, Wageningen, The Netherlands, 2018.
11. Traka, M.; Mithen, R. Glucosinolates, isothiocyanates and human health. *Phytochem. Rev.* **2009**, *8*, 269–282. [CrossRef]
12. Kim, S.-H.; Subramanian, P.; Hahn, B.-S. Glucosinolate diversity analysis in choy sum (*Brassica rapa* subsp. chinensis var. parachinensis) germplasms for functional food breeding. *Foods* **2023**, *12*, 2400. [CrossRef] [PubMed]
13. Miękus, N.; Marszałek, K.; Podlacha, M.; Iqbal, A.; Puchalski, C.; Świergiel, A.H. Health benefits of plant-derived sulfur compounds, glucosinolates, and organosulfur compounds. *Molecules* **2020**, *25*, 3804. [CrossRef]
14. Traka, M. Health benefits of glucosinolates. *Adv. Bot. Res.* **2016**, *80*, 247–279.
15. Krumbein, A.; Schonhof, I.; Schreiner, M. Composition and contents of phytochemicals (glucosinolates, carotenoids and chlorophylls) and ascorbic acid in selected *Brassica* species (*B. juncea*, *B. rapa* subsp. nipposinica var. chinoleifera, *B. rapa* subsp. chinensis and *B. rapa* subsp. rapa). *J. Appl. Bot. Food Qual.* **2005**, *79*, 168–174.

16. Hanson, P.; Yang, R.Y.; Chang, L.C.; Ledesma, L.; Ledesma, D. Contents of carotenoids, ascorbic acid, minerals and total glucosinolates in leafy brassica pakchoi (*Brassica rapa* L. chinensis) as affected by season and variety. *J. Sci. Food Agric.* **2009**, *89*, 906–914. [CrossRef]
17. Klopsch, R.; Witzel, K.; Börner, A.; Schreiner, M.; Hanschen, F.S. Metabolic profiling of glucosinolates and their hydrolysis products in a germplasm collection of Brassica rapa turnips. *Food Res. Int.* **2017**, *100*, 392–403. [CrossRef] [PubMed]
18. Gratacós-Cubarsí, M.; Ribas-Agusti, A.; García-Regueiro, J.A.; Castellari, M. Simultaneous evaluation of intact glucosinolates and phenolic compounds by UPLC-DAD-MS/MS in *Brassica oleracea* L. var. botrytis. *Food Chem.* **2010**, *121*, 257–263. [CrossRef]
19. Ye, L.; Zhang, B.; Yang, X.; Li, X.; Tan, W.; Zhang, X. Ultra-performance liquid chromatography-tandem mass spectrometry revealed the significantly different metabolic profiles of Auricularia cornea growing on weakly acidic and weakly alkaline substrates. *Can. J. Microbiol.* **2023**, *69*, 262–278. [CrossRef]
20. Rathod, R.H.; Chaudhari, S.R.; Patil, A.S.; Shirkhedkar, A.A. Ultra-high performance liquid chromatography-MS/MS (UHPLC-MS/MS) in practice: Analysis of drugs and pharmaceutical formulations. *Future J. Pharm. Sci.* **2019**, *5*, 1–26. [CrossRef]
21. Walter, T.H.; Andrews, R.W. Recent innovations in UHPLC columns and instrumentation. *TrAC Trends Anal. Chem.* **2014**, *63*, 14–20. [CrossRef]
22. Birch, C.S.; Bonwick, G.A. Ensuring the future of functional foods. *Int. J. Food Sci. Technol.* **2019**, *54*, 1467–1485. [CrossRef]
23. Gul, K.; Singh, A.; Jabeen, R. Nutraceuticals and functional foods: The foods for the future world. *Crit. Rev. Food Sci. Nutr.* **2016**, *56*, 2617–2627. [CrossRef] [PubMed]
24. Kim, S.-H.; Lee, G.-A.; Subramanian, P.; Hahn, B.-S. Quantification and diversity analyses of major glucosinolates in conserved chinese cabbage (*Brassica rapa* L. ssp. pekinensis) germplasms. *Foods* **2023**, *12*, 1243. [CrossRef] [PubMed]
25. Sheliya, K.; Shah, K. Ultra performance liquid chromatography (UPLC): A modern chromatography technique. *Pharma Sci. Monit.* **2013**, *4*, 78–99.
26. Wiesner, M.; Zrenner, R.; Krumbein, A.; Glatt, H.; Schreiner, M. Genotypic variation of the glucosinolate profile in pak choi (*Brassica rapa* ssp. chinensis). *J. Agric. Food Chem.* **2013**, *61*, 1943–1953. [CrossRef]
27. Kołodziejski, D.; Koss-Mikołajczyk, I.; Abdin, A.Y.; Jacob, C.; Bartoszek, A. Chemical aspects of biological activity of isothiocyanates and indoles, the products of glucosinolate decomposition. *Curr. Pharm. Des.* **2019**, *25*, 1717–1728. [CrossRef]
28. Rhee, J.-H.; Choi, S.; Lee, J.-E.; Hur, O.-S.; Ro, N.-Y.; Hwang, A.-J.; Ko, H.-C.; Chung, Y.-J.; Noh, J.-J.; Assefa, A.D. Glucosinolate content in Brassica genetic resources and their distribution pattern within and between inner, middle, and outer leaves. *Plants* **2020**, *9*, 1421. [CrossRef]
29. Kim, M.J.; Chiu, Y.-C.; Kim, N.K.; Park, H.M.; Lee, C.H.; Juvik, J.A.; Ku, K.-M. Cultivar-specific changes in primary and secondary metabolites in pak choi (*Brassica rapa*, Chinensis group) by methyl jasmonate. *Int. J. Mol. Sci.* **2017**, *18*, 1004. [CrossRef]
30. He, H.; Ping, L.; Bonnema, G.; Dekker, M.; Verkerk, R. Genetic variation in glucosinolate content within Brassica rapa vegetables. *Acta Hortic.* **2012**, *944*, 129–140.
31. Jang, M.; Hong, E.; Kim, G.H. Evaluation of antibacterial activity of 3-butenyl, 4-pentenyl, 2-phenylethyl, and benzyl isothiocyanate in Brassica vegetables. *J. Food Sci.* **2010**, *75*, M412–M416. [CrossRef]
32. Soundararajan, P.; Park, S.-G.; Won, S.Y.; Moon, M.-S.; Park, H.W.; Ku, K.-M.; Kim, J.S. Influence of genotype on high glucosinolate synthesis lines of *Brassica rapa*. *Int. J. Mol. Sci.* **2021**, *22*, 7301. [CrossRef] [PubMed]
33. Bhandari, S.R.; Jo, J.S.; Lee, J.G. Comparison of glucosinolate profiles in different tissues of nine Brassica crops. *Molecules* **2015**, *20*, 15827–15841. [CrossRef] [PubMed]
34. Kristal, A.R.; Lampe, J.W. *Brassica* vegetables and prostate cancer risk: A review of the epidemiological evidence. *Nutr. Cancer* **2002**, *42*, 1–9. [CrossRef] [PubMed]
35. Van Rheenen, W.; Peyrot, W.J.; Schork, A.J.; Lee, S.H.; Wray, N.R. Genetic correlations of polygenic disease traits: From theory to practice. *Nat. Rev. Genet.* **2019**, *20*, 567–581. [CrossRef] [PubMed]
36. Neyhart, J.L.; Lorenz, A.J.; Smith, K.P. Multi-trait improvement by predicting genetic correlations in breeding crosses. *G3 Genes Genomes Genet.* **2019**, *9*, 3153–3165. [CrossRef] [PubMed]
37. Girdthai, T.; Jogloy, S.; Vorasoot, N.; Akkasaeng, C.; Wongkaew, S.; Patanothai, A.; Holbrook, C. Inheritance of the physiological traits for drought resistance under terminal drought conditions and genotypic correlations with agronomic traits in peanut. *Sabrao J. Breed. Genet.* **2012**, *44*, 240–262.
38. Camacho, D.; De La Fuente, A.; Mendes, P. The origin of correlations in metabolomics data. *Metabolomics* **2005**, *1*, 53–63. [CrossRef]
39. Osman, M.A.; Mahmoud, G.I.; Shoman, S.S. Correlation between total phenols content, antioxidant power and cytotoxicity. *Biointerface Res. Appl. Chem.* **2020**, *11*, 10640–10653.
40. Parveen, S.; Bukhari, N.; Nazir, M.; Qureshi, W.A.; Yaqoob, A.; Shahid, M. Phytochemical analysis, in-vitro biological activities and Pearson correlation of total polyphenolic content with antioxidant activities of Ziziphus mauritiana fruit pulp and seed during different ripening stages. *South Afr. J. Bot.* **2023**, *157*, 346–354. [CrossRef]
41. Bernado, R. *Breeding For Quantitative Traits in Plants*; Stemma Press: Woodbury, MN, USA, 2002; p. 369.
42. Sotelo, T.; Velasco, P.; Soengas, P.; Rodríguez, V.M.; Cartea, M.E. Modification of leaf glucosinolate contents in Brassica oleracea by divergent selection and effect on expression of genes controlling glucosinolate pathway. *Front. Plant Sci.* **2016**, *7*, 1012. [CrossRef]
43. Granato, D.; Santos, J.S.; Escher, G.B.; Ferreira, B.L.; Maggio, R.M. Use of principal component analysis (PCA) and hierarchical cluster analysis (HCA) for multivariate association between bioactive compounds and functional properties in foods: A critical perspective. *Trends Food Sci. Technol.* **2018**, *72*, 83–90. [CrossRef]

44. Ikotun, A.M.; Ezugwu, A.E.; Abualigah, L.; Abuhaija, B.; Heming, J. K-means clustering algorithms: A comprehensive review, variants analysis, and advances in the era of big data. *Inf. Sci.* **2022**, *622*, 178–210. [CrossRef]
45. Barker, M.; Rayens, W. Partial least squares for discrimination. *J. Chemom. A J. Chemom. Soc.* **2003**, *17*, 166–173. [CrossRef]
46. Hamany Djande, C.Y.; Piater, L.A.; Steenkamp, P.A.; Tugizimana, F.; Dubery, I.A. A metabolomics approach and chemometric tools for differentiation of barley cultivars and biomarker discovery. *Metabolites* **2021**, *11*, 578. [CrossRef] [PubMed]
47. Ghisoni, S.; Lucini, L.; Rocchetti, G.; Chiodelli, G.; Farinelli, D.; Tombesi, S.; Trevisan, M. Untargeted metabolomics with multivariate analysis to discriminate hazelnut (*Corylus avellana* L.) cultivars and their geographical origin. *J. Sci. Food Agric.* **2020**, *100*, 500–508. [CrossRef] [PubMed]
48. Wang, M.; Li, Y.; Yang, Y.; Tao, H.; Mustafa, G.; Meng, F.; Sun, B.; Wang, J.; Zhao, Y.; Zhang, F. Biofortification of health-promoting glucosinolates in cruciferous sprouts along the whole agro-food chain. *Trends Food Sci. Technol.* **2023**, *140*, 104164. [CrossRef]
49. Yu, S.; Zhang, F.; Wang, X.; Zhao, X.; Zhang, D.; Yu, Y.; Xu, J. Genetic diversity and marker-trait associations in a collection of Pak-choi (*Brassica rapa* L. ssp. chinensis Makino) Accessions. *Genes Genom.* **2010**, *32*, 419–428. [CrossRef]
50. Park, S.-Y.; Lim, S.-H.; Ha, S.-H.; Yeo, Y.; Park, W.T.; Kwon, D.Y.; Park, S.U.; Kim, J.K. Metabolite profiling approach reveals the interface of primary and secondary metabolism in colored cauliflowers (*Brassica oleracea* L. ssp. botrytis). *J. Agric. Food Chem.* **2013**, *61*, 6999–7007. [CrossRef]

plants

MDPI

Review

Unlocking the Multifaceted Mechanisms of Bud Outgrowth: Advances in Understanding Shoot Branching

Yundong Yuan [1,*], Said Khourchi [2], Shujia Li [3], Yanfang Du [1] and Pierre Delaplace [2]

[1] National Key Laboratory of Wheat Improvement, College of Life Sciences, Shandong Agricultural University, Tai'an 271018, China
[2] Plant Sciences, TERRA Teaching and Research Center, Gembloux Agro-Bio Tech, University of Liège, 5030 Gembloux, Belgium
[3] Institute of Genetics and Developmental Biology, Chinese Academy of Sciences, Beijing 100101, China
* Correspondence: yyd2021@outlook.com; Tel.: +86-13701252493

Abstract: Shoot branching is a complex and tightly regulated developmental process that is essential for determining plant architecture and crop yields. The outgrowth of tiller buds is a crucial step in shoot branching, and it is influenced by a variety of internal and external cues. This review provides an extensive overview of the genetic, plant hormonal, and environmental factors that regulate shoot branching in several plant species, including rice, Arabidopsis, tomato, and wheat. We especially highlight the central role of *TEOSINTE BRANCHED 1* (*TB1*), a key gene in orchestrating bud outgrowth. In addition, we discuss how the phytohormones cytokinins, strigolactones, and auxin interact to regulate tillering/branching. We also shed light on the involvement of sugar, an integral component of plant development, which can impact bud outgrowth in both trophic and signaling ways. Finally, we emphasize the substantial influence of environmental factors, such as light, temperature, water availability, biotic stresses, and nutrients, on shoot branching. In summary, this review offers a comprehensive evaluation of the multifaced regulatory mechanisms that underpin shoot branching and highlights the adaptable nature of plants to survive and persist in fluctuating environmental conditions.

Keywords: bud outgrowth; tillering; branching; plant hormones; axillary meristem (AM); sugars; light; temperature; water; nutrients; biotic stresses

Citation: Yuan, Y.; Khourchi, S.; Li, S.; Du, Y.; Delaplace, P. Unlocking the Multifaceted Mechanisms of Bud Outgrowth: Advances in Understanding Shoot Branching. *Plants* **2023**, *12*, 3628. https://doi.org/10.3390/plants12203628

Academic Editor: Federica Della Rovere

Received: 15 September 2023
Revised: 12 October 2023
Accepted: 18 October 2023
Published: 20 October 2023

1. Introduction

The plasticity exhibited by plants in their shoot development is remarkable, as it allows them to adapt to various harmful external and internal conditions in order to survive and thrive. Shoot architecture in seed plants is primarily determined by factors such as the number, position, orientation, and size of shoot branches. The regulation of shoot branching/tillering constitutes a critical survival and propagation strategy governed by a complex, sophisticated regulatory network.

Initiation of the primary shoot axis can be traced back to the shoot apical meristem (SAM), a group of mitotic cells that forms during embryogenesis. Subsequently, the derivatives of this meristem give rise to all above-ground parts of plants [1]. The SAM produces aerial organs by continuously adding growth units called phytomers, generally comprising three parts: an internode, a leaf, and an axillary meristem (AM) that emerges at the leaf axil [2].

AMs are new stem cell niches derived from the SAM during post-embryonic development. AM activity plays a vital role in generating the intricate branching patterns that contribute to a plant's fractal architecture. Given its significant influence on shoot branching/tillering and panicle branching, the AM has been a focal point in breeding selection for improving crop production and management [3–5]. For the convenience of readers to understand the influence of crop yield, please refer to Figure 1 and

Supplementary Video S1, which dynamically indicates the process of AMs' outgrowth and their impact on grain yield.

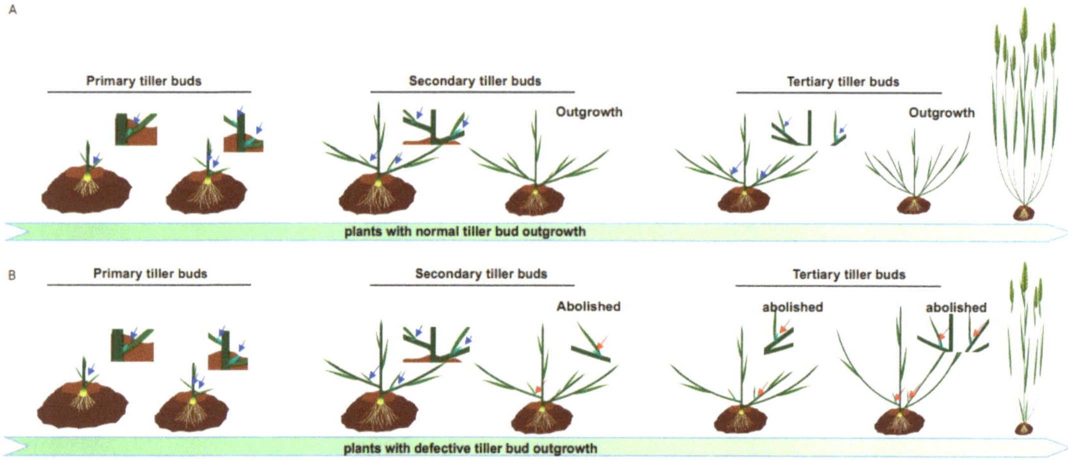

Figure 1. Illustration of the dynamic process of tiller formation and comparisons between plants with normal and defective tiller bud outgrowth. (**A**) portrays the successive processes of tiller bud formation and their outgrowth to generate more panicles than in (**B**). The primary tiller buds (arrows indicated) arise from the leaf axils of the main stem. Secondary tiller buds occur from leaf axils of the primary tillers and so on. We have highlighted with red arrows the abolished tiller buds that cannot grow to form tillers in (**B**).

Numerous studies conducted over the past decades have aimed to unravel the mechanisms underlying shoot branching. The prevailing understanding is that various inputs, such as endogenous factors, developmental cues, and environmental signals, interlock to regulate shoot branching. For instance, among multiple essential genes controlling shoot branching in plants, *TEOSINTE BRANCHED 1* (*TB1*) acts locally in buds to inhibit bud outgrowth and is considered to be an integrator of diverse phytohormonal, trophic, and environmental signaling networks (Figure 2) [6,7]. In addition, the signals defined by phytohormones, nitrogen, light, and sugars have been shown to significantly affect shoot branching (Figure 2) [8–21].

Significant advances have been made in our understanding of shoot branching, and numerous essential genes described in the literature have been demonstrated to influence shoot branching. However, the underlying mechanisms involving these genes are complicated, such as the effect of plant resistance genes on branching/tillering (in addition to resistance) when disrupted [22,23]. This review aims to provide a comprehensive summary of recent advances in our understanding of shoot branching, with a primary focus on Arabidopsis (*Arabidopsis thaliana*), wheat (*Triticum aestivum*), tomato (*Solanum lycopersicum*), and rice (*Oryza sativa*). Elucidating and differentiating the complex mechanisms underlying shoot branching will contribute to the field of crop breeding, as shoot branching/tillering crucially determines plant architecture, which directly influences yield and overall productivity.

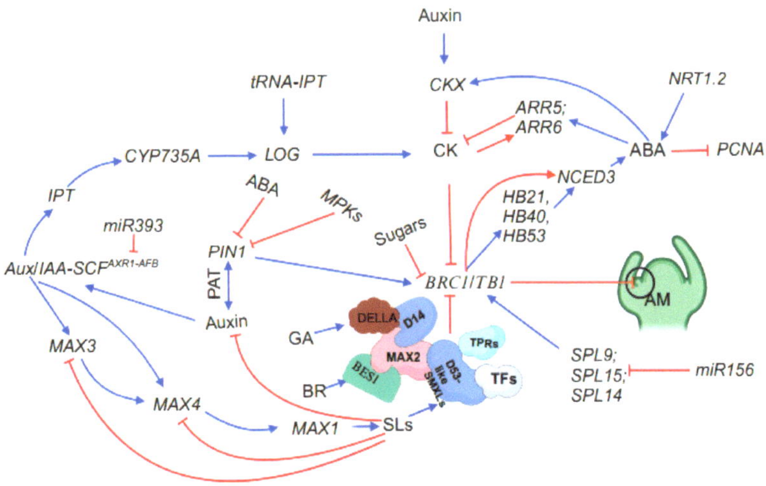

Figure 2. Summary of phytohormones and key genes involved in shoot branching. Blue arrows represent promotion, whilst red flat-ended lines denote inhibition. In this model, *BRC1/TB1* acts as an integrator to interact with other genes, such as *SPL* genes and phytohormones, to mediate shoot branching. Abbreviations: SL, strigolactone; ABA, abscisic acid; CK, cytokinin; GA, gibberellin; *PIN1, PIN-FORMED1; SCF, Skp-Cullin-F-box; Aux/IAA, Auxin/Indole-3-Acetic Acid; IPT, ADENYLATE ISOPENTENYLTRANSFERASE*; PAT, polar auxin transport; *CYP735A, cytochrome P450 monooxygenase 735A; LOG, LONELY GUY; tRNA-IPT, transfer RNA isopentenyltransferase; CKX, cytokinin oxidase; ARR5* and *ARR6, RESPONSE REGULATOR5* and *6; NRT1.2, nitrate transporter 1.2; PCNA1, PROLIFERATING CELL NUCLEAR ANTIGEN1; SPL9, SPL14* and *SPL15, SQUAMOSA PROMOTER BINDIN PROTEIN-LIKE9,14* and *15; BRC1/TB1, BRANCHED1/TEOSINTE BRANCH 1; NCED3, 9-CIS-EPOXYCAROTENOID DIOXYGENASE3; HB21, HOMEOBOX PROTEIN 21; HB40, HOMEOBOX PROTEIN 40; HB53, HOMEOBOX PROTEIN 53*; DELLA, aspartic acid–glutamic acid–leucine–leucine–alanine; D14, DWARF 14; *MAX3, 2, 1,* and *4, more axillary growth 3, 2, 1,* and *4*; TPRs, TOPLESS-RELATED PROTEINs; TFs, transcription factors; D53-like SMXLs, DWARF53-LIKE SMAX1-LIKEs; BES1, bri1-EMS-suppressor 1; AM, axillary meristem.

2. Mechanisms Regulating Shoot Branching

Once a branch/tiller bud is formed, the plant is confronted with a critical decision to stimulate bud sprouting, giving rise to a branch/tiller, or to maintain bud dormancy. Bud outgrowth typically progresses through three discernible stages: dormancy, transition, and sustained growth [24–26]. The fate of buds in the transition stage is influenced by the complex interplay of environmental and endogenous cues, ultimately determining whether buds return to dormancy or enter a sustained growth phase [27]. Consequently, the final count of branches/tillers is not solely determined by the number of axillary buds but is also influenced by the potential of buds to undergo growth [28]. In the following sections, we primarily focus on elucidating shoot branching through the lens of endogenous cues and environmental signals. For a visual representation of the interplay among various components, please refer to the conceptual model of bud outgrowth shown in Figure 2. The genes mentioned in this section are referred to in Table 1.

Table 1. List of genes related to lateral bud outgrowth. The table below lists a number of genes that have been identified and characterized for functions in the mentioned plant species.

Gene Names	Accession Numbers	Reported Species (Homolog)	Functional Annotation	References
OsTB1	*Os03g0706500*	Arabidopsis, rice, maize, pea, tomato	Transcription factor TCP family	[6,29–31]
OsSPL14 (IPA1), OsSPL15	*Os08g0509600, Os08g0513700*	Rice, Arabidopsis	SQUAMOSA promoter binding protein-like transcription factors	[5,32,33]
AXR1	*AT1G05180*	Arabidopsis	A subunit of the RUB1 activating enzyme	[34]
YUCCA	*AT4G32540*	Arabidopsis	A flavin monooxygenase-like enzyme, auxin biosynthesis	[35]
PIN1	*Os02g0743400*	Rice	An auxin transporter	[36]
OsPIN5b	*Os09g0505400*	Rice	An auxin transporter	[37]
MKK7	*AT1G18350*	Arabidopsis	MAP kinase kinase7	[38]
MKK6	*AT5G56580*	Arabidopsis	MAP kinase kinase 6	[39]
TIR1	*Os05g0150500, AT1G72930*	Rice, Arabidopsis	Auxin receptor	[9,40]
IAA12	*AT1G04550*	Arabidopsis	An auxin-responsive gene	[41]
AFB2	*Os04g0395600*	Rice	Auxin signaling f-box 2	[42]
RUB1	*AT1G31340*	Arabidopsis	A ubiquitin-related protein	[9]
D27	*Os11g0587000*	Rice	An iron-containing protein	[12,43]
CCD7/MAX3	*AT2G44990, Os04g0550600*	Arabidopsis, rice	Carotenoid cleavage dioxygenases	[44]
CCD8/MAX4	*AT4G32810, Os01g0746400*	Arabidopsis, rice	Carotenoid cleavage dioxygenases	[45]
MAX1	*AT2G26170*	Arabidopsis	Belonging to the CYP711A cytochrome P450 family	[46]
MAX2/D3	*AT2G42620, Os06g0154200*	Arabidopsis, rice	Belonging to a member of the F-box leucine-rich repeat family	[46]
D14	*Os03g0203200, AT3G03990*	Rice, Arabidopsis	An alpha/beta hydrolase	[47]
D53	*Os11g0104300*	Rice	The substrate of SCF-D3 ubiquitin complex	[48]
SMXL6, SMXL7, SMXL8	*AT1G07200, AT2G29970, AT2G40130*	Arabidopsis	D53-like proteins	[48,49]
IPT	*AT3G23630*	Arabidopsis	An isopentenyl transferase	[50]
SPS	*AT1G16410*	Arabidopsis	Belonging to a member of CYP79F	[51]
AMP1	*AT3G54720*	Arabidopsis	A glutamate carboxypeptidase	[8]
PsCKX2	*LOC127082854*	Pea	Cytokinin dehydrogenase 6-like	[52]
NCED3	*AT3G14440*	Arabidopsis	A 9-cis-epoxycarotenoid dioxygenase	[53]
ABA2	*AT1G52340*	Arabidopsis	A cytosolic short-chain dehydrogenase	[54]
HB21, HB40, HB53	*AT2G02540, AT4G36740, AT5G66700*	Arabidopsis	Homeobox proteins	[14]
SLR1	*Os03g0707600*	Rice	A DELLA protein	[32]
BES1	*AT1G19350*	Arabidopsis	A transcription factor	[55]
MOC2	*Os01g0866400*	Rice	A cytosolic fructose 1,6-bisphosphatase	[56]
OsNPF7.7	*Os10g0579600*	Rice	One nitrate transporter	[57]
TaNAC2-5A	*AY625683*	Wheat	A transcription factor	[58]
OsMADS57	*Os02g0731200*	Rice	A MADS transcription factor 57	[59]

Table 1. *Cont.*

Gene Names	Accession Numbers	Reported Species (Homolog)	Functional Annotation	References
OsBZR1, BES1	*Os07g0580500, AT1G19350*	Rice, Arabidopsis	A key transcription factor involved in brassinosteroid (BS) signaling	[60]
DLT	*Os06g0127800*	Rice	A GRAS protein	[61]
GSK2	*Os05g0207500*	Rice	A conserved glycogen synthase kinase 3-like kinase	[61]
RLA1	*Os05g0389000*	Rice	An APETALA2 (AP2) DNA binding domain protein	[62]
BRI1/D61	*Os01g0718300*	Rice	A BR receptor	[61]

2.1. Internal Inputs Determine Bud Outgrowth

2.1.1. TB1/BRC1 Acts as a Key Integrator of Branching

The expression of *TB1*, encoding a non-canonical basic helix–loop–helix (bHLH) transcription factor of the TCP family, is negatively correlated with bud growth [63,64]. This TCP protein family is represented by four founding members: TB1, CYCLOIDEA (CYC), PROLIFERATING CELL NUCLEAR ANTIGEN FACTOR1 (PCF1), and PCF2. These family members were identified by their functions in plant development or their DNA binding capacity [63,65–68]. The maize (*Zea mays*) *tb1* mutant exhibits an uncontrolled proliferation of tillers, resulting in a bushy architecture reminiscent of its ancestor, teosinte [29]. The inhibitory effect of *TB1* on bud outgrowth is spatially restricted to axillary buds as soon as they become visible [29]. By contrast, in teosinte, *TB1* is not expressed in axillary buds, allowing axillary bud outgrowth [29]. The role of *TB1* in suppressing axillary bud outgrowth is conserved in rice, as ectopic overexpression of its ortholog *OsTB1* under the control of the actin promoter leads to reduced tillering [69]. Conversely, loss-of-function mutants of *OsTB1*, such as *fine culm 1* (*fc1*), show increased tillering [69]. *TB1* and its orthologs (e.g., *OsTB1* or *FC1* in rice, *BRANCHED1* (*BRC1*) in Arabidopsis, *PsBRC1* in pea [*Pisum sativum*], and *SlBRC* in tomato) operate in conjunction with other vital genes and plant hormones to regulate bud outgrowth [6,30,31,70]. These genes and phytohormones are discussed in subsequent sections. The coordinated action of TB1 with these factors has earned it the title "branching integrator" (Figure 2).

It is worth noting that several studies have indicated that inhibition of bud outgrowth can occur independently of *TB1* and its orthologs [2,31,71,72]. Hence, *TB1* may condition bud activation potential, thus contributing to the regulation of branching.

2.1.2. SQUAMOSA Binding Proteins Inhibit Bud Outgrowth

SQUAMOSA promoter binding protein-like (SPL) transcription factors, which are specific to plants, mediate various aspects of plant development, including branching [32]. Different members of the *SPL* gene family in Arabidopsis are post-transcriptionally regulated by miR156 [73]. Overaccumulation of miR156 leads to a considerably bushy phenotype [74,75]. Notably, double mutants of the Arabidopsis paralogs *SPL9* and *SPL15* exhibit an increased branching phenotype, highlighting the crucial role of miR156-targeted *SPL* genes in regulating shoot branching [33] (Figure 2). Accumulation of the SPL9 and SPL15 ortholog OsSPL14 results in fewer tillers and increased yield in rice [5]. These findings highlight the roles of SPL proteins in inhibiting branching. Additionally, OsSPL14, whose encoding transcripts are targeted by miR156, directly activates *OsTB1* expression [5].

2.1.3. Auxin Indirectly Inhibits Sustained Bud Outgrowth

A principal function of auxin, as observed in various studies, is mediating apical dominance, as lateral bud outgrowth is inhibited by auxin. For instance, the *auxin-resistant 1* (*axr1*) mutant of Arabidopsis exhibits a lower sensitivity to auxin than wild-type plants, resulting in weak apical dominance and an increased number of branches, due to the release of axillary buds, highlighting the role of auxin in inhibiting axillary bud outgrowth [34]. Conversely, auxin-overproducing mutants with elevated levels of free auxin, such as the Arabidopsis *yucca* mutants, exhibit stronger apical dominance than wild-type plants [35].

Moreover, apical dominance largely depends on polar auxin transport (PAT) mediated by PIN-FORMED (PIN) proteins in the stem [76]. Studies involving RNA interference (RNAi)-mediated knockdown and overexpression of *OsPIN1* have demonstrated the negative effect of *OsPIN1* on tillering in rice [36]. In addition, overexpression and knockdown experiments with *OsPIN5B* revealed the influence of this gene on tiller numbers [37]. Mitogen-activated protein kinase (MAPK) cascades are essential for transducing external and internal cues into adaptive and programmed responses. The MKK7 (MAPK KINASE 7)-MPK6 signaling pathway regulates PAT by phosphorylating the specific substrate PIN1, thereby modifying shoot branching in Arabidopsis [38,39].

In addition to the crucial role of PAT in regulating branching/tillering, the auxin signaling pathway also influences shoot branching/tillering via SKP1-CULLIN1-F-box (SCF)-mediated protein degradation. Auxin receptors, including TRANSPORT INHIBITOR RESPONSE 1 (TIR1) and closely related family numbers (AUXIN SIGNALING F-BOX [AFB]) [9,40,41], bind to auxin to stabilize the interactions between TIR1/AFBs and members of the Aux/IAA (Auxin/INDOLE-3-ACETIC ACID INDUCIBLE) family of transcriptional repressors [77]. Their interaction with TIR1/AFBs leads to the degradation of Aux/IAA, permitting auxin-mediated upregulation of transcription [78–80]. By contrast, loss of function of *IAA12* in Arabidopsis leads to auxin-resistant stabilization of the SCF complex and, thus, constitutive suppression of target auxin-upregulated genes [41], resulting in a bushy phenotype. Overexpressing *OsMIR393*, whose mature miRNA product OsmiR393 targets and downregulates the transcripts of the auxin receptor genes *OsTIR1* and *OsAFB2*, leads to increased tiller production [42].

Moreover, many *Aux/IAA* genes are rapidly transcriptionally induced by auxin in a $SCF^{TIR/AFB}$-dependent manner [81]. *TIR1/AFB* genes encode F-box proteins that interact with the cullin CUL1 and the SKP1-like proteins ASK1 or ASK2 to form an SCF ubiquitin protein ligase (E3). TIR1/AFB genes function as transcription factors that bind to auxin-response elements (AuxREs) located in the upstream regions of auxin-inducible genes [82]. Auxin resistant 1 (AXR1) is required for proper SCF function as it facilitates conjugation of the ubiquitin-like protein RELATED TO UBIQUITIN 1 (RUB1) to the cullin subunit [9,83,84]. Correspondingly, mutations in *AXR1* result in changes in the expression of $SCF^{TIR1/AFB}$-dependent auxin-responsive genes, leading to defects in downstream auxin responses [9] (Figure 2).

It should be noted that the effect of auxin on branching/tillering is indirect, as apically derived auxin cannot enter buds [85]. Two primary models have been proposed to explain this phenomenon: the canalization model and the second-messenger-based model. According to the canalization model, axillary buds are activated when the amount of auxin initially flowing out of the bud is sufficient to trigger the establishment of polar auxin transport connected to the auxin stream in the stem, thereby promoting bud outgrowth [86,87]. Conversely, the continuous flow of auxin in the stem originating from the apex restricts the export of auxin from the axillary buds on the same axis, thereby maintaining apical dominance [88]. The establishment of auxin transport involves a positive regulatory feedback loop between the polarization of auxin efflux-facilitating PINs at the plasma membrane in the direction of the initial flow and the directional flow [89]. In addition, strigolactones (SLs) act upstream of auxin by stimulating the removal of PIN1 from the plasma membrane, thereby reducing the ability of the bud to create its own polar auxin transport [90]. According to the second-messenger-based model, auxin flow in the main stem negatively modulates cytokinin (Ck) biosynthesis [52] and positively regulates SL levels [91], with these two phytohormones acting antagonistically on buds by inducing and inhibiting their outgrowth, respectively [92,93]. Furthermore, the antagonistic effects of Cks and SLs in buds are integrated by *BRC1*, the Arabidopsis homolog of *TB1* that is mainly expressed in dormant axillary buds [6,31,93].

The transition of a bud from dormancy or quasi-dormancy to more active outgrowth is associated with increased expression of genes involved in the cell cycle [25,94,95]. Expression of cell cycle-related genes is repressed by auxin biosynthesized in the shoot apex, leading to suspension of cell division and bud dormancy [25].

2.1.4. Strigolactones Have an Inhibitory Effect on Bud Outgrowth

SLs are a collection of carotenoid-derived lactones secreted by plants. These phytohormones are primarily known for their roles as rhizosphere signals used by root-parasitic plants to detect their hosts [96] and as cues for mycorrhizal fungi to form symbiotic associations [97]. Importantly, SLs also inhibit the outgrowth of axillary buds. This inhibitory effect was initially observed in *ramosus* (*rms*) mutants in pea, which exhibit excessive branching, as well as *decreased apical dominance* (*dad*) mutants in petunia (*Petunia hybrida*) and *more axillary growth* (*max*) mutants in Arabidopsis [45,98–104].

SL biosynthesis involves several enzymes. DWARF27 (D27) is responsible for isomerizing all-*trans*-β-carotene at the C-9 position to form 9-*cis*-carotene [12,43]. Carotenoid cleavage dioxygenase 7 (CCD7, also named MAX3) and CCD8 (also named MAX4) then cleave 9-*cis*-carotene to produce carlactone, a key endogenous SL precursor [44,45]. The conversion of carlactone to SLs is catalyzed by MAX1, an Arabidopsis cytochrome P450 that acts downstream of MAX4 and MAX3 [46]. MAX1 converts carlactone to carlactonoic acid, which is further converted to methyl carlactonoate, an SL-like compound (Figure 2) [105].

SLs are perceived by MAX2, an ortholog of D3 from rice, which plays a crucial role in regulating plant branching. Disruption of *MAX2* leads to a bushy phenotype in Arabidopsis [46]. Moreover, the perception and signaling roles of SLs in rice require their interaction with D14, a putative SL receptor. D14 interacts with D3, an F-box protein of the SCF E3 ubiquitin ligase complex, to form an SL-induced D14-D3 complex [47]. This complex targets proteins for ubiquitination and degradation, resulting in changes in plant branching/tillering [48] (Figure 2).

The dominant *d53* mutant in rice is characterized by a high-tillering and dwarf phenotype and is resistant to the exogenous application of GR24, a synthetic SL. D53 is targeted for SL-dependent degradation by the SCF^{D3} ubiquitination complex [106]. In addition, SL treatment results in D53 degradation via the ubiquitin–proteasome system in a D14- and D3-dependent manner [48]. D53 interacts with members of the TOPLESS-RELATED PROTEIN (TPR) family of transcriptional co-repressors, which may suppress the activities of their downstream transcription factors (Figure 2) [47,107].

In Arabidopsis, D53-like proteins, including SMAX1-LIKE6 (SMXL6), SMXL7, and SMXL8, are targeted for proteolysis by MAX2-mediated SL signaling [48,49]. Their over-accumulation in SL signaling mutants (such as *max2*) leads to increased branching and constitutively low *BRC1* expression in buds. Conversely, disruption of *SMXL6*, *SMXL7*, and *SMXL8* completely restores branching of *max2* to wild-type levels and leads to very high levels of *BRC1* transcript in inhibited buds [48,49]. This effect highlights the role of SLs in inhibiting bud outgrowth by upregulating *BRC1* transcription. Additionally, these SMXL proteins can form a complex with TPR2 and function as transcriptional repressors. However, D14 interacts with SMXLs and MAX2 in an SL-dependent manner, thereby inhibiting axillary bud outgrowth in Arabidopsis [48].

The relationship between SLs and *BRC1* and its corresponding orthologs suggests that the SL pathway acts upstream of *FC1/OsTB1* in rice. For example, the *fc1* mutant does not respond to application of GR24, and the phenotype of the *fc1 d17* double mutant is similar to that of the SL-deficient mutant *d17*, suggesting the involvement of SLs in regulating *FC1/OsTB1* expression in rice [70]. In line with this notion, *PsBRC1* expression levels are significantly lower in SL-deficient mutants (*rms1*, *rms2*, and *rms4*) than in wild-type pea plants [31] but are upregulated in *rms1* and *rms2* after GR24 application, further supporting the idea that SLs act upstream of *BRC1*. Specifically, *BRC1* is repressed in non-elongated *max2* and *max3* axillary buds but induced in the *smxl6 smxl7 smxl8*, *max2 smxl6 smxl7 smxl8*, and *max3-9 smxl6 smxl7 smxl8* mutants of Arabidopsis, suggesting that SMXL6, SMXL7, and SMXL8 inhibit *BRC1* expression [48].

2.1.5. Other Phytohormones Regulate Tillering/Branching

The inhibitory role of auxin in bud outgrowth is exerted indirectly within buds, as apically derived auxin is not transported into buds [108] and exogenous auxin directly supplied to buds fails to prevent their growth [109]. Cytokinins (CKs) are believed to be crucial in relaying the auxin signal into buds and promoting axillary branching. CKs are an important class of phytohormones that participate in various aspects of plant development, including organ formation, apical dominance, and leaf senescence [110]. Therefore, CKs facilitate the growth and development of axillary buds by serving as a second messenger for the auxin signal.

Several studies have demonstrated that CKs can promote the outgrowth of buds that would otherwise remain inhibited and that CK levels in or near the bud are well correlated with bud fate [50,111]. For example, in chickpea (*Cicer arietinum* L.) plants, CK levels dramatically increased in axillary buds within 24 h of shoot decapitation [112]. Elevated levels of CKs, as achieved by overexpressing *ISOPENTENYL TRANSFERASE* (*IPT*) (encoding a key enzyme in CK biosynthesis), lead to reduced apical dominance (Figure 2) [50]. Continuous treatment of pea plants with synthetic CKs overcomes the inhibition of lateral bud release, turning these into dominant organs [113]. In Arabidopsis, basally applied CKs suppress the inhibitory effects of apically supplied auxin in isolated nodes [114]. Conversely, low local CK levels can limit bud outgrowth, even in auxin- and strigolactone-deficient plants [115]. The *supershoot* (*sps*) mutant of Arabidopsis displays overproliferating branching due to cytokinin-promoted bud initiation as well as bud outgrowth, as endogenous cytokinin levels are elevated in this mutant [51]. The Arabidopsis *altered meristem program 1* (*amp1*) mutant, with increased cytokinin levels, shows enhanced AM formation and bud elongation to generate branches [8]. Interestingly, in this mutant, *BRC1* is slightly downregulated, suggesting that Cks downregulate *BRC1* expression [6].

In addition to their biosynthesis, the metabolism of CKs also determines endogenous cytokinin levels. Cytokinin oxidase (CKX) is the enzyme responsible for inactivating CKs by irreversibly degrading active CKs, thereby regulating endogenous levels of active CKs [116,117]. Various CKXs, such as PsCKX2 in pea, predominantly regulate CK levels in the stem [52]. Notably, the expression pattern of *PsCKX2* in the stem is opposite to that of CK levels before and after decapitation, suggesting that *PsCKX2* induces a decrease in CK levels in the stem. Overall, these findings support the notion that CKs promote bud outgrowth.

While abscisic acid (ABA) is predominantly recognized for its roles in seed dormancy, growth inhibition, and stress responses, emerging evidence indicates that this phytohormone also influences plant branching/tillering. Mutations in key genes involved in ABA biosynthesis, such as *9-CIS-EPOXYCAROTENOID DIOXYGENASE3* (*NCED3*) and *ABA DEFICIENT2* (*ABA2*), enhance bud outgrowth [53,54]. Notably, ABA levels are elevated in buds with delayed outgrowth but are reduced in elongated buds [54]. Exogenous application of ABA partially suppresses branch elongation, suggesting that ABA functions downstream or independently of genes responsible for bud growth [54]. Unlike the IAA biosynthesis gene *TRYPTOPHAN AMINOTRANSFERASE OF ARABIDOPSIS 1* (*TAA1*) and the auxin transporter gene *PIN1*, exogenously supplied ABA does not affect *BRC1* expression, confirming the downstream role of ABA in regulating *BRC1*. Furthermore, BRC1 binds to the promoter of and positively regulates the transcription of three genes encoding related homeodomain leucine zipper proteins (HD-ZIP): *HOMEOBOX PROTEIN 21* (*HB21*), *HB40*, and *HB53*. Together with BRC1, these three proteins promote *NCED3* expression, resulting in ABA accumulation and triggering a phytohormonal response, thereby suppressing bud development [14] (Figure 2). Finally, ABA treatment represses the expression of the cell cycle-related gene *PROLIFERATING CELL NUCLEAR ANTIGEN1*

(*PCNA1*), suggesting that ABA modulates bud outgrowth by regulating cell cycle-related gene expression [54].

Gibberellins (GAs) are renowned for their ability to regulate internode elongation. Dwarfism is often associated with an increase in shoot branching. GA-deficient mutants in Arabidopsis, rice, and pea exhibit higher levels of branching/tillering than their wild-type counterparts [10,118]. Likewise, overexpressing GA catabolism genes decreases GA levels, resulting in increased branching/tillering [10]. Moreover, mutants with disruption of DELLA proteins, the major negative regulators of GA signaling, display reduced shoot branching and/or altered branching patterns [32]. Additionally, in rice, DELLA SLENDER RICE 1 (SLR1) was found to interact with the SL receptor D14 in an SL-dependent manner [32] (Figure 2).

Brassinosteroids (BRs) are important plant hormones that regulate various developmental processes, including stem elongation, leaf development, senescence, and branching [119]. A series of BR-deficient mutants have been used to elucidate the function of BR. For instance, reduced expression of *Brassinazole Resistant 1* (*OsBZR1*) in rice leads to a dwarf phenotype with erected leaves and reduced BR sensitivity [60]. *Dwarf and Low Tillering* (*DLT*), encoding a GRAS family protein, is another BR-related gene. Disruption of *DLT* results in a semi-dwarf mutant with fewer tillers and decreased BR responses. The promoter of *DLT* can be targeted by *OsBZR1* [61]. *GLYCOGEN SYNTHASE KINASE 2* (*GSK2*) encodes a conserved glycogen synthase kinase 3-like kinase. Gain-of-function mutations within the *GSK2* coding sequencing or its overexpression suppress BR signaling, leading to plants with phenotypes resembling BR-deficient mutants [120]. The *reduced leaf angle 1* (*rla1*) mutant encodes a transcription factor containing an APETALA2 (AP2) DNA binding domain that is required for *OsBZR1* function [62]. RLA also can interact with GSK2 [62].

A BR-defective rice mutant displays reduced branching like *dlt* [61], while in Arabidopsis, the *bri1-EMS-suppressor 1* (*bes1*) mutant displays a highly branched phenotype. In contrast, *BES1*-RNAi lines have fewer branches than the wild type [55,119]. Interestingly, the dominant *bes1-D* mutant does not respond to GR24 treatment, and BES1 can interact with MAX2 and act as its substrate for degradation, which is regulated by SLs [15]. Therefore, SL and BR signaling pathways converge on the same transcription factor BES1, an Arabidopsis homolog of OsBZR1, to control branching [15]. However, other components upstream of *AtBES1* in the BR signaling pathway do not alter branching in Arabidopsis [15]. In rice, BRs bind to the receptor *Brassinosteroid insensitive 1* (*BRI1*), activating the receptor complex and inhibiting the ability of OsGSK2 [120] to phosphorylate members of the downstream transcriptional module, including OsBZR1, DLT, and RLA1, and thereby regulate their stability [62]. Furthermore, BRs strongly enhance tillering by promoting bud outgrowth in rice through regulating the stability of D53 and/or the *OsBRZ1-RLA1-DLT* module, a transcriptional complex in the BR signaling pathway [15]. In addition, D53 interacts with OsBZR1 to inhibit the expression of *OsTB1*. This interaction depends on direct DNA binding by OsBZR1, which recruits D53 to the *OsTB1* promoter in axillary buds [15].

2.1.6. Phytohormones Interact Influencing Bud Outgrowth

The regulation of axillary bud outgrowth involves a complex network of phytohormones. Plant hormones, including auxin, CKs, and SLs, play central roles in bud outgrowth. Furthermore, plant hormones can mutually affect each other, ultimately regulating branching/tillering.

Auxin is an indispensable player in plant architecture, particularly branching, and a central component of interacting networks that regulate branching. Auxin exerts its role indirectly in buds [85], whereas SLs are direct components that act on buds via auxin to inhibit bud outgrowth. Auxin inhibits bud outgrowth by regulating SL biosynthesis, as observed in Arabidopsis and rice [92,121,122]. For example, auxin that originates in the shoot apex modulates SL levels in pea by maintaining *RMS1* and *RMS5* transcript abundance [123,124]. The *iaa12* mutant exhibits increased numbers of branches and shows reduced expression of the SL biosynthesis genes *MAX3* and *MAX4* [92], while *MAX3* and *MAX4* transcription is mediated by auxin [125]. Conversely, Arabidopsis mutants defective in SL biosynthesis (such as *max4*) exhibit increased branching and resistance to auxin [45]. Additionally, branching is inhibited by SL treatment in auxin-response mutants such as *axr1* and the *tir1 afb1 afb2 afb3* quadruple mutant [126], providing compelling evidence that loss of auxin signaling promotes bud outgrowth via SL depletion. Moreover, both auxin biosynthetic and auxin signaling mutants respond to SL treatment, indicating that SLs function downstream of auxin [91].

SLs also inversely affect polar auxin transport by limiting the accumulation of the auxin efflux carrier PIN1 in cells involved in polar auxin transport [87]. The Arabidopsis *max* mutants, with defects in the SL pathway, show enhanced polar PIN accumulation and auxin transport [127–129]. Similarly, the rice *d27* mutant (characterized by low SL concentrations) displays enhanced auxin transport [12]. Therefore, auxin and SLs interact via interconnected feedback loops, where each phytohormone regulates the level of the other.

Auxin and CKs play antagonistic roles in regulating bud outgrowth. Auxin inhibits AM outgrowth, whereas CKs counteract auxin activity in Arabidopsis by promoting bud activation [20]. The inhibitory effect of auxin is probably mediated, at least in part, by its ability to reduce both CK export from roots and CK biosynthesis locally at the node [130,131]. Stem girdling in pea, which prevents polar auxin transport via a mechanism similar to decapitation, increases the expression of CK biosynthesis genes and promotes the growth of buds below the girdling site [115]. Decapitated bean (*Phaseolus vulgaris* L.) plants (with a decrease or loss of polar auxin transport) have higher CK concentrations in xylem exudates than control plants. However, applying auxin to the shoots of decapitated plants eliminates the effect of shoot-tip removal on CK concentration, further supporting the antagonistic relationship between auxin and CKs [130]. Auxin inhibits expression of the CK biosynthesis gene *IPT* in the stem [111,115,130,132]. In Arabidopsis, auxin-mediated suppression of CK biosynthesis is dependent on AXR1 [131]. However, in addition to repression by auxin, CKs also enhance auxin production and promote downward auxin transport out of growing buds. This, in turn, suppresses the production of CKs lower within the stem, limiting their accessibility to other buds [52,111,133].

In addition to the inhibitory roles of auxin on CKs, ABA represses CK signaling by inducing the expression of type-A response regulator genes, such as *RESPONSE REGULATOR5* (*ARR5*) and *ARR6*, encoding negative regulators of CK signaling [134] (Figure 2). Moreover, ABA inhibits the accumulation of CKs in the roots and shoots of wheat and promotes CKX activity, contributing to the decrease in CK levels [135]. Conversely, overexpression of *IPT* results in reduced ABA abundance in petunia flowers [136]. ABA suppresses auxin biosynthesis and auxin transport out of axillary buds [54].

2.1.7. Sugars Play an Essential Role in Bud Release

Sugars such as sucrose serve not only as a carbon source for plant metabolism but also as essential signaling compounds [137–139]. Bud outgrowth responds to limiting nutrients and resources, and thus sugars have been attracting increasing interest. Axillary buds are often maintained in a dormant state or their growth is suppressed by the growing shoot apex long after their initial formation. Intriguingly, changes in sugar availability can facilitate the first visible growth of buds, a phenomenon known as axillary bud release [16]. The term "apical dominance" is commonly used to describe shoot branching, referring to the role of the shoot tip in preventing the growth of the axillary buds below it [140]. However, in several plant species, auxin supplementation to the decapitated stump, even at high levels, fails to fully restore apical dominance [109,141]. Therefore, it is unlikely that auxin is the first component inhibiting axillary bud outgrowth. Rather, after decapitation, sugars are rapidly redistributed over a significant distance and accumulate in axillary buds, coinciding with the timing of bud release, before auxin depletion occurs in the corresponding axillary buds [16,142]. This observation suggests that sugars have the potential to promote bud release.

Enhancing the sugar supply alone is sufficient for bud release. Plants employ various mechanisms to maintain apical dominance, one of which is limiting the sugar supply to axillary buds [16]. This effect can be observed in the wheat *tiller inhibition* (*tin*) mutant, whose reduced tillering is associated with a decreased sucrose content in axillary buds [143]. Likewise, reduced tiller formation in the rice *monoculm 2* (*moc2*) mutant is attributed to a disruption in fructose-1,6-bisphosphatase, an enzyme involved in sucrose biosynthesis, resulting in a decline in the sucrose supply [56]. The requirement for sugars for bud outgrowth has been demonstrated in rose (*Rosa hybrida*), where sugar is required for triggering bud outgrowth in single nodes cultivated in vitro [88,144]. Sucrose can also modulate the dynamics of bud outgrowth in a concentration-dependent manner, especially during the transition phase between bud release and sustained bud elongation [88]. Additionally, removal of competing sugar sources or sinks within buds through defoliation further supports the role of sugars in bud release [16,145]. Collectively, these lines of evidence highlight the trophic role of sugars in bud release.

Besides the roles of sugars as nutrients, an effect of sugars on phytohormone homeostasis has been demonstrated in single nodes of *R. hybrida* [88]. For instance, sucrose stimulates CK biosynthesis in bud-bearing stem segments by upregulating the expression of two CK biosynthesis-related genes [88]. Sucrose can also modulate auxin metabolism, as treatment with sucrose or its non-metabolizable analogs increases auxin levels in *R. hybrida* buds in a concentration-dependent manner [88]. Furthermore, elevated sucrose levels in buds promote the export of auxin from the bud to the stem, which is favorable for bud outgrowth according to the auxin canalization model [88]. When exogenously supplied to rose, sucrose reduces the expression of *MAX2*, a gene involved in SL signaling [88]. A dose-dependent inhibitory effect by sucrose has also been detected for *RhBRC1* expression, which is decisive in preventing bud outgrowth [31,146]. Notably, palatinose, a non-metabolizable sucrose analog, can trigger bud outgrowth [144]. These findings collectively demonstrate the crucial role of sugar signaling in regulating bud release.

From a trophic perspective, axillary buds are sink organs that require imported sugars to fulfill their metabolic demands and support their growth. In the context of apical dominance, the supply of sugars to lateral buds is the first signal that releases bud dormancy, preceding detectable auxin depletion. Thus, the growth potential of a bud can be determined by its sink strength, representing its ability to acquire and utilize sugars. The interplay in the demand for sugar between the apical bud and lateral buds is thus crucial for the systemic regulation of shoot branching, encompassing both nutritional support and signaling mechanisms.

2.2. *Effects of Environmental Inputs on Bud Outgrowth*

Bud outgrowth and the transition to dormancy are tightly regulated by various environmental factors, including photoperiod, light intensity, nutrient availability, and stress conditions [26,147]. Notably, ABA, a pivotal phytohormone involved in regulating bud outgrowth, strongly accumulates under stressful conditions such as osmotic stress [148]. Interestingly, ABA levels are elevated in buds with delayed outgrowth but are reduced in elongated buds [54]. This dynamic modulation of ABA levels underscores the remarkable ability of plants to adjust their branching capability in response to diverse environmental cues. Such adaptability represents a successful evolutionary trait that has evolved to accommodate the sessile nature of plants. Here, we focus on environmental cues reported so far. The genes mentioned in this section are referred to in Table 2.

Table 2. List of genes related to lateral bud outgrowth response to environmental inputs. The table below lists a number of genes that have been identified and characterized for functions in the mentioned plant species.

Gene Names	Accession Numbers	Reported Species (Homolog)	Functional Annotation	References
PHYB	*LOC8081072*	*Sorghum bicolor*	Phytochrome B	[149]
SbTB1	*LOC8062930*	*Sorghum bicolor*	Belonging to transcription factor of the TCP family	[149]
NCED3	*AT3G14440*	Arabidopsis	A 9-*cis*-epoxycarotenoid dioxygenase	[53]
ABA2	*AT1G52340*	Arabidopsis	A cytosolic short-chain dehydrogenase/reductase	[53]
OsNPF7.7	*Os10g0579600*	Rice	Belonging to the peptide transporter (PTR) gene family	[57]
OsNR2	*Os02g0770800*	Rice	NADH/NADPH-dependent NO_3^- reductase 2	[150]
NGR5	*Os05g0389000*	Rice	One APETALA2-domain transcription factor	[151]
PRC2	*Os03g0108700*	Rice	A polycomb repressive complex 2-associated coiled-coil protein	[151]
D14	*Os03g0203200*	Rice	A strigolactone receptor	[151]
SPL14	*Os08g0509600*	Rice	A squamosa promoter-binding-like transcription activator	[151]
OsDEP1	*Os09g0441900*	Rice	One unknown phosphatidylethanolamine-binding protein (PEBP)-like domain protein	[152]
OsAFB2	*Os04g0395600*	Rice	An auxin receptor	[153]
OsTIR1	*Os05g0150500*	Rice	A F-Box auxin receptor protein	[153]
OsTCP19	*Os06g0226700*	Rice	A class-I TCP transcription factor	[154]
TaNAC2-5A	*LOC606326*	Wheat	NAC domain-containing protein 2	[58]
OsMADS57	*Os02g0731200*	Rice	A MADS-box transcription factor	[59]
OsPHR2	*Os07g0438800*	Rice	A MYB-CC family protein	[155]
NSP1	*Os03g0408600*	Rice	A GRAS-domain transcription factor	[156]
NSP2	*Os03g0263300*	Rice	A GRAS-domain transcription factor	[156]
OsHAK5	*Os01g0930400*	Rice	A potassium transporter	[157]
OsABCB14	*Os04g0459000*	Rice	An auxin transport	[158]
WOX11	*Os07g0684900*	Rice	A WUSCHEL-related homeobox protein	[159]
OsHAK16	*Os03g0575200*	Rice	A high-affinity potassium transporter	[159]
OsAUX1	*Os01g0856500*	Rice	An auxin transporter	[42]
LRK2	*Os02g0154000*	Rice	A leucine-rich repeat receptor-like kinase	[160]
GHD7	*Os07g0261200*	Rice	A CCT(CONSTANS, CONSTANS-LIKE, and TIMING OF CHLOROPHYLL A/B BINDING1) domain protein	[161]
OsRAN1	*Os01g0611100*	Rice	A small GTPase	[162]

2.2.1. Light Plays a Critical Role in Bud Outgrowth

Light intensity is pivotal for regulating bud outgrowth across numerous plant species. For instance, low-intensity light inhibits tillering in wheat [163]. By contrast, high-intensity light stimulates branching, as observed in *Rosa* species [164].

Photoperiod also has a significant influence on the distribution of bud outgrowth along the plant stem. For example, under short-day conditions, the formation of basal branches is enhanced in pea. Bud outgrowth in the upper nodes often coincides with the onset of flowering and may also be controlled by photoperiod [165,166].

At high density, shade also regulates bud dormancy in cultivated plants [167]. When plants intercept incident light, the light intensity decreases, preferentially in the red part of the light spectrum. Shade is, therefore, characterized by a reduction in the red (R) to far-red (FR) light ratio (R:FR) due to R light absorption and FR reflection by leaves [168]. This decrease in R:FR serves as a signal of shade or competition for light, prompting plants to respond by inhibiting axillary bud outgrowth, elongating their stature, and accelerating flowering to evade the detrimental consequences of shading. This suite of responses, known as shade avoidance syndrome, is mediated by the R- and FR-absorbing photoreceptor phytochrome B (PHYB) [169]. In densely grown sorghum (*Sorghum bicolor*) plants experiencing shade, inhibition of bud outgrowth due to an enriched FR-light environment is associated with activation of the *TB1-like* gene *SbTB1* in buds [13,149]. Conversely, in the absence of shade, a higher proportion of phyB in the active form induces bud outgrowth by downregulating *SbTB1*. Shade signals with their low R:FR ratios decrease the proportion of the active form of phyB, thereby enhancing the expression of *SbTB1* and promoting bud dormancy (Figure 3). Interestingly, ABA also controls bud growth in response to R:FR shifts [53]. Furthermore, branching in response to a low R:FR ratio is defective in the ABA-deficient mutant *aba2-1* and the ABA biosynthetic mutant *nced3-2* [53].

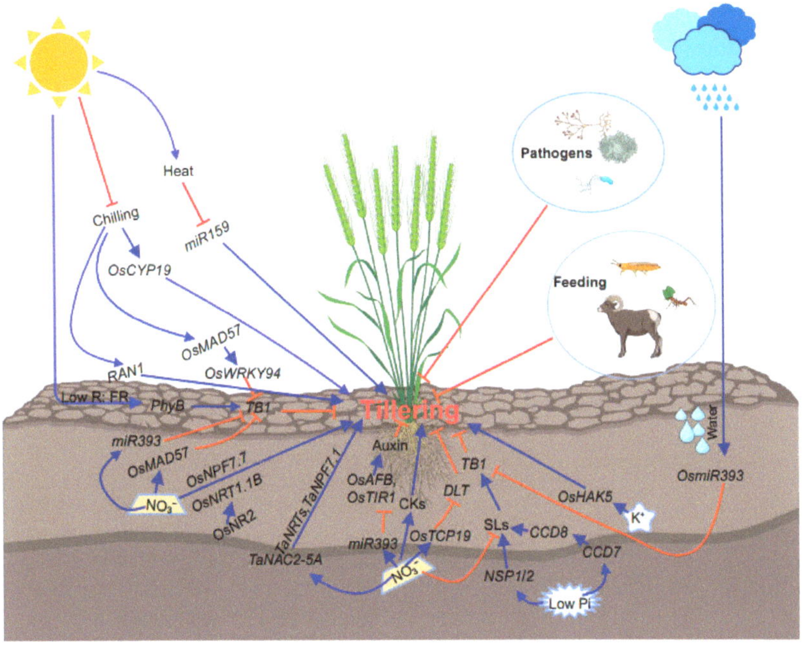

Figure 3. Effects of environmental inputs on bud outgrowth. The roles of various environmental factors are indicated, and genes involved are depicted. Blue arrows represent promotion, while red-flat-ended lines denote inhibition. In this model, the abiotic factors like nutrients, light, temperature, and biotic stresses impact tillering significantly. Abbreviations: *OsCYP19*, *Oryza sativa cytochrome P450 family 19; OsMAD57, Oryza sativa MADS-box protein 57; miR159, microRNA 159; PhyB, phytochrome B; TB1, TEOSINTE BRANCHED 1; OsNPF7.7, peptide transporters family 7.7; OsNRT1.1B.* Oryza sativa *Nitrate Transporter 1.1B; OsNR2, Oryza sativa NADH/NADPH-DEPENDENT NITRATE REDUCTASE 2; OsWRKY94, Oryza sativa WRKY transcription factor 94; TaNAC2-5A, Triticum aestivum NAC transcription factor 2-5A; miR393, microRNA 393;* CK, cytokinin; SL, strigolactone; *OsAFB, Oryza sativa AUXIN SIGNALING F-BOX 2; OsTIR1, Oryza sativa TRANSPORT INHIBITOR RESPONSE; OsTCP19, TEOSINTE BRANCHED1/CYCLOIDEA/PROLIFERATING CELL FACTOR 19; DLT, DWARF AND LOW-TILLERING; NSP1/2, Nodulation Signaling Pathway 1/2;* Pi, phosphorus; *OsHAK5, Oryza sativa High-Affinity* K^+ *Transporter 5;* K^+, potassium; *CCD7/8, Carotenoid cleavage dioxygenase 7/8.*

2.2.2. Impacts of Nutrients on Bud Outgrowth

Nitrogen (N) is an essential macronutrient that dominates plant growth and plant productivity [170,171]. N limitation can significantly limit the tiller numbers in rice [170]. Furthermore, high branching ability is positively correlated with the capacity for N uptake. Several N transporters have been identified that regulate shoot branching in rice. One such transporter is peptide transporters family 7.7 (OsNPF7.7), whose increased abundance facilitates the influx of NO_3^- and NH_4^+, thereby promoting the outgrowth of axillary buds [57]. *Indica NADH/NADPH-DEPENDENT NITRATE REDUCTASE 2* (*OsNR2*) promotes NO_3^- uptake through interaction with *OsNRT1.1B*, a low-affinity NO_3^- transport gene, and increases effective tillers in *japonica* rice. Similarly, the *japonica* allele of *OsNR2* also promotes tillering but not to the extent observed in *indica OsNR2*-overexpression lines [150,172]. The rice APETALA2-domain transcription factor encoded by a *NITROGEN MEDIATED TILLER GROWTH RESPONSE 5* (*NGR5*) allele is upregulated under conditions of increased nitrogen availability. NGR5 interacts with a component of the polycomb repressive complex 2 (PRC2) to regulate the expression of *D14* and *OsSPL14* by mediating levels of histone methylation (H3K27me3) modification, thereby regulating rice tillering [151,172]. Overexpression of *DENSE AND ERECT PANICLE1* (*OsDEP1*) can increase tiller numbers under high N supply [152]. OsSPL14 can directly activate the expression of *OsDEP1* and *OsTB1* to regulate tiller bud outgrowth [173]. A microRNA, *miR393*, can target and repress the expression of *OsTB1* and the two auxin receptors *OsAFB2* and *OsTIR1* under NO_3^- conditions, which influences the transport of auxin and eventually regulates tillering [153]. The rice *TEOSINTE BRANCHED1/CYCLOIDEA/PROLIFERATING CELL FACTOR 19* (*OsTCP19*) transcription factor can directly bind and repress the activity of the tiller-promoting *DLT* gene, thereby negatively regulating tillering in the presence of nitrogen (N). Further investigation revealed a 29-bp insertion/deletion polymorphism in the *OsTCP19* promoter that confers differential transcriptional response to N among rice varieties [154]. In wheat, the NO_3^--inducible *CEREAL-SPECIFIC NAM, ATAF*, and *CUC* (*NAC*) transcription factor *TaNAC2-5A*, whose gene expression is induced by a limited NO_3^- supply, directly binds to the promoters of genes encoding NO_3^- transporters and glutamine synthetase, thereby enhancing N acquisition and assimilation. Notably, *TaNAC2-5A* overexpression leads to enhanced tiller numbers [58]. OsMADS57, a NO_3^--inducible MADS-box transcription factor, interacts with OsTB1 and targets D14 to control the outgrowth of axillary buds in rice [59] (Figure 3).

Following decapitation, the primary forms of available and transported N are amino acids through the phloem, which might be easy for buds to obtain [17]. This notion is supported by the finding that the levels of three key amino acids, aspartate, asparagine, and glutamine, increase in axillary buds after decapitation, coinciding with the initiation of bud outgrowth. Interestingly, sucrose fails to trigger bud outgrowth in excised rose nodes in the absence of asparagine in the growth medium.

Given that N availability influences CK and SL levels, it is plausible that N functions as a second messenger to mediate branching/tillering [174–176]. In several species, limited N availability promotes SL production, subsequently inhibiting branching/tillering. N limitation also leads to a reduction in CK production. N fertilization can suppress the expression of SL biosynthesis genes [171,177]. Collectively, these findings highlight the important role of N (as with sugars) in influencing branching.

Phosphorus (Pi) is an essential macronutrient for plant growth and metabolism. However, the availability of Pi in soils is often low due to chemical fixation and poor diffusion, resulting in low-Pi environments that can limit plant development and processes like tillering [178]. For example, Pi deficiency has been shown to reduce tiller numbers in rice [179], with the transcription factor *Oryza sativa PHOSPHATE STARVATION RESPONSE2* (*OsPHR2*) implicated in the repression of tillering under low Pi conditions [155]. Pi deficiency also induced SL biosynthesis by increasing transcription of SL biosynthetic genes like the β-carotene *cis-trans isomerase DWARF27 (D27)*, the *carotenoid cleavage dioxygenase 7 (CCD7)/D17*, and *CCD8/D10* [2,179,180], which play a key role in regulating tillering, as described above. Additionally, nodulation signaling pathway 1 (NSP1) and NSP2, two GRAS family transcription factors, have been shown to promote SL production in rice under low-Pi conditions by directly binding to SL biosynthetic gene promoters as a complex [156]. Moreover, Pi deficiency represses tiller numbers by promoting the degradation of D53 and the expression of *OsTB1*. Further examination of the mechanisms of genes in response to Pi availability will provide a deeper understanding of nutrient-mediated tillering.

Potassium (K^+) is the most abundant cation in plants and an essential macronutrient [157,181]. Adequate K^+ availability can increase tillering in plants, as evidenced by enhanced tillering in rice overexpressing *Oryza sativa High-Affinity* K^+ *Transporter 5* (*OsHAK5*). In contrast, knockout of *OsHAK5* reduces tillers in rice [157], producing a phenotype resembling loss-of-function mutants of the auxin transporter *OsABCB14* [158], implying a potential interaction between K^+ and auxin. Furthermore, driving expression of the *WUSCHEL-related homeobox* transcription factor gene *WOX11* using the promoter of *OsHAK16*, which encodes a low K^+-induced K^+ transporter, leads to increased effective tiller numbers in rice [159]. Together, these findings indicate that K^+ availability modulates tillering at least through effects on K^+ transporters like *OsHAK5* and associated transcriptional networks [159]. However, the specific molecular mechanisms by which K^+ influences tiller development remain unclear. Further research is needed to elucidate the signaling pathways and gene regulatory networks through which K^+ is perceived and transduced in axillary buds, promoting bud outgrowth.

2.2.3. Water Availability Influences Bud Outgrowth

Tillering/branching processes are susceptible to drought stress, one of the most limiting factors affecting agricultural yields. Optimal water availability is critical for normal plant growth and development, including tillering. Tolerance to this stress is multigenic and complex in nature. Drought stress triggers specific alterations of gene-expression patterns in plant tissues [182]. For instance, the drought-inducible microRNA *miR393* was shown to be upregulated in Arabidopsis [183], and *miR393* also regulates tiller number increases in rice by modulating auxin signaling through auxin receptor genes like *OsAUX1* and *OsTIR1*. In addition, overexpression of *OsmiR393* downregulates the rice tillering inhibitor *OsTB1*, leading to increased tiller numbers [42] (Figure 3). Overexpression of *LRK2*, which encodes a leucine-rich receptor-like kinase gene, increases drought tolerance and tiller numbers in rice [160]. *Grain number*, *plant height*, and *heading date7* (*GHD7*) encoding a *CCT* (*CONSTANS, CONSTANS-LIKE, and TIMING OF CHLOROPHYLLA/B BINDING1*) domain protein regulates the rice flowering pathway and also contributes to rice yield potential. Overexpression of *GHD7* increases drought sensitivity, while knock-down of *GHD7* raises drought tolerance. Moreover, *GHD7* also regulates the plasticity of tillering by mediating the *PHYTOCHROME B-TEOSINTE BRANCHED1* pathway [161]. In summary, water availability or drought stress regulates tillering/branching in plants through effects on gene expression, microRNA levels, and modulation of hormonal signaling pathways like auxin signaling.

2.2.4. Effects of Temperature on Tillering

Temperature is a critical factor influencing tillering in crops [184]. High temperatures caused by extreme weather events can reduce tiller numbers, as evidenced in *B. distachyon*. Tiller numbers declined linearly in B. *distachyon* from 24 to 36 °C at a rate of approximately one tiller for every 1.7 °C increase in temperature [185]. Genes involved in heat stress play an important role in tillering. For instance, in rice, *miR159* is downregulated by heat stress, and its overexpression increases heat sensitivity and significantly reduces tillering [186]. At the other extreme, chilling also detrimentally impacts tillering. Chilling tolerance is a complex agronomic trait governed by intricate genetic networks and signal transduction cascades. Mechanistic insights into cold-stress effects on tillering are emerging. For example, overexpression of *OsMADS57* maintains rice tiller growth under chilling stress. OsMADS57 directly binds and activates the defense gene *OsWRKY94* for cold-stress responses while suppressing its activity under normal temperatures [187]. Additionally, OsWRKY94 is directly targeted and repressed by the tillering inhibitor OsTB1 during chilling. *D14* transcription was directly promoted by OsMADS57 for suppressing tillering under chilling treatment, whereas *D14* was repressed for enhancing tillering under normal conditions [187] (Figure 3). Likewise, overexpression of *OsCYP19-4* results in cold-resistance phenotypes with significantly increased tiller numbers [188]. Ran is a small GTPase that involves various developments like nuclear assembly and cell-cycle control [189]. The levels of mRNA encoding OsRAN1 were greatly increased by chilling, and *OsRAN1* overexpression in *Arabidopsis* increased tiller numbers [162]. Elevated expression of *miR393* also improves chilling tolerance and tillering [190]. In summary, modulation of chilling-tolerance genes may benefit crop-breeding efforts to sustain tiller development under temperature extremes caused by climate change.

2.2.5. Biotic Stresses Impact Tiller Bud Outgrowth

Biotic stress affects plant development, including tillering, fundamentally disrupting and depriving plants of the nutrients they rely on for survival. More specifically, biotic stresses caused by plant pathogens, insect pests, and parasitic organisms can impair growth and developmental processes such as tillering and branching (Figure 3). These biotic agents injure plant tissues both directly through feeding/infection and indirectly by inhibiting the uptake and utilization of water, nutrients, and photoassimilates required for plant growth. Pathogens, insects, and parasites disrupt key physiological processes like metabolism, resource allocation, and energy balance, ultimately reducing the plant's capacity for producing new tillers or branches. Here, we take some examples to elucidate these processes. For instance, *Striga* is an obligate parasitic plant that can attach to host roots to deplete them of nutrients [191]. The rice cultivar Azucena, belonging to the japonica subspecies, exudes high SL levels and induces high germination of the root-parasitic plant *Striga hermonthica*. In contrast, Bala, an indica cultivar, is a low-SL producer, stimulates less *Striga* germination, and is highly tillered [192]. Plants relocate resources while fighting against pathogens and exhibit reduced tillering/branching. For example, the *UNI* gene encodes a coiled-coil nucleotide-binding leucine-rich repeat protein that belongs to the disease-resistance (R) protein family involved in pathogen recognition. The *uni-1D* mutation induces the upregulation of the pathogenesis-related gene while evoking some morphological defects like increased branches [22]. Further research into mitigation strategies against prevalent biotic agents would benefit efforts to secure plant growth and crop yields.

3. Concluding Remarks

Shoot branching is a highly intricate regulatory developmental program that involves the complex interplay of multiple genes, plant hormones, and environmental cues. A thorough understanding of the underlying mechanisms governing shoot branching/tillering is crucial for crop breeding and improving productivity (Video S1 and Figure 1). In this review, we provide a comprehensive overview of current advances in understanding the intricate regulatory mechanisms of shoot branching. We have highlighted key genes such as *TB1*, several phytohormones such as auxin, and environmental and internal inputs such as N, Pi, K^+, light, water availability, and biotic stresses, underscoring the interplay of these components.

However, the complex regulation of shoot branching/tillering requires further investigation. The interplay of the various underlying internal and external cues is still not entirely understood. For example, further research is needed to fully understand how sugars collaborate with other factors to regulate shoot branching/tillering. Notably, plants with strong resistance to some pathogens often display fewer branches/tillers, but the exact mechanisms at play require further study. Importantly, AMs, which give rise to shoot branches/tillers, can also influence the patterns of panicles, hence affecting crop yields, an important factor in crop production and management [3–5]. With new achievements in understanding shoot branching, a comprehensive landscape of factors controlling shoot branching will be established. A better understanding of shoot branching will ultimately facilitate the control of the process.

Supplementary Materials: The following supporting information can be downloaded at: https://www.mdpi.com/article/10.3390/plants12203628/s1, Video S1: Illustration of the dynamic process of tiller formation and comparisons between plants with normal and defective tiller bud outgrowth.

Author Contributions: Y.Y. conceived the study; Y.Y. wrote and edited the manuscript; Y.Y. and Y.D. created and edited the figures. S.L. and S.K. read through this manuscript and gave critical advice; P.D. conceived, read through, and gave critical decisions. All authors have read and agreed to the published version of the manuscript.

Funding: This work was funded by the National Key Research and Development Program of China (2021YFD1200601-08).

Data Availability Statement: No new data were created or analyzed in this study. Data sharing is not applicable to this article.

Acknowledgments: We thank Yonghong Wang for the critical review of this manuscript.

Conflicts of Interest: The authors declare no conflict of interest.

References

1. Bowman, J.L.; Eshed, Y. Formation and maintenance of the shoot apical meristem. *Trends Plant Sci.* **2000**, *5*, 110–115. [CrossRef] [PubMed]
2. Wang, B.; Smith, S.M.; Li, J. Genetic Regulation of Shoot Architecture. *Annu. Rev. Plant Biol.* **2018**, *69*, 437–468. [CrossRef]
3. Springer, N. Shaping a better rice plant. *Nat. Genet.* **2010**, *42*, 475–476. [CrossRef]
4. Wang, Y.; Li, J. Molecular Basis of Plant Architecture. *Annu. Rev. Plant Biol.* **2008**, *59*, 253–279. [CrossRef] [PubMed]
5. Jiao, Y.; Wang, Y.; Xue, D.; Wang, J.; Yan, M.; Liu, G.; Dong, G.; Zeng, D.; Lu, Z.; Zhu, X.; et al. Regulation of *OsSPL14* by OsmiR156 defines ideal plant architecture in rice. *Nat. Genet.* **2010**, *42*, 541–544. [CrossRef]
6. Aguilar-Martínez, J.A.; Poza-Carrión, C.; Cubas, P. Arabidopsis BRANCHED1 acts as an integrator of branching signals within axillary buds. *Plant Cell* **2007**, *19*, 458–472. [CrossRef]
7. Wang, M.; Le Moigne, M.-A.; Bertheloot, J.; Crespel, L.; Perez-Garcia, M.-D.; Ogé, L.; Demotes-Mainard, S.; Hamama, L.; Davière, J.-M.; Sakr, S. BRANCHED1: A key hub of shoot branching. *Front. Plant Sci.* **2019**, *10*, 76. [CrossRef]
8. Helliwell, C.A.; Chin-Atkins, A.N.; Wilson, I.W.; Chapple, R.; Dennis, E.S.; Chaudhury, A. The *Arabidopsis AMP1* Gene Encodes a Putative Glutamate Carboxypeptidase. *Plant Cell* **2001**, *13*, 2115–2125. [CrossRef]
9. Dharmasiri, N.; Dharmasiri, S.; Estelle, M. The F-box protein TIR1 is an auxin receptor. *Nature* **2005**, *435*, 441–445. [CrossRef]
10. Lo, S.-F.; Yang, S.-Y.; Chen, K.-T.; Hsing, Y.-I.; Zeevaart, J.A.D.; Chen, L.-J.; Yu, S.-M. A Novel Class of Gibberellin 2-Oxidases Control Semidwarfism, Tillering, and Root Development in Rice. *Plant Cell* **2008**, *20*, 2603–2618. [CrossRef] [PubMed]
11. Bayer, E.M.; Smith, R.S.; Mandel, T.; Nakayama, N.; Sauer, M.; Prusinkiewicz, P.; Kuhlemeier, C. Integration of transport-based models for phyllotaxis and midvein formation. *Genes Dev.* **2009**, *23*, 373–384. [CrossRef]
12. Lin, H.; Wang, R.; Qian, Q.; Yan, M.; Meng, X.; Fu, Z.; Yan, C.; Jiang, B.; Su, Z.; Li, J.; et al. DWARF27, an iron-containing protein required for the biosynthesis of strigolactones, regulates rice tiller bud outgrowth. *Plant Cell* **2009**, *21*, 1512–1525. [CrossRef] [PubMed]
13. Finlayson, S.A.; Krishnareddy, S.R.; Kebrom, T.H.; Casal, J.J. Phytochrome Regulation of Branching in *Arabidopsis*. *Plant Physiol.* **2010**, *152*, 1914–1927. [CrossRef] [PubMed]
14. González-Grandío, E.; Pajoro, A.; Franco-Zorrilla, J.M.; Tarancón, C.; Immink, R.G.H.; Cubas, P. Abscisic acid signaling is controlled by a BRANCHED1/HD-ZIP/cascade in *Arabidopsis* axillary buds. *Proc. Natl. Acad. Sci. USA* **2017**, *114*, E245–E254. [CrossRef]
15. Wang, Y.; Sun, S.; Zhu, W.; Jia, K.; Yang, H.; Wang, X. Strigolactone/MAX2-Induced Degradation of Brassinosteroid Transcriptional Effector BES1 Regulates Shoot Branching. *Dev. Cell* **2013**, *27*, 681–688. [CrossRef]
16. Mason, M.G.; Ross, J.J.; Babst, B.A.; Wienclaw, B.N.; Beveridge, C.A. Sugar demand, not auxin, is the initial regulator of apical dominance. *Proc. Natl. Acad. Sci. USA* **2014**, *111*, 6092–6097. [CrossRef] [PubMed]
17. Tegeder, M. Transporters involved in source to sink partitioning of amino acids and ureides: Opportunities for crop improvement. *J. Exp. Bot.* **2014**, *65*, 1865–1878. [CrossRef] [PubMed]
18. Tian, C.; Zhang, X.; He, J.; Yu, H.; Wang, Y.; Shi, B.; Han, Y.; Wang, G.; Feng, X.; Zhang, C.; et al. An organ boundary-enriched gene regulatory network uncovers regulatory hierarchies underlying axillary meristem initiation. *Mol. Syst. Biol.* **2014**, *10*, 755. [CrossRef]
19. Wang, Q.; Kohlen, W.; Rossmann, S.; Vernoux, T.; Theres, K. Auxin Depletion from the Leaf Axil Conditions Competence for Axillary Meristem Formation in *Arabidopsis* and Tomato. *Plant Cell* **2014**, *26*, 2068–2079. [CrossRef]
20. Wang, Y.; Wang, J.; Shi, B.; Yu, T.; Qi, J.; Meyerowitz, E.M.; Jiao, Y. The Stem Cell Niche in Leaf Axils Is Established by Auxin and Cytokinin in *Arabidopsis*. *Plant Cell* **2014**, *26*, 2055–2067. [CrossRef]
21. Zhang, Q.Q.; Wang, J.G.; Wang, L.Y.; Wang, J.F.; Wang, Q.; Yu, P.; Bai, M.Y.; Fan, M. Gibberellin repression of axillary bud formation in *Arabidopsis* by modulation of DELLA-SPL9 complex activity. *J. Integr. Plant Biol.* **2020**, *62*, 421–432. [CrossRef] [PubMed]
22. Igari, K.; Endo, S.; Hibara, K.; Aida, M.; Sakakibara, H.; Kawasaki, T.; Tasaka, M. Constitutive activation of a CC-NB-LRR protein alters morphogenesis through the cytokinin pathway in Arabidopsis. *Plant J. Cell Mol. Biol.* **2008**, *55*, 14–27. [CrossRef] [PubMed]

23. Pan, Y.H.; Chen, L.; Guo, H.F.; Feng, R.; Lou, Q.J.; Rashid, M.A.R.; Zhu, X.Y.; Qing, D.J.; Liang, H.F.; Gao, L.J.; et al. Systematic Analysis of NB-ARC Gene Family in Rice and Functional Characterization of GNP12. *Front Genet* **2022**, *13*, 887217. [CrossRef]
24. Stafstrom, J.P.; Sussex, I.M. Expression of a Ribosomal Protein Gene in Axillary Buds of Pea Seedlings. *Plant Physiol.* **1992**, *100*, 1494–1502. [CrossRef]
25. Devitt, M.L.; Stafstrom, J.P. Cell cycle regulation during growth-dormancy cycles in pea axillary buds. *Plant Mol. Biol.* **1995**, *29*, 255–265. [CrossRef] [PubMed]
26. Dun, E.A.; Ferguson, B.J.; Beveridge, C.A. Apical dominance and shoot branching. Divergent opinions or divergent mechanisms? *Plant Physiol.* **2006**, *142*, 812–819. [CrossRef] [PubMed]
27. Waldie, T.; Hayward, A.; Beveridge, C.A. Axillary bud outgrowth in herbaceous shoots: How do strigolactones fit into the picture? *Plant Mol. Biol.* **2010**, *73*, 27–36. [CrossRef]
28. Wang, Y.; Li, J. Branching in rice. *Curr. Opin. Plant Biol.* **2011**, *14*, 94–99. [CrossRef]
29. Hubbard, L.; McSteen, P.; Doebley, J.; Hake, S. Expression patterns and mutant phenotype of *teosinte branched1* correlate with growth suppression in maize and teosinte. *Genetics* **2002**, *162*, 1927–1935. [CrossRef]
30. Martín-Trillo, M.; Grandío, E.G.; Serra, F.; Marcel, F.; Rodríguez-Buey, M.L.; Schmitz, G.; Theres, K.; Bendahmane, A.; Dopazo, H.; Cubas, P. Role of tomato *BRANCHED1*-like genes in the control of shoot branching. *Plant J.* **2011**, *67*, 701–714. [CrossRef]
31. Braun, N.; de Saint Germain, A.; Pillot, J.-P.; Boutet-Mercey, S.; Dalmais, M.; Antoniadi, I.; Li, X.; Maia-Grondard, A.; Le Signor, C.; Bouteiller, N. The pea TCP transcription factor PsBRC1 acts downstream of strigolactones to control shoot branching. *Plant Physiol.* **2012**, *158*, 225–238. [CrossRef]
32. Rameau, C.; Bertheloot, J.; Leduc, N.; Andrieu, B.; Foucher, F.; Sakr, S. Multiple pathways regulate shoot branching. *Front. Plant Sci.* **2015**, *5*, 741–755. [CrossRef]
33. Schwarz, S.; Grande, A.V.; Bujdoso, N.; Saedler, H.; Huijser, P. The microRNA regulated SBP-box genes *SPL9* and *SPL15* control shoot maturation in *Arabidopsis*. *Plant Mol. Biol.* **2008**, *67*, 183–195. [CrossRef]
34. Lincoln, C.; Britton, J.H.; Estelle, M. Growth and development of the *axr1* mutants of *Arabidopsis*. *Plant Cell* **1990**, *2*, 1071–1080. [CrossRef] [PubMed]
35. Zhao, Y.; Christensen, S.K.; Fankhauser, C.; Cashman, J.R.; Cohen, J.D.; Weigel, D.; Chory, J. A role for flavin monooxygenase-like enzymes in auxin biosynthesis. *Science* **2001**, *291*, 306–309. [CrossRef] [PubMed]
36. Xu, M.; Zhu, L.; Shou, H.; Wu, P. A *PIN1* family gene, *OsPIN1*, involved in auxin-dependent adventitious root emergence and tillering in rice. *Plant Cell Physiol.* **2005**, *46*, 1674–1681. [CrossRef] [PubMed]
37. Lu, G.; Coneva, V.; Casaretto, J.A.; Ying, S.; Mahmood, K.; Liu, F.; Nambara, E.; Bi, Y.M.; Rothstein, S.J. *OsPIN5b* modulates rice (*Oryza sativa*) plant architecture and yield by changing auxin homeostasis, transport and distribution. *Plant J.* **2015**, *83*, 913–925. [CrossRef]
38. Dai, Y.; Wang, H.; Li, B.; Huang, J.; Liu, X.; Zhou, Y.; Mou, Z.; Li, J. Increased expression of MAP KINASE KINASE7 causes deficiency in polar auxin transport and leads to plant architectural abnormality in Arabidopsis. *Plant Cell* **2006**, *18*, 308–320. [CrossRef]
39. Jia, W.; Li, B.; Li, S.; Liang, Y.; Wu, X.; Ma, M.; Wang, J.; Gao, J.; Cai, Y.; Zhang, Y.; et al. Mitogen-Activated Protein Kinase Cascade MKK7-MPK6 Plays Important Roles in Plant Development and Regulates Shoot Branching by Phosphorylating PIN1 in Arabidopsis. *PLoS Biol.* **2016**, *14*, e1002550. [CrossRef]
40. Kepinski, S.; Leyser, O. The *Arabidopsis* F-box protein TIR1 is an auxin receptor. *Nature* **2005**, *435*, 446–451. [CrossRef]
41. Dharmasiri, N.; Dharmasiri, S.; Weijers, D.; Lechner, E.; Yamada, M.; Hobbie, L.; Ehrismann, J.S.; Jürgens, G.; Estelle, M. Plant Development Is Regulated by a Family of Auxin Receptor F Box Proteins. *Dev. Cell* **2005**, *9*, 109–119. [CrossRef]
42. Xia, K.; Wang, R.; Ou, X.; Fang, Z.; Tian, C.; Duan, J.; Wang, Y.; Zhang, M. OsTIR1 and OsAFB2 downregulation via OsmiR393 overexpression leads to more tillers, early flowering and less tolerance to salt and drought in rice. *PLoS ONE* **2012**, *7*, e30039. [CrossRef]
43. Alder, A.; Jamil, M.; Marzorati, M.; Bruno, M.; Vermathen, M.; Bigler, P.; Ghisla, S.; Bouwmeester, H.; Beyer, P.; Al-Babili, S. The Path from β-Carotene to Carlactone, a Strigolactone-Like Plant Hormone. *Science* **2012**, *335*, 1348–1351. [CrossRef] [PubMed]
44. Booker, J.; Auldridge, M.; Wills, S.; McCarty, D.; Klee, H.; Leyser, O. MAX3/CCD7 Is a Carotenoid Cleavage Dioxygenase Required for the Synthesis of a Novel Plant Signaling Molecule. *Curr. Biol.* **2004**, *14*, 1232–1238. [CrossRef]
45. Sorefan, K. *MAX4* and *RMS1* are orthologous dioxygenase-like genes that regulate shoot branching in *Arabidopsis* and pea. *Genes Dev.* **2003**, *17*, 1469–1474. [CrossRef]
46. Booker, J.; Sieberer, T.; Wright, W.; Williamson, L.; Willett, B.; Stirnberg, P.; Turnbull, C.; Srinivasan, M.; Goddard, P.; Leyser, O. *MAX1* Encodes a Cytochrome P450 Family Member that Acts Downstream of *MAX3/4* to Produce a Carotenoid-Derived Branch-Inhibiting Hormone. *Dev. Cell* **2005**, *8*, 443–449. [CrossRef]
47. Smith, S.M.; Li, J. Signalling and responses to strigolactones and karrikins. *Curr. Opin. Plant Biol.* **2014**, *21*, 23–29. [CrossRef] [PubMed]
48. Wang, L.; Wang, B.; Jiang, L.; Liu, X.; Li, X.; Lu, Z.; Meng, X.; Wang, Y.; Smith, S.M.; Li, J. Strigolactone Signaling in Arabidopsis Regulates Shoot Development by Targeting D53-Like SMXL Repressor Proteins for Ubiquitination and Degradation. *Plant Cell* **2015**, *27*, 3128–3142. [CrossRef] [PubMed]

49. Soundappan, I.; Bennett, T.; Morffy, N.; Liang, Y.; Stanga, J.P.; Abbas, A.; Leyser, O.; Nelson, D.C. SMAX1-LIKE/D53 Family Members Enable Distinct MAX2-Dependent Responses to Strigolactones and Karrikins in Arabidopsis. *Plant Cell* **2015**, *27*, 3143–3159. [CrossRef] [PubMed]
50. Medford, J.I.; Horgan, R.; El-Sawi, Z.; Klee, H.J. Alterations of Endogenous Cytokinins in Transgenic Plants Using a Chimeric Isopentenyl Transferase Gene. *Plant Cell* **1989**, *1*, 403–413. [CrossRef]
51. Tantikanjana, T.; Yong, J.W.H.; Letham, D.S.; Griffith, M.; Hussain, M.; Ljung, K.; Sandberg, G.; Sundaresan, V. Control of axillary bud initiation and shoot architecture in *Arabidopsis* through the *SUPERSHOOT* gene. *Genes Dev.* **2001**, *15*, 1577–1588. [CrossRef] [PubMed]
52. Shimizu-Sato, S.; Tanaka, M.; Mori, H. Auxin–cytokinin interactions in the control of shoot branching. *Plant Mol. Biol.* **2009**, *69*, 429–435. [CrossRef] [PubMed]
53. Reddy, S.K.; Holalu, S.V.; Casal, J.J.; Finlayson, S.A. Abscisic acid regulates axillary bud outgrowth responses to the ratio of red to far-red light. *Plant Physiol.* **2013**, *163*, 1047–1058. [CrossRef]
54. Yao, C.; Finlayson, S.A. Abscisic Acid Is a General Negative Regulator of *Arabidopsis* Axillary Bud Growth. *Plant Physiol.* **2015**, *169*, 611–626. [CrossRef] [PubMed]
55. Yin, Y.; Vafeados, D.; Tao, Y.; Yoshida, S.; Asami, T.; Chory, J. A New Class of Transcription Factors Mediates Brassinosteroid-Regulated Gene Expression in *Arabidopsis*. *Cell* **2005**, *120*, 249–259. [CrossRef] [PubMed]
56. Koumoto, T.; Shimada, H.; Kusano, H.; She, K.-C.; Iwamoto, M.; Takano, M. Rice monoculm mutation *moc2*, which inhibits outgrowth of the second tillers, is ascribed to lack of a fructose-1, 6-bisphosphatase. *Plant Biotechnol.* **2013**, *30*, 47–56. [CrossRef]
57. Huang, W.; Bai, G.; Wang, J.; Zhu, W.; Zeng, Q.; Lu, K.; Sun, S.; Fang, Z. Two splicing variants of *OsNPF7.7* regulate shoot branching and nitrogen utilization efficiency in rice. *Front. Plant Sci.* **2018**, *9*, 300. [CrossRef]
58. He, X.; Qu, B.; Li, W.; Zhao, X.; Teng, W.; Ma, W.; Ren, Y.; Li, B.; Li, Z.; Tong, Y. The nitrate-inducible NAC transcription factor TaNAC2-5A controls nitrate response and increases wheat yield. *Plant Physiol.* **2015**, *169*, 1991–2005. [CrossRef]
59. Guo, S.; Xu, Y.; Liu, H.; Mao, Z.; Zhang, C.; Ma, Y.; Zhang, Q.; Meng, Z.; Chong, K. The interaction between OsMADS57 and OsTB1 modulates rice tillering via DWARF14. *Nat. Commun.* **2013**, *4*, 1566. [CrossRef]
60. Bai, M.Y.; Zhang, L.Y.; Gampala, S.S.; Zhu, S.W.; Song, W.Y.; Chong, K.; Wang, Z.Y. Functions of OsBZR1 and 14-3-3 proteins in brassinosteroid signaling in rice. *Proc. Natl. Acad. Sci. USA* **2007**, *104*, 13839–13844. [CrossRef]
61. Tong, H.; Jin, Y.; Liu, W.; Li, F.; Fang, J.; Yin, Y.; Qian, Q.; Zhu, L.; Chu, C. DWARF AND LOW-TILLERING, a new member of the GRAS family, plays positive roles in brassinosteroid signaling in rice. *Plant J.* **2009**, *58*, 803–816. [CrossRef]
62. Qiao, S.; Sun, S.; Wang, L.; Wu, Z.; Li, C.; Li, X.; Wang, T.; Leng, L.; Tian, W.; Lu, T.; et al. The RLA1/SMOS1 Transcription Factor Functions with OsBZR1 to Regulate Brassinosteroid Signaling and Rice Architecture. *Plant Cell* **2017**, *29*, 292–309. [CrossRef] [PubMed]
63. Doebley, J.; Stec, A.; Hubbard, L. The evolution of apical dominance in maize. *Nature* **1997**, *386*, 485–488. [CrossRef] [PubMed]
64. Cubas, P.; Lauter, N.; Doebley, J.; Coen, E. The TCP domain: A motif found in proteins regulating plant growth and development. *Plant J.* **1999**, *18*, 215–222. [CrossRef]
65. Doebley, J.; Stec, A.; Gustus, C. teosinte branched1 and the origin of maize: Evidence for epistasis and the evolution of dominance. *Genetics* **1995**, *141*, 333–346. [CrossRef]
66. Doebley, J. The genetics of maize evolution. *Annu. Rev. Genet.* **2004**, *38*, 37–59. [CrossRef] [PubMed]
67. Luo, D.; Carpenter, R.; Vincent, C.; Copsey, L.; Coen, E. Origin of floral asymmetry in Antirrhinum. *Nature* **1996**, *383*, 794–799. [CrossRef]
68. Kosugi, S.; Ohashi, Y. PCF1 and PCF2 specifically bind to cis elements in the rice proliferating cell nuclear antigen gene. *Plant Cell* **1997**, *9*, 1607–1619. [CrossRef]
69. Takeda, T.; Suwa, Y.; Suzuki, M.; Kitano, H.; Ueguchi-Tanaka, M.; Ashikari, M.; Matsuoka, M.; Ueguchi, C. The *OsTB1* gene negatively regulates lateral branching in rice. *Plant J.* **2003**, *33*, 513–520. [CrossRef] [PubMed]
70. Minakuchi, K.; Kameoka, H.; Yasuno, N.; Umehara, M.; Luo, L.; Kobayashi, K.; Hanada, A.; Ueno, K.; Asami, T.; Yamaguchi, S.; et al. *FINE CULM1* (*FC1*) Works Downstream of Strigolactones to Inhibit the Outgrowth of Axillary Buds in Rice. *Plant Cell Physiol.* **2010**, *51*, 1127–1135. [CrossRef]
71. González-Grandío, E.; Poza-Carrión, C.; Sorzano, C.O.S.; Cubas, P. *BRANCHED1* promotes axillary bud dormancy in response to shade in *Arabidopsis*. *Plant Cell* **2013**, *25*, 834–850. [CrossRef]
72. Seale, M.; Bennett, T.; Leyser, O. *BRC1* expression regulates bud activation potential, but is not necessary or sufficient for bud growth inhibition in *Arabidopsis*. *Development* **2017**, *144*, 1661–1673. [CrossRef] [PubMed]
73. Rhoades, M.W.; Reinhart, B.J.; Lim, L.P.; Burge, C.B.; Bartel, B.; Bartel, D.P. Prediction of Plant MicroRNA Targets. *Cell* **2002**, *110*, 513–520. [CrossRef] [PubMed]
74. Schwab, R.; Palatnik, J.F.; Riester, M.; Schommer, C.; Schmid, M.; Weigel, D. Specific Effects of MicroRNAs on the Plant Transcriptome. *Dev. Cell* **2005**, *8*, 517–527. [CrossRef] [PubMed]
75. Wei, S.; Gruber, M.Y.; Yu, B.; Gao, M.-J.; Khachatourians, G.G.; Hegedus, D.D.; Parkin, I.A.; Hannoufa, A. *Arabidopsis* mutant *sk156* reveals complex regulation of *SPL15* in a *miR156*-controlled gene network. *BMC Plant Biol.* **2012**, *12*, 169. [CrossRef]
76. Matthes, M.S.; Best, N.B.; Robil, J.M.; Malcomber, S.; Gallavotti, A.; McSteen, P. Auxin EvoDevo: Conservation and Diversification of Genes Regulating Auxin Biosynthesis, Transport, and Signaling. *Mol. Plant* **2019**, *12*, 298–320. [CrossRef]

77. Tan, X.; Calderon-Villalobos, L.I.A.; Sharon, M.; Zheng, C.; Robinson, C.V.; Estelle, M.; Zheng, N. Mechanism of auxin perception by the TIR1 ubiquitin ligase. *Nature* **2007**, *446*, 640–645. [CrossRef]
78. Zenser, N.; Ellsmore, A.; Leasure, C.; Callis, J. Auxin modulates the degradation rate of Aux/IAA proteins. *Proc. Natl. Acad. Sci. USA* **2001**, *98*, 11795–11800. [CrossRef]
79. Tiwari, S.B.; Wang, X.-J.; Hagen, G.; Guilfoyle, T.J. AUX/IAA proteins are active repressors, and their stability and activity are modulated by auxin. *Plant Cell* **2001**, *13*, 2809–2822. [CrossRef]
80. Gray, W.M.; Kepinski, S.; Rouse, D.; Leyser, O.; Estelle, M. Auxin regulates SCFT^{IR1}-dependent degradation of AUX/IAA proteins. *Nature* **2001**, *414*, 271–276. [CrossRef]
81. Park, J.-Y.; Kim, H.-J.; Kim, J. Mutation in domain II of IAA1 confers diverse auxin-related phenotypes and represses auxin-activated expression of *Aux/IAA* genes in steroid regulator-inducible system. *Plant J.* **2002**, *32*, 669–683. [CrossRef]
82. Ulmasov, T.; Hagen, G.; Guilfoyle, T.J. ARF1, a transcription factor that binds to auxin response elements. *Science* **1997**, *276*, 1865–1868. [CrossRef]
83. Wu, K.; Chen, A.; Pan, Z.-Q. Conjugation of Nedd8 to CUL1 Enhances the Ability of the ROC1-CUL1 Complex to Promote Ubiquitin Polymerization. *J. Biol. Chem.* **2000**, *275*, 32317–32324. [CrossRef] [PubMed]
84. Del Pozo, J.C.; Dharmasiri, S.; Hellmann, H.; Walker, L.; Gray, W.M.; Estelle, M. AXR1-ECR1–Dependent Conjugation of RUB1 to the *Arabidopsis* Cullin AtCUL1 Is Required for Auxin Response. *Plant Cell* **2002**, *14*, 421–433. [CrossRef] [PubMed]
85. Prasad, T.; Li, X.; Abdel-Rahman, A.; Hosokawa, Z.; Cloud, N.; Lamotte, C.; Cline, M. Does auxin play a role in the release of apical dominance by shoot inversion in Ipomoea nil? *Ann. Bot.* **1993**, *71*, 223–229. [CrossRef]
86. Li, C.J.; Bangerth, F. Autoinhibition of indoleacetic acid transport in the shoots of two-branched pea (*Pisum sativum*) plants and its relationship to correlative dominance. *Physiol. Plant.* **1999**, *106*, 415–420. [CrossRef]
87. Domagalska, M.A.; Leyser, O. Signal integration in the control of shoot branching. *Nat. Rev. Mol. Cell Biol.* **2011**, *12*, 211–221. [CrossRef]
88. Barbier, F.; Péron, T.; Lecerf, M.; Perez-Garcia, M.-D.; Barrière, Q.; Rolčík, J.; Boutet-Mercey, S.; Citerne, S.; Lemoine, R.; Porcheron, B. Sucrose is an early modulator of the key hormonal mechanisms controlling bud outgrowth in *Rosa hybrida*. *J. Exp. Bot.* **2015**, *66*, 2569–2582. [CrossRef]
89. Bennett, T.; Hines, G.; Leyser, O. Canalization: What the flux? *Trends Genet.* **2014**, *30*, 41–48. [CrossRef]
90. Shinohara, N.; Taylor, C.; Leyser, O. Strigolactone Can Promote or Inhibit Shoot Branching by Triggering Rapid Depletion of the Auxin Efflux Protein PIN1 from the Plasma Membrane. *PLoS Biol.* **2013**, *11*, e1001474. [CrossRef]
91. Brewer, P.B.; Dun, E.A.; Ferguson, B.J.; Rameau, C.; Beveridge, C.A. Strigolactone acts downstream of auxin to regulate bud outgrowth in pea and *Arabidopsis*. *Plant Physiol.* **2009**, *150*, 482–493. [CrossRef]
92. Hayward, A.; Stirnberg, P.; Beveridge, C.; Leyser, O. Interactions between Auxin and Strigolactone in Shoot Branching Control. *Plant Physiol.* **2009**, *151*, 400–412. [CrossRef] [PubMed]
93. Dun, E.A.; de Saint Germain, A.; Rameau, C.; Beveridge, C.A. Antagonistic action of strigolactone and cytokinin in bud outgrowth control. *Plant Physiol.* **2012**, *158*, 487–498. [CrossRef]
94. Shimizu, S.; Mori, H. Analysis of Cycles of Dormancy and Growth in Pea Axillary Buds Based on mRNA Accumulation Patterns of Cell Cycle-Related Genes. *Plant Cell Physiol.* **1998**, *39*, 255–262. [CrossRef]
95. Kebrom, T.H.; Brutnell, T.P.; Finlayson, S.A. Suppression of sorghum axillary bud outgrowth by shade, phyB and defoliation signalling pathways. *Plant Cell Environ.* **2009**, *33*, 48–58. [CrossRef]
96. Cook, C.; Whichard, L.P.; Turner, B.; Wall, M.E.; Egley, G.H. Germination of witchweed (*Striga lutea* Lour.): Isolation and properties of a potent stimulant. *Science* **1966**, *154*, 1189–1190. [CrossRef]
97. Akiyama, K.; Matsuzaki, K.-I.; Hayashi, H. Plant sesquiterpenes induce hyphal branching in arbuscular mycorrhizal fungi. *Nature* **2005**, *435*, 824–827. [CrossRef] [PubMed]
98. Napoli, C.; Ruehle, J. New mutations affecting meristem growth and potential in Petunia hybrida Vilm. *J. Hered.* **1996**, *87*, 371–377. [CrossRef]
99. Beveridge, C.A.; Symons, G.M.; Murfet, I.C.; Ross, J.J.; Rameau, C. The *rms1* mutant of pea has elevated indole-3-acetic acid levels and reduced root-sap zeatin riboside content but increased branching controlled by graft-transmissible signal (s). *Plant Physiol.* **1997**, *115*, 1251. [CrossRef]
100. Beveridge, C.A. Long-distance signalling and a mutational analysis of branching in pea. *Plant Growth Regul.* **2000**, *32*, 193–203. [CrossRef]
101. Beveridge, C.A. Axillary bud outgrowth: Sending a message. *Curr. Opin. Plant Biol.* **2006**, *9*, 35–40. [CrossRef]
102. Turnbull, C.G.N.; Booker, J.P.; Leyser, H.M.O. Micrografting techniques for testing long-distance signalling in *Arabidopsis*. *Plant J.* **2002**, *32*, 255–262. [CrossRef] [PubMed]
103. Foo, E.; Turnbull, C.G.N.; Beveridge, C.A. Long-Distance Signaling and the Control of Branching in therms1 Mutant of Pea. *Plant Physiol.* **2001**, *126*, 203–209. [CrossRef]
104. Morris, S.E.; Turnbull, C.G.; Murfet, I.C.; Beveridge, C.A. Mutational analysis of branching in pea. Evidence that *Rms1* and *Rms5* regulate the same novel signal. *Plant Physiol.* **2001**, *126*, 1205–1213. [CrossRef] [PubMed]
105. Abe, S.; Sado, A.; Tanaka, K.; Kisugi, T.; Asami, K.; Ota, S.; Kim, H.I.; Yoneyama, K.; Xie, X.; Ohnishi, T.; et al. Carlactone is converted to carlactonoic acid by MAX1 in *Arabidopsis* and its methyl ester can directly interact with AtD14 in vitro. *Proc. Natl. Acad. Sci.* **2014**, *111*, 18084–18089. [CrossRef]

106. Jiang, L.; Liu, X.; Xiong, G.; Liu, H.; Chen, F.; Wang, L.; Meng, X.; Liu, G.; Yu, H.; Yuan, Y.; et al. DWARF 53 acts as a repressor of strigolactone signalling in rice. *Nature* **2013**, *504*, 401–405. [CrossRef] [PubMed]
107. Xiong, G.; Wang, Y.; Li, J. Action of strigolactones in plants. *Enzymes* **2014**, *35*, 57–84. [CrossRef] [PubMed]
108. Morris, D.A. Transport of exogenous auxin in two-branched dwarf pea seedlings (*Pisum sativum* L.). *Planta* **1977**, *136*, 91–96. [CrossRef]
109. CLINE, M.G. Exogenous auxin effects on lateral bud outgrowth in decapitated shoots. *Ann. Bot.* **1996**, *78*, 255–266. [CrossRef]
110. El-Showk, S.; Ruonala, R.; Helariutta, Y. Crossing paths: Cytokinin signalling and crosstalk. *Development* **2013**, *140*, 1373–1383. [CrossRef]
111. Tanaka, M.; Takei, K.; Kojima, M.; Sakakibara, H.; Mori, H. Auxin controls local cytokinin biosynthesis in the nodal stem in apical dominance. *Plant J.* **2006**, *45*, 1028–1036. [CrossRef]
112. Turnbull, C.G.; Raymond, M.A.; Dodd, I.C.; Morris, S.E. Rapid increases in cytokinin concentration in lateral buds of chickpea (*Cicer arietinum* L.) during release of apical dominance. *Planta* **1997**, *202*, 271–276. [CrossRef]
113. Li, C.; Bangerth, F. The possible role of cytokinins, ethylene and indoleacetic acid in apical dominance. In *Progress in Plant Growth Regulation. Current Plant Science and Biotechnology in Agriculture*; Springer: Dordrecht, The Netherlands, 1992; pp. 431–436. [CrossRef]
114. Chatfield, S.P.; Stirnberg, P.; Forde, B.G.; Leyser, O. The hormonal regulation of axillary bud growth in *Arabidopsis*. *Plant J.* **2000**, *24*, 159–169. [CrossRef] [PubMed]
115. Ferguson, B.J.; Beveridge, C.A. Roles for auxin, cytokinin, and strigolactone in regulating shoot branching. *Plant Physiol.* **2009**, *149*, 1929–1944. [CrossRef] [PubMed]
116. Jones, R.J.; Schreiber, B. Role and function of cytokinin oxidase in plants. *Plant Growth Regul.* **1997**, *23*, 123–134. [CrossRef]
117. Werner, T.; Motyka, V.; Strnad, M.; Schmülling, T. Regulation of plant growth by cytokinin. *Proc. Natl. Acad. Sci. USA* **2001**, *98*, 10487–10492. [CrossRef]
118. Silverstone, A.L.; Mak, P.Y.A.; Martinez, E.C.; Sun, T.-P. The New *RGA* Locus Encodes a Negative Regulator of Gibberellin Response in *Arabidopsis thaliana*. *Genetics* **1997**, *146*, 1087–1099. [CrossRef]
119. Yin, Y.; Wang, Z.-Y.; Mora-Garcia, S.; Li, J.; Yoshida, S.; Asami, T.; Chory, J. BES1 Accumulates in the Nucleus in Response to Brassinosteroids to Regulate Gene Expression and Promote Stem Elongation. *Cell* **2002**, *109*, 181–191. [CrossRef]
120. Li, J.; Nam, K.H. Regulation of brassinosteroid signaling by a GSK3/SHAGGY-like kinase. *Science* **2002**, *295*, 1299–1301. [CrossRef]
121. Arite, T.; Iwata, H.; Ohshima, K.; Maekawa, M.; Nakajima, M.; Kojima, M.; Sakakibara, H.; Kyozuka, J. *DWARF10*, an *RMS1/MAX4/DAD1* ortholog, controls lateral bud outgrowth in rice. *Plant J.* **2007**, *51*, 1019–1029. [CrossRef] [PubMed]
122. Zou, J.; Zhang, S.; Zhang, W.; Li, G.; Chen, Z.; Zhai, W.; Zhao, X.; Pan, X.; Xie, Q.; Zhu, L. The rice *HIGH-TILLERING DWARF1* encoding an ortholog of Arabidopsis MAX3 is required for negative regulation of the outgrowth of axillary buds. *Plant J.* **2006**, *48*, 687–698. [CrossRef]
123. Johnson, X.; Brcich, T.; Dun, E.A.; Goussot, M.; Haurogné, K.; Beveridge, C.A.; Rameau, C. Branching Genes Are Conserved across Species. Genes Controlling a Novel Signal in Pea Are Coregulated by Other Long-Distance Signals. *Plant Physiol.* **2006**, *142*, 1014–1026. [CrossRef]
124. Foo, E.; Bullier, E.; Goussot, M.; Foucher, F.; Rameau, C.; Beveridge, C.A. The branching gene *RAMOSUS1* mediates interactions among two novel signals and auxin in pea. *Plant Cell* **2005**, *17*, 464–474. [CrossRef] [PubMed]
125. Bainbridge, K.; Sorefan, K.; Ward, S.; Leyser, O. Hormonally controlled expression of the *Arabidopsis MAX4* shoot branching regulatory gene. *Plant J.* **2005**, *44*, 569–580. [CrossRef]
126. Beveridge, C.A.; Dun, E.A.; Rameau, C. Pea has its tendrils in branching discoveries spanning a century from auxin to strigolactones. *Plant Physiol.* **2009**, *151*, 985–990. [CrossRef]
127. Bennett, T.; Leyser, O. Something on the Side: Axillary Meristems and Plant Development. *Plant Mol. Biol.* **2006**, *60*, 843–854. [CrossRef]
128. Prusinkiewicz, P.; Crawford, S.; Smith, R.S.; Ljung, K.; Bennett, T.; Ongaro, V.; Leyser, O. Control of bud activation by an auxin transport switch. *Proc. Natl. Acad. Sci. USA* **2009**, *106*, 17431–17436. [CrossRef]
129. Crawford, S.; Shinohara, N.; Sieberer, T.; Williamson, L.; George, G.; Hepworth, J.; Müller, D.; Domagalska, M.A.; Leyser, O. Strigolactones enhance competition between shoot branches by dampening auxin transport. *Development* **2010**, *137*, 2905–2913. [CrossRef]
130. Bangerth, F. Response of cytokinin concentration in the xylem exudate of bean (*Phaseolus vulgaris* L.) plants to decapitation and auxin treatment, and relationship to apical dominance. *Planta* **1994**, *194*, 439–442. [CrossRef]
131. Nordström, A.; Tarkowski, P.; Tarkowska, D.; Norbaek, R.; Åstot, C.; Dolezal, K.; Sandberg, G. Auxin regulation of cytokinin biosynthesis in *Arabidopsis thaliana*: A factor of potential importance for auxin–cytokinin-regulated development. *Proc. Natl. Acad. Sci. USA* **2004**, *101*, 8039–8044. [CrossRef]
132. Li, C.J.; Guevara, E.; Herrera, J.; Bangerth, F. Effect of apex excision and replacement by 1-naphthylacetic acid on cytokinin concentration and apical dominance in pea plants. *Physiol. Plant.* **1995**, *94*, 465–469. [CrossRef]
133. Bangerth, F.; Li, C.-J.; Gruber, J. Mutual interaction of auxin and cytokinins in regulating correlative dominance. *Plant Growth Regul.* **2000**, *32*, 205–217. [CrossRef]

134. To, J.P.C.; Haberer, G.; Ferreira, F.J.; DerueRe, J.; Mason, M.G.; Schaller, G.E.; Alonso, J.M.; Ecker, J.R.; Kieber, J.J. Type-A Arabidopsis Response Regulators Are Partially Redundant Negative Regulators of Cytokinin Signaling. *Plant Cell* **2004**, *16*, 658–671. [CrossRef] [PubMed]
135. Vysotskaya, L.B.; Korobova, A.V.; Veselov, S.Y.; Dodd, I.C.; Kudoyarova, G.R. ABA mediation of shoot cytokinin oxidase activity: Assessing its impacts on cytokinin status and biomass allocation of nutrient-deprived durum wheat. *Funct. Plant Biol.* **2009**, *36*, 66. [CrossRef]
136. Chang, H.; Jones, M.L.; Banowetz, G.M.; Clark, D.G. Overproduction of Cytokinins in Petunia Flowers Transformed with P_{SAG12}-IPT Delays Corolla Senescence and Decreases Sensitivity to Ethylene. *Plant Physiol.* **2003**, *132*, 2174–2183. [CrossRef]
137. Rolland, F.; Baena-Gonzalez, E.; Sheen, J. Sugar sensing and signaling in plants: Conserved and novel mechanisms. *Annu. Rev. Plant Biol.* **2006**, *57*, 675–709. [CrossRef]
138. Smeekens, S.; Ma, J.; Hanson, J.; Rolland, F. Sugar signals and molecular networks controlling plant growth. *Curr. Opin. Plant Biol.* **2010**, *13*, 273–278. [CrossRef]
139. Wind, J.; Smeekens, S.; Hanson, J. Sucrose: Metabolite and signaling molecule. *Phytochemistry* **2010**, *71*, 1610–1614. [CrossRef]
140. Barbier, F.F.; Lunn, J.E.; Beveridge, C.A. Ready, steady, go! A sugar hit starts the race to shoot branching. *Curr. Opin. Plant Biol.* **2015**, *25*, 39–45. [CrossRef]
141. Morris, S.E.; Cox, M.C.; Ross, J.J.; Krisantini, S.; Beveridge, C.A. Auxin dynamics after decapitation are not correlated with the initial growth of axillary buds. *Plant Physiol.* **2005**, *138*, 1665–1672. [CrossRef]
142. Renton, M.; Hanan, J.; Ferguson, B.J.; Beveridge, C.A. Models of long-distance transport: How is carrier-dependent auxin transport regulated in the stem? *New Phytol.* **2012**, *194*, 704–715. [CrossRef] [PubMed]
143. Kebrom, T.H.; Chandler, P.M.; Swain, S.M.; King, R.W.; Richards, R.A.; Spielmeyer, W. Inhibition of tiller bud outgrowth in the *tin* mutant of wheat is associated with precocious internode development. *Plant Physiol.* **2012**, *160*, 308–318. [CrossRef] [PubMed]
144. Rabot, A.; Henry, C.; Ben Baaziz, K.; Mortreau, E.; Azri, W.; Lothier, J.; Hamama, L.; Boummaza, R.; Leduc, N.; Pelleschi-Travier, S.; et al. Insight into the Role of Sugars in Bud Burst Under Light in the Rose. *Plant Cell Physiol.* **2012**, *53*, 1068–1082. [CrossRef] [PubMed]
145. Kebrom, T.H.; Mullet, J.E. Photosynthetic leaf area modulates tiller bud outgrowth in sorghum. *Plant Cell Environ.* **2015**, *38*, 1471–1478. [CrossRef]
146. Brewer, P.B.; Koltai, H.; Beveridge, C.A. Diverse Roles of Strigolactones in Plant Development. *Mol. Plant* **2013**, *6*, 18–28. [CrossRef]
147. Ongaro, V.; Leyser, O. Hormonal control of shoot branching. *J. Exp. Bot.* **2007**, *59*, 67–74. [CrossRef] [PubMed]
148. Yoshida, T.; Mogami, J.; Yamaguchi-Shinozaki, K. ABA-dependent and ABA-independent signaling in response to osmotic stress in plants. *Curr. Opin. Plant Biol.* **2014**, *21*, 133–139. [CrossRef] [PubMed]
149. Kebrom, T.H.; Burson, B.L.; Finlayson, S.A. Phytochrome B Represses *Teosinte Branched1* Expression and Induces Sorghum Axillary Bud Outgrowth in Response to Light Signals. *Plant Physiol.* **2006**, *140*, 1109–1117. [CrossRef]
150. Gao, Z.; Wang, Y.; Chen, G.; Zhang, A.; Yang, S.; Shang, L.; Wang, D.; Ruan, B.; Liu, C.; Jiang, H.; et al. The indica nitrate reductase gene OsNR2 allele enhances rice yield potential and nitrogen use efficiency. *Nat. Commun.* **2019**, *10*, 5207. [CrossRef]
151. Wu, K.; Wang, S.; Song, W.; Zhang, J.; Wang, Y.; Liu, Q.; Yu, J.; Ye, Y.; Li, S.; Chen, J.; et al. Enhanced sustainable green revolution yield via nitrogen-responsive chromatin modulation in rice. *Science* **2020**, *367*, eaaz2046. [CrossRef]
152. Sun, H.; Qian, Q.; Wu, K.; Luo, J.; Wang, S.; Zhang, C.; Ma, Y.; Liu, Q.; Huang, X.; Yuan, Q.; et al. Heterotrimeric G proteins regulate nitrogen-use efficiency in rice. *Nat. Genet.* **2014**, *46*, 652–656. [CrossRef]
153. Li, X.; Xia, K.; Liang, Z.; Chen, K.; Gao, C.; Zhang, M. MicroRNA393 is involved in nitrogen-promoted rice tillering through regulation of auxin signal transduction in axillary buds. *Sci. Rep.* **2016**, *6*, 32158. [CrossRef] [PubMed]
154. Liu, Y.; Wang, H.; Jiang, Z.; Wang, W.; Xu, R.; Wang, Q.; Zhang, Z.; Li, A.; Liang, Y.; Ou, S.; et al. Genomic basis of geographical adaptation to soil nitrogen in rice. *Nature* **2021**, *590*, 600–605. [CrossRef]
155. Fioreze, S.L.; Castoldi, G.; Pivetta, L.A.; Pivetta, L.G.; Fernandes, D.M.; Büll, L.T. Tillering of two wheat genotypes as affected by phosphorus levels. *Acta Sci. Agron.* **2012**, *34*, 331–338. [CrossRef]
156. Li, X.R.; Sun, J.; Albinsky, D.; Zarrabian, D.; Hull, R.; Lee, T.; Jarratt-Barnham, E.; Chiu, C.H.; Jacobsen, A.; Soumpourou, E.; et al. Nutrient regulation of lipochitooligosaccharide recognition in plants via NSP1 and NSP2. *Nat. Commun.* **2022**, *13*, 6421. [CrossRef]
157. Chérel, I. Regulation of K+ channel activities in plants: From physiological to molecular aspects. *J. Exp. Bot.* **2004**, *55*, 337–351. [CrossRef] [PubMed]
158. Xu, Y.; Zhang, S.; Guo, H.; Wang, S.; Xu, L.; Li, C.; Qian, Q.; Chen, F.; Geisler, M.; Qi, Y.; et al. OsABCB14 functions in auxin transport and iron homeostasis in rice (*Oryza sativa* L.). *Plant J. Cell Mol. Biol.* **2014**, *79*, 106–117. [CrossRef] [PubMed]
159. Chen, G.; Feng, H.; Hu, Q.; Qu, H.; Chen, A.; Yu, L.; Xu, G. Improving rice tolerance to potassium deficiency by enhancing OsHAK16p:WOX11-controlled root development. *Plant Biotechnol. J.* **2015**, *13*, 833–848. [CrossRef]
160. Kang, J.; Li, J.; Gao, S.; Tian, C.; Zha, X. Overexpression of the leucine-rich receptor-like kinase gene LRK 2 increases drought tolerance and tiller number in rice. *Plant Biotechnol. J.* **2017**, *15*, 1175–1185. [CrossRef]
161. Weng, X.; Wang, L.; Wang, J.; Hu, Y.; Du, H.; Xu, C.; Xing, Y.; Li, X.; Xiao, J.; Zhang, Q. Grain number, plant height, and heading date7 is a central regulator of growth, development, and stress response. *Plant Physiol.* **2014**, *164*, 735–747. [CrossRef] [PubMed]

162. Xu, P.; Cai, W. RAN1 is involved in plant cold resistance and development in rice (*Oryza sativa*). *J. Exp. Bot.* **2014**, *65*, 3277–3287. [CrossRef]
163. Evers, J.B.; Vos, J.; Andrieu, B.; Struik, P.C. Cessation of Tillering in Spring Wheat in Relation to Light Interception and Red: Far-red Ratio. *Ann. Bot.* **2006**, *97*, 649–658. [CrossRef]
164. Girault, T.; Bergougnoux, V.; Combes, D.; Viemont, J.-D.; Leduc, N. Light controls shoot meristem organogenic activity and leaf primordia growth during bud burst in *Rosa* sp. *Plant Cell Environ.* **2008**, *31*, 1534–1544. [CrossRef]
165. Stirnberg, P.; Van De Sande, K.; Leyser, H.M.O. *MAX1* and *MAX2* control shoot lateral branching in *Arabidopsis*. *Development* **2002**, *129*, 1131–1141. [CrossRef]
166. Beveridge, C.A.; Weller, J.L.; Singer, S.R.; Hofer, J.M.I. Axillary Meristem Development. Budding Relationships between Networks Controlling Flowering, Branching, and Photoperiod Responsiveness. *Plant Physiol.* **2003**, *131*, 927–934. [CrossRef] [PubMed]
167. Kebrom, T.H.; Brutnell, T.P. The molecular analysis of the shade avoidance syndrome in the grasses has begun. *J. Exp. Bot.* **2007**, *58*, 3079–3089. [CrossRef]
168. Ballaré, C.L.; Scopel, A.L.; Sánchez, R.A. Far-red radiation reflected from adjacent leaves: An early signal of competition in plant canopies. *Science* **1990**, *247*, 329–332. [CrossRef]
169. Smith, H.; Whitelam, G.C. The shade avoidance syndrome: Multiple responses mediated by multiple phytochromes. *Plant Cell Environ.* **1997**, *20*, 840–844. [CrossRef]
170. Luo, L.; Zhang, Y.; Xu, G. How does nitrogen shape plant architecture? *J. Exp. Bot.* **2020**, *71*, 4415–4427. [CrossRef]
171. Luo, Z.; Janssen, B.J.; Snowden, K.C. The molecular and genetic regulation of shoot branching. *Plant Physiol.* **2021**, *187*, 1033–1044. [CrossRef] [PubMed]
172. Gao, Y.; Qi, S.; Wang, Y. Nitrate signaling and use efficiency in crops. *Plant Commun.* **2022**, *3*, 100353. [CrossRef] [PubMed]
173. Duan, E.; Wang, Y.; Li, X.; Lin, Q.; Zhang, T.; Wang, Y.; Zhou, C.; Zhang, H.; Jiang, L.; Wang, J.; et al. OsSHI1 Regulates Plant Architecture Through Modulating the Transcriptional Activity of IPA1 in Rice. *Plant Cell* **2019**, *31*, 1026–1042. [CrossRef] [PubMed]
174. Drummond, R.S.M.; Janssen, B.J.; Luo, Z.; Oplaat, C.; Ledger, S.E.; Wohlers, M.W.; Snowden, K.C. Environmental Control of Branching in Petunia. *Plant Physiol.* **2015**, *168*, 735–751. [CrossRef] [PubMed]
175. Yoneyama, K.; Xie, X.; Kim, H.I.; Kisugi, T.; Nomura, T.; Sekimoto, H.; Yokota, T.; Yoneyama, K. How do nitrogen and phosphorus deficiencies affect strigolactone production and exudation? *Planta* **2012**, *235*, 1197–1207. [CrossRef]
176. Kamada-Nobusada, T.; Makita, N.; Kojima, M.; Sakakibara, H. Nitrogen-Dependent Regulation of De Novo Cytokinin Biosynthesis in Rice: The Role of Glutamine Metabolism as an Additional Signal. *Plant Cell Physiol.* **2013**, *54*, 1881–1893. [CrossRef]
177. Yoneyama, K.; Xie, X.; Kisugi, T.; Nomura, T.; Yoneyama, K. Nitrogen and phosphorus fertilization negatively affects strigolactone production and exudation in sorghum. *Planta* **2013**, *238*, 885–894. [CrossRef]
178. Paz-Ares, J.; Puga, M.I.; Rojas-Triana, M.; Martinez-Hevia, I.; Diaz, S.; Poza-Carrión, C.; Miñambres, M.; Leyva, A. Plant adaptation to low phosphorus availability: Core signaling, crosstalks, and applied implications. *Mol. Plant* **2022**, *15*, 104–124. [CrossRef]
179. Yuan, K.; Zhang, H.; Yu, C.; Luo, N.; Yan, J.; Zheng, S.; Hu, Q.; Zhang, D.; Kou, L.; Meng, X.; et al. Low phosphorus promotes NSP1-NSP2 heterodimerization to enhance strigolactone biosynthesis and regulate shoot and root architectures in rice. *Mol. Plant* **2023**. [CrossRef]
180. Umehara, M.; Hanada, A.; Magome, H.; Takeda-Kamiya, N.; Yamaguchi, S. Contribution of strigolactones to the inhibition of tiller bud outgrowth under phosphate deficiency in rice. *Plant Cell Physiol.* **2010**, *51*, 1118–1126. [CrossRef]
181. Yang, T.; Feng, H.; Zhang, S.; Xiao, H.; Hu, Q.; Chen, G.; Xuan, W.; Moran, N.; Murphy, A.; Yu, L.; et al. The Potassium Transporter OsHAK5 Alters Rice Architecture via ATP-Dependent Transmembrane Auxin Fluxes. *Plant Commun.* **2020**, *1*, 100052. [CrossRef] [PubMed]
182. Panda, D.; Mishra, S.S.; Behera, P.K. Drought tolerance in rice: Focus on recent mechanisms and approaches. *Rice Sci.* **2021**, *28*, 119–132. [CrossRef]
183. Sunkar, R.; Zhu, J.-K. Novel and stress-regulated microRNAs and other small RNAs from Arabidopsis. *Plant Cell* **2004**, *16*, 2001–2019. [CrossRef] [PubMed]
184. Prasanth, V.V.; Babu, M.S.; Basava, R.K.; Tripura Venkata, V.G.N.; Mangrauthia, S.K.; Voleti, S.R.; Neelamraju, S. Trait and Marker Associations in Oryza nivara and O. rufipogon Derived Rice Lines under Two Different Heat Stress Conditions. *Front. Plant Sci.* **2017**, *8*, 1819. [CrossRef]
185. Harsant, J.; Pavlovic, L.; Chiu, G.; Sultmanis, S.; Sage, T.L. High temperature stress and its effect on pollen development and morphological components of harvest index in the C3 model grass Brachypodium distachyon. *J. Exp. Bot.* **2013**, *64*, 2971–2983. [CrossRef] [PubMed]
186. Wang, Y.; Sun, F.; Cao, H.; Peng, H.; Ni, Z.; Sun, Q.; Yao, Y. TamiR159 directed wheat TaGAMYB cleavage and its involvement in anther development and heat response. *PLoS ONE* **2012**, *7*, e48445. [CrossRef]
187. Chen, L.; Zhao, Y.; Xu, S.; Zhang, Z.; Xu, Y.; Zhang, J.; Chong, K. OsMADS57 together with OsTB1 coordinates transcription of its target OsWRKY94 and D14 to switch its organogenesis to defense for cold adaptation in rice. *New Phytol.* **2018**, *218*, 219–231. [CrossRef] [PubMed]
188. Yoon, D.H.; Lee, S.S.; Park, H.J.; Lyu, J.I.; Chong, W.S.; Liu, J.R.; Kim, B.G.; Ahn, J.C.; Cho, H.S. Overexpression of OsCYP19-4 increases tolerance to cold stress and enhances grain yield in rice (*Oryza sativa*). *J. Exp. Bot.* **2016**, *67*, 69–82. [CrossRef]

189. Clarke, P.R.; Zhang, C. Spatial and temporal coordination of mitosis by Ran GTPase. *Nat. Rev. Mol. Cell Biol.* **2008**, *9*, 464–477. [CrossRef]
190. Liu, Y.; Wang, K.; Li, D.; Yan, J.; Zhang, W. Enhanced Cold Tolerance and Tillering in Switchgrass (*Panicum virgatum* L.) by Heterologous Expression of Osa-miR393a. *Plant Cell Physiol.* **2017**, *58*, 2226–2240. [CrossRef]
191. Holbrook-Smith, D.; Toh, S.; Tsuchiya, Y.; McCourt, P. Small-molecule antagonists of germination of the parasitic plant Striga hermonthica. *Nat. Chem. Biol.* **2016**, *12*, 724–729. [CrossRef]
192. Cardoso, C.; Zhang, Y.; Jamil, M.; Hepworth, J.; Charnikhova, T.; Dimkpa, S.O.; Meharg, C.; Wright, M.H.; Liu, J.; Meng, X.; et al. Natural variation of rice strigolactone biosynthesis is associated with the deletion of two MAX1 orthologs. *Proc. Natl. Acad. Sci. USA* **2014**, *111*, 2379–2384. [CrossRef] [PubMed]

plants

MDPI

Article

Genome Survey and Chromosome-Level Draft Genome Assembly of *Glycine max* var. Dongfudou 3: Insights into Genome Characteristics and Protein Deficiencies

Yajuan Duan [1,†], Yue Li [1,†], Jing Zhang [1], Yongze Song [1], Yan Jiang [1], Xiaohong Tong [1], Yingdong Bi [2,*], Shaodong Wang [1,*] and Sui Wang [1,*]

1 Key Laboratory of Soybean Biology of Chinese Education Ministry, Northeast Agricultural University, 600 Changjiang Road, Harbin 150030, China; duanyajuan2902@163.com (Y.D.); zhly200026@163.com (Y.L.)

2 Institute of Crop Cultivation and Tillage, Heilongjiang Academy of Agricultural Sciences, Harbin 150028, China

* Correspondence: yingdongbi@haas.cn (Y.B.); wsdhlj@neau.edu.cn (S.W.); wangsui.ws@163.com (S.W.)

† These authors contributed equally to this work.

Abstract: Dongfudou 3 is a highly sought-after soybean variety due to its lack of beany flavor. To support molecular breeding efforts, we conducted a genomic survey using next-generation sequencing. We determined the genome size, complexity, and characteristics of Dongfudou 3. Furthermore, we constructed a chromosome-level draft genome and speculated on the molecular basis of protein deficiency in GmLOX1, GmLOX2, and GmLOX3. These findings set the stage for high-quality genome analysis using third-generation sequencing. The estimated genome size is approximately 1.07 Gb, with repetitive sequences accounting for 72.50%. The genome is homozygous and devoid of microbial contamination. The draft genome consists of 916.00 Mb anchored onto 20 chromosomes, with annotations of 46,446 genes and 77,391 transcripts, achieving Benchmarking Single-Copy Orthologue (BUSCO) completeness of 99.5% for genome completeness and 99.1% for annotation. Deletions and substitutions were identified in the three *GmLox* genes, and they also lack corresponding active proteins. Our proposed approach, involving *k*-mer analysis after filtering out organellar DNA sequences, is applicable to genome surveys of all plant species, allowing for accurate assessments of size and complexity. Moreover, the process of constructing chromosome-level draft genomes using closely related reference genomes offers cost-effective access to valuable information, maximizing data utilization.

Keywords: genome survey; next-generation sequencing; *Glycine max*; Dongfudou 3; *k*-mer analysis; organellar DNA; *GmLox* genes

check for updates

Citation: Duan, Y.; Li, Y.; Zhang, J.; Song, Y.; Jiang, Y.; Tong, X.; Bi, Y.; Wang, S.; Wang, S. Genome Survey and Chromosome-Level Draft Genome Assembly of *Glycine max* var. Dongfudou 3: Insights into Genome Characteristics and Protein Deficiencies. *Plants* **2023**, *12*, 2994. https://doi.org/10.3390/plants12162994

Academic Editor: Alex Troitsky

Received: 5 July 2023
Revised: 11 August 2023
Accepted: 17 August 2023
Published: 19 August 2023

1. Introduction

Soybean (*Glycine max* (Linn.) Merr.), a member of the family Fabaceae, is a crucial crop that provides a major source of proteins and oils for human consumption and livestock feed worldwide [1]. In addition to its nutritional value, soybean plays a pivotal role in sustainable agriculture by fixing atmospheric nitrogen through a symbiotic relationship with microorganisms [2,3]. Moreover, soybean is also a crop that is highly sensitive to photoperiod and temperature, and there may be significant variations in regional adaptability and yield among different soybean varieties [4,5]. It is generally believed to have originated in China and was domesticated from wild soybean (*Glycine soja* Siebold and Zucc.) [6]. Research on the soybean genome began relatively early. As early as 2010, the soybean cultivar Williams 82's genome was sequenced and has been consistently updated since then [7]. Over the past decade, several high-quality soybean genomes have been obtained, and a soybean pan-genome has been constructed, ushering in the era of crop pan-genomics [8–10]. The abundant genomic information has greatly advanced various studies on soybean. However, compared to the vast number of soybean varieties, the number of varieties with completed whole-genome sequencing remains rare.

With the continuous maturation of third-generation sequencing technologies, particularly PacBio HiFi and ONT ultra-long sequencing, soybean genomes at T2T or even gap-free levels are gradually becoming achievable [11]. However, even with the decrease in the cost of third-generation sequencing, it is still not feasible to sequence thousands of soybean varieties. As a result, next-generation sequencing remains the mainstream approach, with hundreds or thousands of completed sequencing data already stored in public databases. The challenge remains how to better utilize these data. Genome survey sequencing is a technique used to analyze the overall structure of a genome without sequencing it completely. It is often used as a first step in the process of genome sequencing as it provides a broad overview of the genome, such as the genome size, heterozygosity, and the proportion of repetitive sequences, and can help guide subsequent sequencing efforts. Further detailed analysis of these data with a depth of over 50× may provide additional information, including species ploidy, symbiotic microorganisms, historical effective population size, and even a chromosome-level genome assembly. In fact, this already satisfies the needs of a considerable portion of research studies.

Soybean and its products are rich and balanced in nutrients, but a significant portion of the population is sensitive to the "beany flavor" in soybean, which directly reduces their desire to consume soybean. Research shows that lipoxygenases (LOXs) in mature soybean seeds can catalyze the oxidation of unsaturated fatty acids, such as linoleic acid and linolenic acid, producing conjugated unsaturated fatty acid hydroperoxides, which are then converted into volatile compounds, resulting in the beany flavor [12,13]. Dongfudou 3 is a special-purpose soybean variety bred by Northeast Agricultural University and Fumin Seed Group Co. Ltd. (Wudalianchi, China) in Wudalianchi, Heilongjiang Province. The approval number is Heishendou 20190045. The main characteristic of this variety is its lack of "beany flavor". Protein analysis shows that its seeds do not contain three main types of lipoxygenases: GmLOX1, GmLOX2, and GmLOX3. Due to its excellent characteristics, Dongfudou 3 has been favored by many soybean farmers and downstream processing enterprises. To further understand its genetic features and utilize them in breeding programs, we initiated high-quality whole-genome sequencing. Prior to third-generation sequencing, we conducted a comprehensive genome survey of Dongfudou 3 using next-generation sequencing data. This involved analyzing its genome characteristics, constructing a chromosome-level draft genome assembly, annotating genes, and investigating the molecular mechanism behind the deficiency of LOX proteins. In this study, we attempted to improve the workflow of plant genome *k*-mer analysis by removing the interference of organellar-derived DNA. This approach has broader applicability to genome surveys of various plant species, facilitating more precise assessments of genome size and complexity. At the same time, we also attempted to establish a simple and fast plant genome survey pipeline, which could quickly obtain chromosome-level draft genomes and annotation files, to assist more researchers in obtaining more beneficial information at a lower cost.

2. Results

2.1. Sequencing and Data Cleaning

To minimize the risk of contamination by endophytic microorganisms, pests, or diseases, strict management practices were implemented throughout the sequencing process. Measures such as seed and substrate sterilization, isolated cultivation, and pest and disease control were employed. After removing adapters and primers, a total of approximately 133.62 Gb of raw data was obtained. Subsequently, trimming was performed to remove low-quality bases at the beginning and end of reads, resulting in clean reads with a length of 140 bp and a total size of 116.28 Gb. The average percentages of Q20 and Q30 were 97.94% and 91.47%, respectively. The filtering rate was approximately 12.98%, and the insert size peak was approximately 260 bp.

2.2. Genome Size Estimation by k-mer Analysis

To accurately assess the genome size of Dongfudou 3, it was necessary to remove reads originating from organelles. By carefully controlling the alignment parameters, we aimed to eliminate organelle-derived reads while minimizing the erroneous removal of similar sequences from the nuclear genome. As a result, approximately 11.52% of the reads were discarded, leaving us with 102.88 Gb of clean reads derived from the nuclear genome. Based on the published genome information of several soybean varieties, we estimated the genome size of Dongfudou 3 to be around 1 Gb. Therefore, we set $k = 17$ for k-mer analysis. Figure 1 illustrates a main peak at a depth of 83×, while on the right side of the peak, around 165× depth, there is a minor peak formed due to repetitive sequences. Based on the results from GCE, we estimated the genome size of Dongfudou 3 to be approximately 1.07 Gb, with repetitive sequences accounting for approximately 72.50% of the whole genome. Based on these preliminary assessments, the genome of Dongfudou 3 reached a state of near purity, with no significant heterozygous peaks observed.

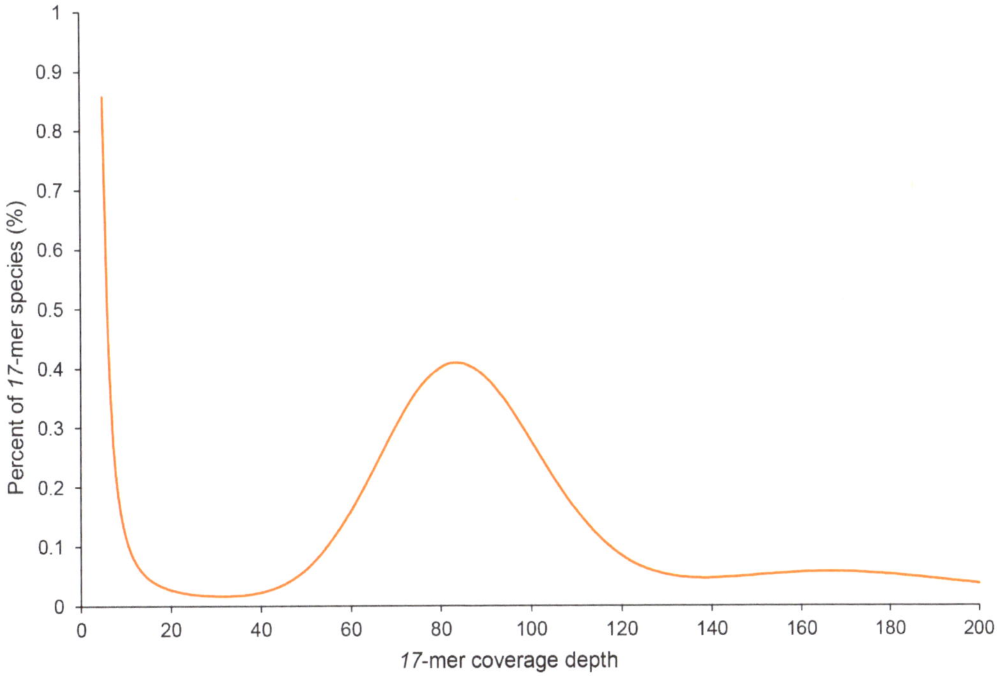

Figure 1. Distribution of 17-mer frequency of *Glycine max* var. Dongfudou 3.

2.3. Preliminary Genome Assembly and GC Content Assessment

In order to obtain higher-quality genome scaffolds, we assessed the optimal assembly k-mer. A value of 117 was chosen according to the k-mer estimation of KmerGenie. Using SOAPdenovo, we combined short reads to generate a draft assembly, which consisted of 2,196,169 contigs and 2,080,395 scaffolds and covered 1.17 Gb. The overall GC content of the whole contigs was 35.70%. To assess the GC content and sequencing coverage of contigs, and to check for potential contamination such as bacterial sequences, we aligned the clean reads back to the contigs assembled by SOAPdenovo. We selected 49,090 contigs with a minimum length of 5000 bp and calculated their coverage and GC content. As shown in Figure 2, most of the contigs were concentrated in the 30–40% GC content and 70–90× average depth area. In areas over 150× depth, there was also a small number of contigs, which may be repetitive sequences in the genome. We did not observe any clustering of contigs in regions with GC content exceeding 60%, indicating a lack of contamination by bacteria or other organisms in the sample.

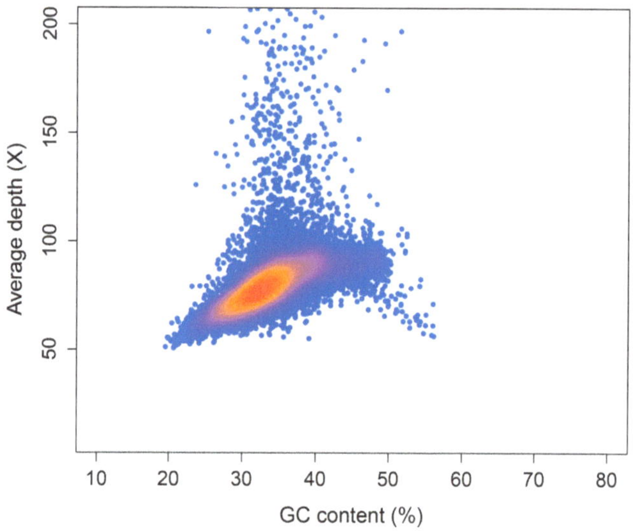

Figure 2. GC depth distribution of the *Glycine max* var. Dongfudou 3 genome. The X-axis is the GC content, and the Y-axis is the average depth. Each dot represents a contig. Color from red to orange and then to blue indicates dot density from high to low.

2.4. Reference-Guided Chromosome Anchoring and Gene Annotation

Thanks to the utilization of the RagTag and Liftoff tools, along with the availability of a high-quality reference genome for soybean, we were able to efficiently perform chromosome-level assembly of the soybean genome. A total of 175,378 scaffolds was successfully anchored onto 20 chromosomes, resulting in a cumulative size of 848.26 Mb, which represents approximately 72.89% of the total genome size. After closing the gaps, the final assembly yielded a chromosome-level genome with a size of 916.00 Mb. The longest contig obtained was 298.19 kb, and the contig N50 was 39.37 kb. We assessed the completeness of our assembly by remapping the BGI short reads, and found that 99.94% of the DNA reads could align properly and the coverage ratio reached 96.04%. The completeness of the assembly was assessed using BUSCO, with a score of 99.5% (S: 43.1%, D: 56.4%, F: 0.3%, and M: 0.2%). Using Liftoff, we mapped the reference genome's GFF file onto the assembled genome of Dongfudou 3 and performed some corrections to the CDS structures. Initially, we identified 47,794 genes, and after further filtering out genes with incomplete structures, we obtained a final set of 46,446 annotated genes and 77,391 transcripts. We extracted the longest transcript as the primary transcript for each gene and re-evaluated the completeness of the annotated genes using BUSCO. The results showed the Benchmarking Single-Copy Orthologue (BUSCO) completeness was 99.1% (S: 45.7%, D: 53.4%, F: 0.4%, and M: 0.5%). Subsequently, we used the GFF file and the file containing the primary transcripts to extract genes, cDNA, CDS, and protein sequences. For gene functional annotation, we focused on the widely used GO and KEGG databases. In the GO database, a total of 30,797 genes were successfully annotated, accounting for 66.31% of all genes. In the KEGG database, 19,737 genes were annotated, representing 42.49% of all genes. Table 1 presents key statistics for genome assembly and annotation.

Table 1. Statistics for the genome assembly and annotation.

Item	Value
Assembled genome size (Mb)	916.00
Number of contigs	60,936
Number of Scaffolds	20
N50 of contigs (bp)	39,370
N90 of contigs (bp)	9078
Longest contig (bp)	298,193
GC content (%)	34.3
BUSCO complete of the genome (%)	99.5
BUSCO complete of the genes (%)	99.1
Number of protein-coding genes	46,446
Number of genes annotated to GO terms	30,797
Number of genes annotated to KEGG terms	19,737

2.5. Mutation Analysis of Lipoxygenase Deficiency

During the previous variety approval process, Dongfudou 3 was identified as a variety with complete deletions of GmLOX1, GmLOX2, and GmLOX3. Here, we first examined the variations in these three genes at the DNA level. Previous studies have indicated that the three proteins are qualitative traits controlled by two linked genes located on chromosome 13 (*SoyZH13_13G319810*, *GmLox1*; *SoyZH13_13G319800*, *GmLox2*) and one gene located on chromosome 15 (*SoyZH13_15G025600*, *GmLox3*). Based on the principles of collinearity and sequence similarity, we mapped the positions of the allelic genes *GmLox1*, *GmLox2*, and *GmLox3* within the Dongfudou 3 genome and examined their potential variations. Specifically, we initially identified the genomic locations of these genes as follows: *GmLox1* on chromosome 13, spanning from 40,034,891 to 40,038,646 bp; *GmLox2* on chromosome 13, spanning from 40,027,569 to 40,031,512 bp; and *GmLox3* on chromosome 15, spanning from 2,052,600 to 2,056,693 bp. During subsequent gene structure annotation, we focused on the variations observed in the coding sequences (CDSs) of the three genes. As shown in Figure 3, compared to the corresponding alleles in Zhonghuang 13, *GmLox1* in Dongfudou 3 exhibited three mutations: a T to A mutation at position 114, which did not alter the amino acid residue; a C to A mutation at position 157, resulting in the substitution of asparagine with histidine; and a fragment deletion from position 1575 to 1648. *GmLox2* had one mutation: a T to A mutation at position 1596, leading to the substitution of histidine with glutamine. *GmLox3* exhibited one mutation: a deletion of a G base at position 101, resulting in premature termination of the encoded protein. These variations were also confirmed by Sanger sequencing (Figure 4).

2.6. Impact of GmLox Genes Variations on Transcription and Translation Processes

To verify the expression status of these three mutated genes in Dongfudou 3 at the RNA level, we conducted qRT-PCR analysis. The results in Figure 5 demonstrate that compared to the control Zhonghuang 13, the expression levels of *GmLox1* and *GmLox3* in Dongfudou 3 are close to negligible, while the expression level of *GmLox2* is only approximately 1/8 of the control. These findings indicate that these genes in Dongfudou 3 exhibit either near-zero expression or extremely low expression levels at the RNA level.

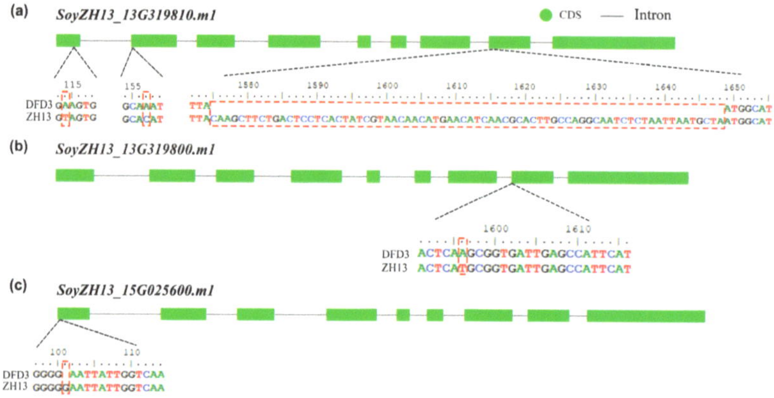

Figure 3. Schematic representation of sequence variations in the *GmLox1*, *GmLox2*, and *GmLox3* genes in Dongfudou 3. The structural diagram represents the corrected structure of the longest transcript corresponding to *GmLox1*, *GmLox2*, and *GmLox3* in Zhonghuang 13. The positions of variations between Zhonghuang 13 and Dongfudou 3 are highlighted in red dashed boxes. The labels "DFD3" and "ZH13" represent Dongfudou 3 and Zhonghuang 13, respectively. Please note that the coordinates in the figure are based on the CDS file of the modified Zhonghuang 13 reference gene, where the first base of the start codon is designated as coordinate 1. The coordinates do not include intronic or other non-coding sequences. The coordinates in the figure are depicted in a 5′ to 3′ orientation, progressing from left to right. (**a**) Schematic representation of the variant sites in *GmLox1*. (**b**) Schematic representation of the variant sites in *GmLox2*. (**c**) Schematic representation of the variant sites in *GmLox3*.

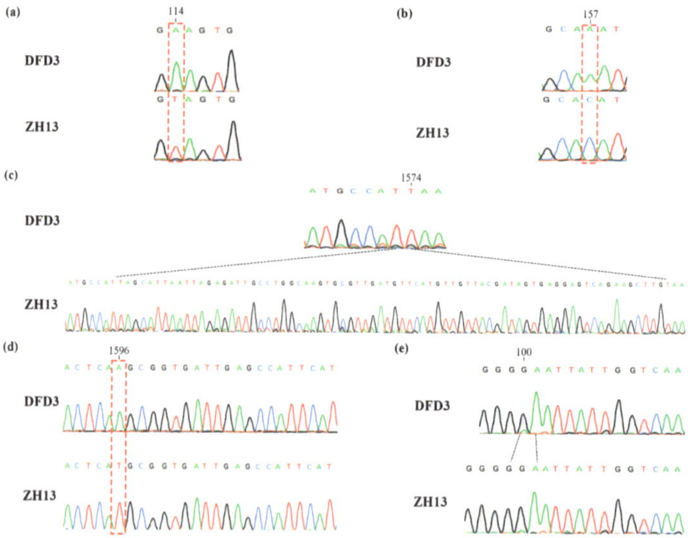

Figure 4. Validation of mutations in the *GmLox1*, *GmLox2*, and *GmLox3* genes in Dongfudou 3 using Sanger sequencing. The labels "DFD3" and "ZH13" represent Dongfudou 3 and Zhonghuang 13, respectively. Mutations are indicated by red and black dashed lines at their respective positions. (**a**) Schematic representation of the mutation at position 114 in the CDS of *GmLox1*. (**b**) Schematic representation of the mutation at position 157 in the CDS of *GmLox1*. (**c**) Schematic representation of the base deletion in the CDS of *GmLox1*. (**d**) Schematic representation of the mutation at position 1596 in the CDS of *GmLox2*. (**e**) Schematic representation of the base deletion in the CDS of *GmLox3*.

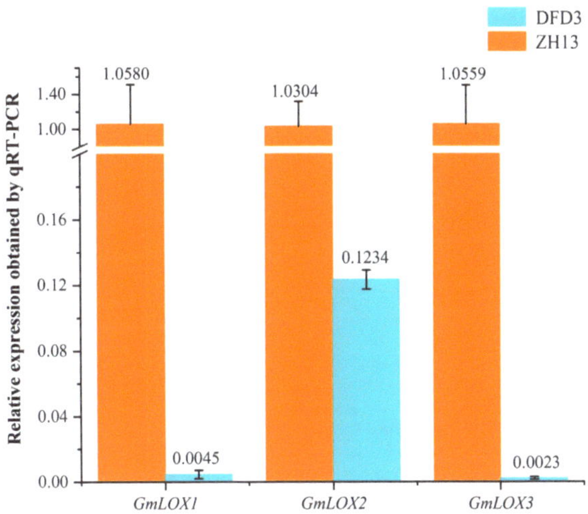

Figure 5. qRT-PCR measurement of 3 *GmLox* genes in Zhonghuang 13 and Dongfudou 3 seeds. In the experiments, we assumed an amplification efficiency (E) of 100%, and each sample was subjected to 3 biological replicates. The internal reference gene chosen was *GmCYP2* (*SoyZH13_12G024400.m2*). The labels "DFD3" and "ZH13" represent Dongfudou 3 and Zhonghuang 13, respectively.

To determine the presence or absence of the three lipoxygenase enzymes (GmLOX1, GmLOX2, and GmLOX3) in Dongfudou 3 seeds at the protein level, we employed a colorimetric assay. As controls, we included the non-deficient variety Zhonghuang 13, which is known to possess three lipoxygenase enzymes. Figure 6 shows that in comparison to the controls, Dongfudou 3 maintains the color of the substrate solution in all treatments, indicating the absence of GmLOX1 (solution remains blue), GmLOX2 (solution remains blue), and GmLOX3 (solution remains yellow). These results align with the SDS-PAGE gel electrophoresis conducted during the variety assessment of Dongfudou 3, providing further confirmation of the protein-level deficiency of GmLOX1, GmLOX2, and GmLOX3.

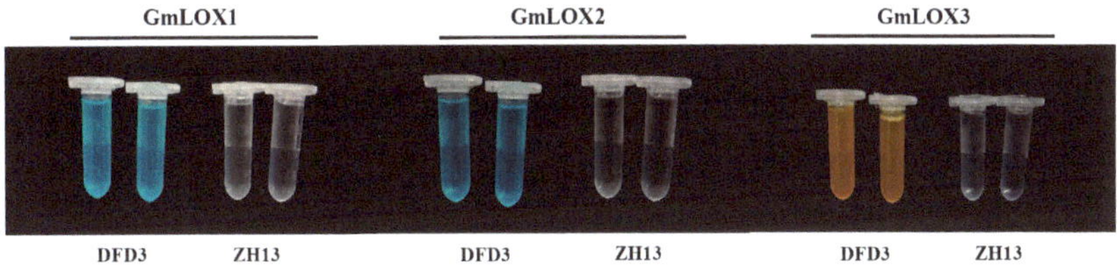

Figure 6. Determination of soybean GmLOX1, GmLOX2, and GmLOX3 activities using a colorimetric assay. The labels "DFD3" and "ZH13" represent Dongfudou 3 and Zhonghuang 13, respectively.

To further analyze whether a single nucleotide variation in the *GmLox2* gene CDS of Dongfudou 3 affects its protein structure, we utilized AlphaFold2 to predict and compare the GmLOX2 proteins from Zhonghuang 13 and Dongfudou 3. Figure 7 illustrates the structural representation of the two proteins overall and in the vicinity of the mutation site. The results reveal that the substitution of asparagine for histidine at position 532 of the GmLOX2 protein does not alter the overall spatial structure of the protein. However, it leads to the absence of an imidazole group at the mutation site.

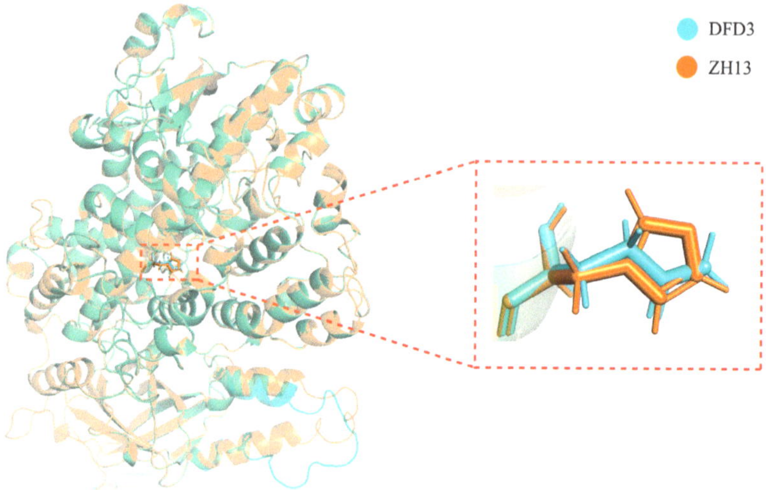

Figure 7. Structural representation of soybean LOX2 protein and illustration of the mutation site.

3. Discussion

The estimation of genome size is a crucial task in genome surveys. Determining the genome size is of significant importance for selecting sequencing strategies, sequencing depth, and evaluating assembly outcomes. Initially, flow cytometry was predominantly used to assess genome size, but the results varied widely, making it challenging to establish a unified standard [14,15]. Therefore, it is currently employed as a rough estimation method for genome size. Currently, the most prevalent method for genome size estimation is primarily based on *k*-mer analysis [16]. This approach is widely regarded as highly accurate and enables the assessment of genome heterozygosity and repetitive sequence content. As a result, it has been extensively applied and has given rise to various evaluation software tools. Currently, this strategy has demonstrated relatively accurate performance in genome size estimation for the majority of microorganisms and animals. However, its performance in plant genome surveys has been less satisfactory. In practice, there are often significant differences in the results obtained from sequencing different individuals or different tissue types, or when using different evaluation software. These discrepancies are difficult to explain solely based on measurement errors or other factors. This discrepancy may be attributed to the fact that plant cells generally contain a larger number of organelles, such as mitochondria and plastids, the genomes of which are significantly larger than those of organelles in animal cells. Due to the current use of sequencing data that include the entire DNA content of plant cells in most plant genome surveys, the inevitable presence of organelle DNA can affect the total count and distribution of *k*-mers, thereby leading to inaccurate assessment results. Additionally, the variation in the number of organelles in plant cells among different individuals, tissues, or growth stages further contributes to varying proportions of organelle-derived DNA in sequencing results, which can have a significant impact on the final outcomes [17]. In fact, a few researchers have already recognized this issue and incorporated strategies to remove high-frequency *k*-mers that may originate from organelle sequences into relevant software tools [18]. However, it is evident that this strategy of filtering out organelle sequences can only serve as a temporary solution. High-frequency *k*-mers derived from nuclear sources, such as telomeres, centromeres, or other repetitive regions, are often inadvertently filtered out as well. Moreover, if plant samples are contaminated with microorganisms or harbor endophytic bacteria, the indiscriminate removal of high-frequency *k*-mers may compromise the subsequent assessment of sample purity. In this study, we employed a direct mapping approach, aligning reads to a reference genome of soybean organelles, to filter out sequences originating from organelles. This approach is primarily based on the highly conserved nature of plant organelles. On the one hand, plastid genomes exhibit a high degree of sequence and structural conservation across different species [19]. On the other hand, although plant mitochondrial genomes may display higher recombination rates among different varieties or individuals, their overall mutation rate is extremely low, resulting in a high level of sequence conservation [20]. Therefore, the impact of this filtering method on the accurate retrieval of most short reads is minimal. In this study, we selected the plastid and mitochondrial genomes of the Williams 82 soybean as references. There is an important consideration to note here. During the course of extensive evolution, certain sequences from organelle genomes originating from endosymbiotic events have been transferred to the nucleus [21,22]. Although accumulated mutations have resulted in sequence differences between the two, to minimize the excessive removal of similar sequences from the nucleus during the alignment process, we also controlled the alignment parameters and modified the scoring matrix accordingly. Considering the error rate of next-generation sequencing and the sequence variations among different varieties, we made several attempts and examined the BAM files. Eventually, we allowed a maximum of three base mismatches or two base deletions, or one base deletion combined with one base mismatch in each aligned read. In fact, during the examination of the BAM files, we focused on the detection of heterozygous variant sites and coverage. Since most angiosperm organelles are maternally inherited, the likelihood of multiple variants occurring at the same position in a single sample is extremely low. Therefore, an excessive number of detected heterozygous sites after alignment is likely due to mismatches. Additionally, the conservation of organelle genomes among closely related species means that low coverage may indicate overly stringent alignment parameters or significant differences between the selected organelle reference genome and the sample. In such cases, parameter adjustments or the selection of a more suitable organelle reference genome are necessary. In order to better determine the sequence differences between the organelle genomes of Dongfudou 3 and Williams 82, we also attempted a preliminary assembly of the organelle genome of Dongfudou 3 using GetOrganelle. We compared it roughly with the sequence of Williams 82 to evaluate our parameter settings and scoring matrix. This approach can also be applied to samples without closely related reference organelle genomes or samples with significant differences from the reference organelle genome. During the assembly process, we did not concern ourselves with the completeness of the organelle genome. In fact, the results of assembly software, such as GetOrganelle, typically include most of the organelle genome, which is usually sufficient for filtering out organelle-derived DNA. The impact of a small amount of remaining or excessively filtered sequences on the results is minimal and can be ignored. We also attempted an evaluation without removing the organelle DNA. With all other parameters unchanged, the resulting size was approximately 1.21 Gb, which is about 140 Mb larger than the result obtained after removing organelle DNA. This size difference is significantly larger than several published soybean genomes and deviates from the general understanding. However, after filtering out organelle-derived DNA, our results were more consistent with the recently assembled high-quality soybean genome using third-generation sequencing, demonstrating the effectiveness and accuracy of our strategy for estimating plant genome size.

GC content evaluation is also an important component of plant genome surveys. It involves plotting the GC content and coverage distribution of preliminary assembled contigs to determine if there are multiple enrichment centers, which can indicate heterozygosity and contamination levels. Although not absolute, there is generally a difference in GC content between plant genomes and microbial genomes [23,24]. Researchers often use scatter plots on the GC content axis to preliminarily assess the presence of microbial contamination in plant samples. The existence of multiple enrichment centers on the GC content axis can indicate the potential presence of microorganisms. Additionally, highly heterozygous regions of the genome are often assembled, which can result in lower coverage during the alignment of reads. Researchers also use scatter plots on the coverage axis to preliminarily assess the homogeneity of the sample genome based on the presence of multiple enrichment centers. Indeed, we had concerns about the heterozygosity of the selected Dongfudou 3 sample prior to sequencing. Despite soybean being a self-pollinating crop, the creation of new varieties often involves hybridization. Dongfudou 3 is a variety obtained through hybridization, and although it has undergone ten consecutive generations of self-pollination, theoretically approaching homozygosity, we still had doubts. Therefore, we conducted an analysis of its heterozygosity using *k*-mer analysis and GC content distribution to gain more insights. Another concern of ours was bacterial contamination, particularly from endophytic bacteria. During the genome assembly of previous species, we often encountered cases where samples were contaminated with endophytic bacteria, leading to significant issues during subsequent assembly and annotation processes [25]. Given the symbiotic nitrogen fixation and other characteristics of soybean, the likelihood of microbial presence, including endophytic bacteria, in its tissues is relatively high. Some preliminary studies have also confirmed the abundance of symbiotic microorganisms in soybean [26,27]. While we took certain disinfection measures during the experiments, the entire process was conducted in an open environment, making it difficult to completely avoid the introduction of microorganisms. However, it was crucial to ensure that their presence in the DNA was minimal and would not significantly impact the genome survey. According to the results of this study, the GC content and coverage of contigs assembled from the Dongfudou 3 sample were concentrated around a single center, indicating a low level of genome heterozygosity and no apparent contaminants. This finding demonstrates that the sample fully meets the requirements for subsequent deep third-generation sequencing.

In this study, we also attempted to construct chromosome-level genome scaffolds and annotate gene structures and functions using survey sequencing data. This clearly expanded the application scope of the data and improved their utility. Clearly, this low-cost sequencing and assembly approach may not suffice for the detection and analysis of repeats, heterozygosity, or complex regions. However, the BUSCO assessment results indicate that the majority of the gene regions have been well assembled and annotated. This approach may be particularly effective for species or varieties that lack genome information but have high-quality reference genomes from closely related species. It is especially suitable for experiments that do not require a high level of chromosome-level genome assembly, such as gene cloning, gene editing, transcriptomics, SNP or indel variant mining, and other studies with lower demands for chromosome-level genome resolution. This simple and cost-effective approach for draft genome construction provides a convenient solution for researchers working with understudied species, those with lower requirements for genome quality, or those facing budget constraints.

Dongfudou 3, a commercially available non-beany soybean cultivar, has gained wide market acceptance. However, the molecular basis underlying the absence of proteins, such as GmLOX1, GmLOX2, and GmLOX3, in this cultivar has remained unclear for a long time due to the lack of genomic information. Here, based on the survey genome draft, we aimed to preliminarily speculate the molecular mechanisms underlying the absence of lipoxygenase enzymes and demonstrate that such assembly results can fulfill the requirements for detecting small fragment variations in most gene regions. Leveraging collinearity and sequence similarity, we conveniently mapped the positions of the *GmLox1*, *GmLox2*, and *GmLox3* genes in the Dongfudou 3 genome and identified their variation sites. The reliability of the sequencing assembly was also confirmed through Sanger sequencing results. Additionally, we employed fluorescent quantitative PCR to detect that the expression levels of GmLox mutant genes at the RNA level were reduced to nearly zero or only about 1/8 of the control. We also utilized colorimetric assays to detect the absence or inactivation of these three proteins at the protein level.

Clearly, it is easy to understand that the DNA-level base deletions in *GmLox1* and *GmLox3* lead to incomplete CDS structures. Concurrently, the expression of the two genes was also close to zero at the RNA level, and protein-level detection indicated the absence of these two proteins. The nucleotide substitution in the *GmLox2* gene, however, did not alter the CDS structure. Although its expression is significantly reduced at the RNA level, we still harbor concerns regarding the potential function of the protein it may translate into. Therefore, we focused on whether a single nucleotide mutation in the *GmLox2* gene affects protein activity. Protein structure prediction revealed that the overall structure of the GmLOX2 protein remained largely unchanged before and after the mutation. However, the mutation caused the amino acid residue to change from histidine to asparagine, resulting in the loss of an imidazole moiety at the mutation site. Previous studies have underscored the critical importance of histidine residues in protein–metal ion interactions. The imidazole side chain of histidine often serves as a ligand in metalloproteins, engaging in coordination bonds with metal ions. This is facilitated by the imidazole side chain's pair of nitrogen atoms, with one acting as a positively charged ligand. Histidine's ability to bind metal ions, such as iron, copper, and zinc, enables its involvement in a range of essential biological functions [28]. Among these, six histidine residues are conserved across all lipoxygenase sequences and are considered potential iron ligands [29]. Further investigations on soybean GmLOX1 and GmLOX3 proteins have provided additional evidence for the crucial role of these six histidines in maintaining Fe binding and protein activity [30,31]. Indeed, preliminary analysis of soybean varieties with a deletion in the GmLOX2 protein revealed that the substitution of histidine with asparagine at position 532 in GmLOX2, aligned with histidine at position 504 in GmLOX1, is a common occurrence. It is likely to disrupt the binding of the protein to iron, leading to the loss of GmLOX2 activity [32]. However, it is evident that we cannot definitively confirm, at this stage, that these mutations are responsible for the deficiency or inactivation of these three proteins. This is because gene transcription and translation involve multiple processes influenced by complex factors, including, but not limited to, variations in the coding region, promoter region, upstream transcription factors, and various epigenetic regulations. Another point of interest for us is the significant reduction in the expression level of the *GmLox2* gene in Dongfudou 3. Given its close linkage with *GmLox1*, we investigated whether mutations in *GmLox1* could potentially impact the transcription of *GmLox2*. The preliminary assembly of the genome only assists us in preliminarily identifying gene variations; the accurate determination of their causes will rely on further molecular experiments for validation in the future.

In summary, this study utilized deep second-generation sequencing to perform a genome survey analysis of the soybean variety Dongfudou 3. It accurately assessed its genome size, complexity, and contamination status. A particular focus was placed on optimizing the *k*-mer analysis process and filtering out potential influences from organelle DNA sequences on genome size estimation. This strategy is applicable not only to soybeans but also to all plant species. It significantly reduces the risk of over-assembly or collapse in the later stages of genome assembly caused by erroneous genome size estimation. However, it should be noted that the reference genome and alignment parameters used in this study may not be optimal for other species and would require further optimization and adjustment. Furthermore, the study also attempted to construct chromosome-level draft genomes using only deep second-generation sequencing data combined with information from closely related reference genomes. The exploration of the molecular mechanisms behind the loss of three GmLOX proteins demonstrated that the draft genomes obtained through this strategy could meet the needs of a substantial portion of research and have practical value. This strategy is also applicable to other species, with the assembly quality highly dependent on the quality of the selected reference genome and its phylogenetic relationship with the study species. In the future, this information will be used to design and optimize sequencing and assembly protocols for the Dongfudou 3 genome, aiming to obtain a high-quality genome.

4. Materials and Methods

4.1. Plant Materials and Sequencing

Healthy and plump seeds of the self-pollinated 10th generation of Dongfudou 3 were surface sterilized with 75% ethanol for 30 s, washed 3 times with sterile water, and soaked in the dark for 4 h. The soaked seeds were placed in a Petri dish with moistened filter paper and germinated in the dark at 24 °C for 5 d, then transferred to a hydroponic tank and grown in 1/2 Hoagland nutrient solution at 24 °C, 12 h light/12 h dark cycle. When the plantlets reached the V3 stage, a fully expanded trifoliate leaf was taken and DNA was extracted using a genomic DNA kit (DP3111, Biotake Corporation, Beijing, China). The DNA was transported to Frasergen Bioinformatics Co. Ltd. (Wuhan, China) on dry ice for library construction and sequencing. A paired-end library with reads of 150 bp and an average insert size of ~300 bp was constructed and sequenced on the MGISEQ-2000RS platform (BGI, Shenzhen, China). Figure 8 illustrates the morphological characteristics of the plants and seeds of Dongfudou 3.

Figure 8. The morphological characteristics of *Glycine max* var. Dongfudou 3. (**a**) The plant. (**b**) The seeds.

4.2. Quality Control and Data Cleaning

To obtain high-quality and vector/adaptor-free reads, raw reads were filtered using fastp (v.0.23.2, main options: -f 8, -t 2, -n 0, -l 140, -q 20, -u 20), with quality checks by FastQC (v.0.11.9) before and after filtering.

4.3. Genome Size and Repeat Rate Estimation by k-mer Analysis

To remove the sequences derived from organelles in the DNA, we downloaded the plastid (MZ964145.1) and mitochondrial (MW331583.1 and MW331584.1) genomes of soybean Williams 82 from NCBI as references. We then used Bowtie2 (v.2.4.4) to align the clean reads to the reference sequences. We strictly controlled the alignment parameters and allowed, at most, one alignment with no more than 3 mismatches or 2 gaps, or 1 gap plus 1 mismatch. Then, samtools (v.1.15.1) was used to convert the format, and bedtools (v.2.29.2) was used to extract the unmapped fastq files, which only came from the nuclear genome of soybean. Finally, genome size, repeat rate, and heterozygous rate were calculated using KmerFreq (v.4.0) and GCE (v.1.0.2) based on a 17-mer distribution [16]. The *k*-mer distribution was drawn using Excel 2019.

4.4. Preliminary Genome Assembly and GC Content Analysis

Before assembly, KmerGenie (v.1.7051) was used to select optimal *k*-mer sizes for assembly [33]. Then, SOAPdenovo2 (v.2.04) was used to build the initial contigs and scaffolds using the selected optimal *k*-mer. Next, we remapped the clean reads to the assembled contigs using bwa-mem2 (v.2.0pre2) and then converted the SAM file to BAM file using samtools. Subsequently, we utilized a self-written script to extract the GC content and coverage depth of contigs longer than 5000 bp. Finally, the densCols function in R (v.4.1.3) was used to plot the GC depth distribution of the contigs.

4.5. Fast Reference-Guided Chromosome Anchoring and Gene Annotation

Draft scaffolds were ordered and oriented by aligning them to the reference genome of *Glycine max* var. Williams 82-ISU-01 (v.2.0, downloaded from Phytozome V13 website) using RagTag v2.1.0 to generate a chromosome-level draft genome [34]. Subsequently, we removed the fragments that did not successfully anchor to the chromosomes and applied GapCloser (v.1.12) to fill the gaps, resulting in the final chromosome-level genome assembly. Next, the gff3 file of the reference genome Williams 82-ISU-01 was transferred into our new draft genome utilizing Liftoff (v.1.6.3) to perform gene annotation with default parameters [35]. We further filtered the results by protein integrity and obtained the final gff3 annotation file. Subsequently, we extracted the longest transcript to generate the primary-transcript-only gff3 file and used the two gff3 files to generate corresponding gene, cDNA, CDS, and protein fasta sequence files.

For gene function annotation, we only performed the primary GO and KEGG annotations on the predicted genes. GO terms were assigned using Interproscan (v.5.54-87.0) and eggNOG database (v.6.0, http://eggnog6.embl.de/, accessed on 12 April 2023). We integrated the two annotation results and filtered them based on go-basic.obo (v.1.2, released on1 April 2023, downloaded from http://geneontology.org/, accessed on 12 April 2023) to ensure that the results were acyclic and non-redundant. KEGG ortholog (KO) annotation was performed using KofamKOALA (v.1.3.0, released on 1 April 2023) [36].

To assess the quality of genome assembly and annotation, QUAST (v.5.2.0) was used to calculate some basic statistics such as the number of contigs, contig N50, and GC content [37]. Then, bwa-mem2 (v2.0pre2) was utilized to remap the DNA short reads to the draft genome; then, the mapping rate and coverage ratio were calculated with samtools to assess the correctness and completeness of the draft genome. In addition, BUSCO (v.5.3.2) analysis was used to assess the completeness, redundancy, and accuracy of the draft genome and the predicted genes with the embryophyte_odb10 dataset, which contains 1614 Embryophyta single-copy orthologs [38].

4.6. Sequence Variations of Seed Lipoxygenases and Their Detection at the Transcriptional and Translational Levels

At the DNA level, we used the modified *GmLox1*, *GmLox2*, and *GmLox3* gene sequences from *Glycine max* var. Zhonghuang 13 to locate the corresponding alleles on the Dongfudou 3 genome, utilizing blastn (v.2.12.0+) and chromosomal collinearity information [39]. Sequences were aligned using MUSCLE in BioEdit (v.7.2.5), followed by manual identification of the variation sites. We also designed primers (Table 2) based on the variant site information for PCR amplification, and sent the products to Shanghai Sangon Biotechnology Co. Ltd. (Shanghai, China) for Sanger sequencing to confirm the accuracy of variant detection. At the transcriptional level, total RNA was extracted from Zhonghuang 13 and Dongfudou 3 seeds using a Plant RNA Extraction Kit (RP3301, BioTeke, China), followed by reverse transcription into cDNA using a ReverTra Ace® qPCR RT Kit (FSQ-101, Toyobo, Japan). *GmCYP2* (*SoyZH13_12G024400*) served as the internal reference, and quantitative PCR primers were designed for the three *GmLox* genes in non-mutated regions. These primers were also synthesized by Shanghai Sangon Biotechnology Co. Ltd (Shanghai, China). And are listed in Table 2. Three technical replicate qRT-PCRs were performed for each biological replicate using SYBR Green Realtime PCR Master Mix (QPK-201, Toyobo, China) in an AriaMx real-time PCR system (Agilent Technologies, Santa Clara, CA, USA). The amplification conditions were as follows: pre-denaturation at 95 °C for 1 min, denaturation at 95 °C for 15 s, 45 cycles, extension at 60 °C for 1 min, with Rox as the reference dye. After the last polymerase chain reaction cycle, the melting curve was generated under the conditions of 95 °C for 15 s, 65 °C for 1 min, and 95 °C for 15 s. The $2^{-\Delta\Delta C_T}$ method was employed to calculate the gene expression data [40], with Zhonghuang 13 genes used as the control. At the protein level, as previously described, we used a colorimetric assay method to measure the activity of lipoxygenases GmLOX1, GmLOX2, and GmLOX3 in soybean seeds [13,41,42]. Finally, AlphaFold (v.2.3.2) was used to predict the potential impact of point mutations in GmLOX2 on protein structure [43]. Visualization and comparison of protein structures were performed using PyMOL (v.2.5).

Table 2. Primers for amplification of variant sites.

Primer Name [1]	Primer
GmLox1-snp-F	AGCTTTGGTTGATTTTCTCACAGGT
GmLox1-snp-R	CGGGGATTCCCATGCTTCCG
GmLox1-del-F	CGAGGTAAACATGCGAAGCG
GmLox1-del-R	GCAGCCCATATCTCCAGTCC
GmLox2-snp-F	TGCCACATCCTGCTGGGGA
GmLox2-snp-R	CCGCTGAAGACATCTCAACGGAA
GmLox3-del-F	TACCACCAGGGGCTGTGCTT
GmLox3-del-R	GAAGACACACAAGGAGGACACGC
GmLox1-qPCR-F	ATTGGTTAAATACTCATGCGGC
GmLox1-qPCR-R	CCGAAGACATCTCCACAGAATA
GmLox2-qPCR-F	TCCTGAACAGAGGAGGAGGG
GmLox2-qPCR-R	GTGCCTATGAGTCCCCCAAC
GmLox3-qPCR-F	GCTTGGGGGTCTTCTCCATAG
GmLox3-qPCR-R	GCTGGAGAGACACGGATCG
GmCYP2-qPCR-F	CAAAAACCCTGTCACGCAGT
GmCYP2-qPCR-R	CACTTTCTCTCAAGGGCACCA

[1] "snp", "del", and "qPCR" in primer names indicate that the primer is designed to detect single nucleotide polymorphisms, base deletions, or qRT-PCR, respectively. "F" and "R" represent the forward and reverse primers, respectively.

5. Conclusions

In this study, we performed a genome survey analysis of the soybean variety Dongfudou 3 using next-generation sequencing, accurately assessing its genome size and complexity. The estimated genome size of Dongfudou 3 was approximately 1.07 Gb, with repetitive sequences accounting for about 72.50% of the genome. The sample appeared to be nearly homozygous, and no significant microbial contamination was detected. Leveraging the soybean reference genome, we assembled a chromosome-level draft genome of Dongfudou 3, with a total of 916.00 Mb sequences anchored onto 20 chromosomes. Through genome annotation, we identified 46,446 genes and 77,391 transcripts. The genome and annotation completeness were evaluated using BUSCO, with assessment scores of 99.5% and 99.1%, respectively. Based on the draft genome, we also conducted a preliminary analysis of the molecular mechanisms underlying the protein loss of GmLOX1, GmLOX2, and GmLOX3 in Dongfudou 3. We identified several base deletions and frameshift mutations in the coding regions of *GmLox1* and *GmLox3*, which may lead to protein inactivation. Additionally, a single-base mutation in the coding region of *GmLox2* resulted in the substitution of glutamine with histidine in the iron-binding active site, which may lead to loss of enzymatic activity. This study not only helps us design further strategies for constructing a high-quality genome of Dongfudou 3 using third-generation sequencing, but also provides a preliminary understanding of certain characteristics of Dongfudou 3 through the assembled draft genome. Moreover, this research has improved the accuracy of estimating plant genome size based on k-mer analysis and tested the process of constructing draft genomes using next-generation data and closely related reference genomes. These findings serve as a valuable reference for conducting plant genome survey analyses on other species.

Author Contributions: Conceptualization, S.W. (Sui Wang), S.W. (Shaodong Wang) and Y.B.; data curation, S.W. (Sui Wang); formal analysis, Y.D., Y.L. and S.W. (Sui Wang); funding acquisition, S.W. (Shaodong Wang), Y.B. and S.W. (Sui Wang); investigation, Y.D. and Y.L.; visualization, Y.D., Y.L. and S.W. (Sui Wang); writing—original draft, Y.D., Y.L. and S.W. (Sui Wang); writing—review and editing, J.Z., Y.S., Y.J. and X.T. All authors have read and agreed to the published version of this manuscript.

Funding: This study was funded by the Heilongjiang Provincial Key Research and Development Plan Project (2022ZX02B05), the National Science and Technology Ministry's "14th Five-Year Plan" project "Creation and Application of High-quality Soybean New Varieties" (2021YFD1201103-01-04), and the Young Talents Program of Northeast Agricultural University (19QC01).

Data Availability Statement: The datasets presented in this study can be found in online repositories. The second-generation sequencing data of Dongfudou 3 have been submitted to the NCBI Sequence Read Archive (SRA) (accession number: SRR22692844). All the draft genome information of Dongfudou 3 was stored at http://www.wangsui.net.cn/resource/database/public/plant/Glycine_max/Gmax_DFD3_Survey/Gmax_DFD3_Survey.tar.gz, accessed on 24 May 2023.

Conflicts of Interest: The authors declare no conflict of interest.

References

1. Bilyeu, K.; Ratnaparkhe, M.B.; Kole, C. *Genetics, Genomics, and Breeding of Soybean*, 1st ed.; CRC Press: Boca Raton, FL, USA, 2010; pp. 1–2. [CrossRef]
2. Roy, S.; Liu, W.; Nandety, R.S.; Crook, A.; Mysore, K.S.; Pislariu, C.I.; Frugoli, J.; Dickstein, R.; Udvardi, M.K. Celebrating 20 years of genetic discoveries in legume nodulation and symbiotic nitrogen fixation. *Plant Cell* **2020**, *32*, 15–41. [CrossRef] [PubMed]
3. Han, Q.; Ma, Q.; Chen, Y.; Tian, B.; Xu, L.; Bai, Y.; Chen, W.-F.; Li, X. Variation in rhizosphere microbial communities and its asso-ciation with the symbiotic efficiency of rhizobia in soybean. *ISME J.* **2020**, *14*, 1915–1928. [CrossRef] [PubMed]
4. Novikova, L.; Seferova, I.; Matvienko, I.; Shchedrina, Z.; Vishnyakova, M. Photoperiod and temperature sensitivity in early soybean accessions from the vir collection in leningrad province of the Russian Federation. *Turk. J. Agric. For.* **2022**, *46*, 947–954. [CrossRef]
5. Lin, X.-Y.; Liu, B.-H.; Weller, J.L.; Abe, J.; Kong, F.-J. Molecular mechanisms for the photoperiodic regulation of flowering in soybean. *J. Integr. Plant Biol.* **2021**, *63*, 981–994. [CrossRef] [PubMed]
6. Li, Y.-H.; Qin, C.; Wang, L.; Jiao, C.-Z.; Hong, H.-L.; Tian, Y.; Li, Y.-F.; Xing, G.-N.; Wang, J.; Gu, Y.-Z.; et al. Genome-wide signatures of the geographic expansion and breeding of soybean. *Sci. China Life Sci.* **2023**, *66*, 350–365. [CrossRef] [PubMed]
7. Schmutz, J.; Cannon, S.B.; Schlueter, J.; Ma, J.-X.; Mitros, T.; Nelson, W.; Hyten, D.L.; Song, Q.-J.; Thelen, J.J.; Cheng, J.-L.; et al. Genome sequence of the palaeopolyploid soybean. *Nature* **2010**, *463*, 178–183. [CrossRef] [PubMed]
8. Shen, Y.-T.; Liu, J.; Geng, H.-Y.; Zhang, J.-X.; Liu, Y.-C.; Zhang, H.-K.; Xing, S.-L.; Du, J.-C.; Ma, S.-S.; Tian, Z.-X. De novo assembly of a Chinese soybean genome. *Sci. China Life Sci.* **2018**, *61*, 871–884. [CrossRef]
9. Shen, Y.-T.; Du, H.-L.; Liu, Y.-C.; Ni, L.-B.; Wang, Z.; Liang, C.-Z.; Tian, Z.-X. Update soybean Zhonghuang 13 genome to a golden reference. *Sci. China Life Sci.* **2019**, *62*, 1257–1260. [CrossRef]
10. Liu, Y.-C.; Du, H.-L.; Li, P.-C.; Shen, Y.-T.; Peng, H.; Liu, S.-L.; Zhou, G.-A.; Zhang, H.-K.; Liu, Z.; Shi, M.; et al. Pan-genome of wild and cultivated soybeans. *Cell* **2020**, *182*, 162–176. [CrossRef]
11. Lang, D.-D.; Zhang, S.-L.; Ren, P.-P.; Liang, F.; Sun, Z.-Y.; Meng, G.-L.; Tan, Y.-T.; Li, X.-K.; Lai, Q.-H.; Han, L.-L.; et al. Comparison of the two up-to-date sequencing technologies for genome assembly: HiFi reads of Pacific Biosciences Sequel II system and ultralong reads of Oxford Nanopore. *Gigascience* **2020**, *9*, 1–7. [CrossRef]
12. Davies, C.S.; Nielsen, S.S.; Nielsen, N.C. Flavor improvement of soybean preparations by genetic removal of lipoxygenase-2. *J. Am. Oil Chem. Soc.* **1987**, *64*, 1428–1433. [CrossRef]
13. Wang, J.; Kuang, H.-Q.; Zhang, Z.-H.; Yang, Y.-Q.; Yan, L.; Zhang, M.-C.; Song, S.-K.; Guan, Y.-F. Generation of seed lipoxygenase-free soybean using CRISPR-Cas9. *Crop J.* **2020**, *8*, 432–439. [CrossRef]
14. Doležel, J.; Bartoš, J. Plant DNA flow cytometry and estimation of nuclear genome size. *Ann. Bot.* **2005**, *95*, 99–110. [CrossRef]
15. Doležel, J.; Greilhuber, J.; Lucretti, S.; Meister, A.; Lysák, M.A.; Nardi, L.; Obermayer, R. Plant genome size estimation by flow cytometry: Inter-laboratory comparison. *Ann. Bot.* **1998**, *82*, 17–26. [CrossRef]
16. Liu, B.-H.; Shi, Y.-J.; Yuan, J.-Y.; Hu, X.-S.; Zhang, H.; Li, N.; Li, Z.-Y.; Chen, Y.-X.; Mu, D.-S.; Fan, W. Estimation of genomic characteristics by analyzing *k*-mer frequency in de novo genome projects. *arXiv* **2013**, arXiv:1308.2012. [CrossRef]
17. Lutz, K.A.; Wang, W.-Q.; Zdepski, A.; Michael, T.P. Isolation and analysis of high quality nuclear DNA with reduced organellar DNA for plant genome sequencing and resequencing. *BMC Biotechnol.* **2011**, *11*, 54. [CrossRef]
18. Vurture, G.W.; Sedlazeck, F.J.; Nattestad, M.; Underwood, C.J.; Fang, H.; Gurtowski, J.; Schatz, M.C. GenomeScope: Fast reference-free genome profiling from short reads. *Bioinformatics* **2017**, *33*, 2202–2204. [CrossRef]
19. Daniell, H.; Lin, C.-S.; Yu, M.; Chang, W.-J. Chloroplast genomes: Diversity, evolution, and applications in genetic engineering. *Genome Biol.* **2016**, *17*, 1–29. [CrossRef]

20. Chevigny, N.; Schatz-Daas, D.; Lotfi, F.; Gualberto, J.M. DNA repair and the stability of the plant mitochondrial genome. *Int. J. Mol. Sci.* **2020**, *21*, 328. [CrossRef]
21. Martin, W. Gene transfer from organelles to the nucleus: Frequent and in big chunks. *Proc. Natl. Acad. Sci. USA* **2003**, *100*, 8612–8614. [CrossRef]
22. Shih, P.M.; Matzke, N.J. Primary endosymbiosis events date to the later Proterozoic with cross-calibrated phylogenetic dating of duplicated ATPase proteins. *Proc. Natl. Acad. Sci. USA* **2013**, *110*, 12355–12360. [CrossRef]
23. Hildebrand, F.; Meyer, A.; Eyre-Walker, A. Evidence of selection upon genomic GC-Content in bacteria. *PLoS Genet.* **2010**, *6*, e1001107. [CrossRef]
24. Singh, R.; Ming, R.; Yu, Q. Comparative Analysis of GC content variations in plant genomes. *Trop. Plant Biol.* **2016**, *9*, 136–149. [CrossRef]
25. Wang, S.; Chen, S.; Liu, C.-X.; Liu, Y.; Zhao, X.-Y.; Yang, C.-P.; Qu, G.-Z. Genome survey sequencing of *Betula platyphylla*. *Forests* **2019**, *10*, 826. [CrossRef]
26. Almeida Lopes, K.B.; Carpentieri Pipolo, V.; Oro, T.H.; Stefani Pagliosa, E.; Degrassi, G. Culturable endophytic bacterial communities associated with field-grown soybean. *J. Appl. Microbiol.* **2016**, *120*, 740–755. [CrossRef]
27. Hung, P.Q.; Kumar, S.M.; Govindsamy, V.; Annapurna, K. Isolation and characterization of endophytic bacteria from wild and cultivated soybean varieties. *Biol. Fertil. Soils* **2007**, *44*, 155–162. [CrossRef]
28. Liao, S.-M.; Du, Q.-S.; Meng, J.-Z.; Pang, Z.-W.; Huang, R.-B. The multiple roles of histidine in protein interactions. *Chem. Cent. J.* **2013**, *7*, 44. [CrossRef] [PubMed]
29. Boyington, J.C.; Gaffney, B.J.; Amzel, L.M. The three-dimensional structure of an arachidonic acid 15-lipoxygenase. *Science* **1993**, *260*, 1482–1486. [CrossRef]
30. Steczko, J.; Donoho, G.P.; Clemens, J.C.; Dixon, J.E.; Axelrod, B. Conserved histidine residues in soybean lipoxygenase: Functional consequences of their replacement. *Biochemistry* **1992**, *31*, 4053. [CrossRef]
31. Steczko, J.; Axelrod, B. Identification of the iron-binding histidine residues in soybean lipoxygenase L-1. *Biochem. Biophys. Res. Commun.* **1992**, *186*, 686–689. [CrossRef]
32. Wang, W.H.; Takano, T.; Shibata, D.; Kitamura, K.; Takeda, G. Molecular basis of a null mutation in soybean lipoxygenase 2: Substitution of glutamine for an iron-ligand histidine. *Proc. Natl. Acad. Sci. USA* **1994**, *91*, 5828–5832. [CrossRef] [PubMed]
33. Chikhi, R.; Medvedev, P. Informed and automated *k*-mer size selection for genome assembly. *Bioinformatics* **2014**, *30*, 31–37. [CrossRef] [PubMed]
34. Alonge, M.; Lebeigle, L.; Kirsche, M.; Jenike, K.; Ou, S.; Aganezov, S.; Wang, X.-G.; Lippman, Z.B.; Schatz, M.C.; Soyk, S. Automated assembly scaffolding using RagTag elevates a new tomato system for high-throughput genome editing. *Genome Biol.* **2022**, *23*, 258. [CrossRef] [PubMed]
35. Shumate, A.; Salzberg, S.L. Liftoff: Accurate mapping of gene annotations. *Bioinformatics* **2021**, *37*, 1639–1643. [CrossRef] [PubMed]
36. Aramaki, T.; Blanc-Mathieu, R.; Endo, H.; Ohkubo, K.; Kanehisa, M.; Goto, S.; Ogata, H. KofamKOALA: KEGG Ortholog assignment based on profile HMM and adaptive score threshold. *Bioinformatics* **2020**, *36*, 2251–2252. [CrossRef] [PubMed]
37. Gurevich, A.; Saveliev, V.; Vyahhi, N.; Tesler, G. QUAST: Quality assessment tool for genome assemblies. *Bioinformatics* **2013**, *29*, 1072–1075. [CrossRef] [PubMed]
38. Manni, M.; Berkeley, M.R.; Seppey, M.; Simão, F.A.; Zdobnov, E.M. BUSCO update: Novel and streamlined workflows along with broader and deeper phylogenetic coverage for scoring of eukaryotic, prokaryotic, and viral genomes. *Mol. Biol. Evol.* **2021**, *38*, 4647–4654. [CrossRef]
39. Zhang, J.; Ng, C.; Jiang, Y.; Wang, X.-X.; Wang, S.-D.; Wang, S. Genome-wide identification and analysis of LOX genes in soybean cultivar "Zhonghuang 13". *Front. Genet.* **2022**, *13*, 1020554. [CrossRef]
40. Livak, K.J.; Schmittgen, T.D. Analysis of relative gene expression data using real-time quantitative PCR and the $2^{-\Delta\Delta C_T}$ method. *Methods* **2001**, *25*, 402–408. [CrossRef]
41. Narvel, J.M.; Fehr, W.R.; Weldon, L.C. Analysis of soybean seed lipoxygenases. *Crop Sci.* **2000**, *40*, 838–840. [CrossRef]
42. Suda, I.; Hajika, M.; Nishiba, Y.; Furuta, S.; Igita, K. Simple and rapid method for the selective detection of individual lipoxygenase isoenzymes in soybean seeds. *J. Agric. Food. Chem.* **1995**, *43*, 742–747. [CrossRef]
43. Jumper, J.; Evans, R.; Pritzel, A.; Green, T.; Figurnov, M.; Ronneberger, O.; Tunyasuvunakool, K.; Bates, R.; Žídek, A.; Potapenko, A.; et al. Highly accurate protein structure prediction with AlphaFold. *Nature* **2021**, *596*, 583–589. [CrossRef]

plants

Article

MBD3 Regulates Male Germ Cell Division and Sperm Fertility in *Arabidopsis thaliana*

Jia Shu [1], Xiaochang Yin [2], Yannan Liu [3], Yingjie Mi [2], Bin Zhang [1], Aoyuan Zhang [1], Hongbo Guo [4,*] and Juane Dong [1,*]

1 College of Life Sciences, Northwest A&F University, Yangling 712100, China
2 School of Advanced Agricultural Sciences, Peking University, Beijing 100871, China
3 Academy for Advanced Interdisciplinary Studies, Peking University, Beijing 100871, China
4 College of Chemistry & Pharmacy, Northwest A&F University, Yangling 712100, China
* Correspondence: hbguo@nwsuaf.edu.cn (H.G.); dzsys@nwsuaf.edu.cn (J.D.)

Abstract: DNA methylation plays important roles through the methyl-CpG-binding domain (MBD) to realize epigenetic modifications. Thirteen AtMBD proteins have been identified from the *Arabidopsis thaliana* genome, but the functions of some members are unclear. AtMBD3 was found to be highly expressed in pollen and seeds and it preferably binds methylated CG, CHG, and unmethylated DNA sequences. Then, two mutant alleles at the AtMBD3 locus were obtained in order to further explore its function using CRISPR/Cas9. When compared with 92.17% mature pollen production in the wild type, significantly lower percentages of 84.31% and 78.91% were observed in the *mbd3-1* and *mbd3-2* mutants, respectively. About 16–21% of pollen from the *mbd3* mutants suffered a collapse in reproductive transmission, whereas the other pollen was found to be normal. After pollination, about 16% and 24% of *mbd3-1* and *mbd3-2* mutant seeds underwent early or late abortion, respectively. Among all the late abortion seeds in *mbd3-2* plants, 25% of the abnormal seeds were at the globular stage, 31.25% were at the transition stage, and 43.75% were at the heart stage. A transcriptome analysis of the seeds found 950 upregulated genes and 1128 downregulated genes between wild type and *mbd3-2* mutants. Some transcriptional factors involved in embryo development were selected to be expressed, and we found significant differences between wild type and *mbd3* mutants, such as *WOXs*, *CUC1*, *AIB4*, and *RGL3*. Furthermore, we found a gene that is specifically expressed in pollen, named *PBL6*. PBL6 was found to directly interact with AtMBD3. Our results provide insights into the function of AtMBD3 in plants, especially in sperm fertility.

Keywords: MBD protein; DNA methylation; pollen; embryo development; male germ cell division

Citation: Shu, J.; Yin, X.; Liu, Y.; Mi, Y.; Zhang, B.; Zhang, A.; Guo, H.; Dong, J. MBD3 Regulates Male Germ Cell Division and Sperm Fertility in *Arabidopsis thaliana*. *Plants* **2023**, *12*, 2654. https://doi.org/10.3390/plants12142654

Academic Editor: Attila Fehér

Received: 31 May 2023
Revised: 3 July 2023
Accepted: 7 July 2023
Published: 15 July 2023

1. Introduction

DNA methylation is an older epigenetic modification in eukaryotes that plays many important roles in the regulation of gene expression and embryo development [1]. It is essential for mammalian embryo development, as well as in plants [2]. DNA methylation can occur in all sequence contexts in plants, including CG, CHG (which are symmetrical sites), and CHH (H = A, T, or C), which is considered asymmetrical. Two *met1b* null mutants of OsMET1b in rice showed abnormal seed phenotypes [3]. The *Oscmt3a* mutants exhibited a severe decrease in CHG methylation and pleiotropic developmental abnormalities [4]. Double mutants of maize (*Zmet2*/*Zmet5* and *Chr101*/*Chr106*) result in seed problems [5].

Plant DNA methylation mainly occurs at transposons and repetitive DNA elements [6,7]. Its process involves writer, reader, and editor proteins. The reader proteins can specifically bind methylated CpG dinucleotides through methyl-CpG-binding domains (MBDs) or SET- and ring-finger-associated domains [8–10]. MBDs are the earliest proteins found to specifically read and bind methylated sequences, playing important roles in the regulation of DNA methylation, chromatin remodeling, and histone modification [11,12].

The MBD protein family includes 13 members in *Arabidopsis thaliana*, 17 in rice, 14 in *Populus*, 14 in maize, and 6 in wheat [13]. In *A. thaliana*, only AtMBD5, AtMBD6, and AtMBD7 have been found to specifically bind to methylated CG sites, whereas AtMBD4 binds to unmethylated DNA [14,15]. *AtMBD4* negatively regulates the phosphate starvation response, altering root morphology [16]. *AtMBD5* performs an important function in maintaining chromatin structure and mitosis [17]. The *AtMBD7* gene is associated with active DNA demethylation and transcriptional gene silencing [18,19]. Other *AtMBDs*, such as *AtMBD8*, result in a delay in flowering time during both long and short days [20]. The mutation of *AtMBD9* results in a significantly earlier flowering than that of wild-type plants [21]. *AtMBD11* displays a variety of phenotypic effects, including abnormal flower position, fertility problems, and late flowering [22]. However, to this day, the functions of the other *AtMBD* members are poorly understood.

In mammals, the *MBD3* knockout is embryonically lethal in mice [23], suggesting the decisive role of *mbd3* in early embryonic development and late embryonic stages. To further explore the function of *AtMBD3* in plants, especially throughout embryo development, CRISPR/Cas9 technology was used to form the *mbd3* mutants of *A. thaliana* and to obtain developmental phenotypes. These mutants will help us understand the regulation of gene expression, particularly in relation to essential and cell cycle genes from the gametophytic to sporophytic phases. Our results provide insights into the function of MBD3 in pollen and embryo development.

2. Results

2.1. AtMBD3 Was Highly Expressed in Pollen and Seeds

A high degree of amino acid sequence conservation was found among MBD1-13 of *A. thaliana*. The distinct expression profiles of ten MBD proteins were presented from the *A. thaliana* eFP Browser, and the results showed that these proteins had different expression patterns but similar evolutionary characteristics (Figure S1).

A quantitative RT-PCR analysis was performed to evaluate the gene expression of 13 *AtMBDs* (Table S1). Except for *AtMBD4* and *AtMBD11*, most of the *AtMBDs* were highly expressed in seeds. *AtMBD3* was also found to be highly expressed in pollen among all the 13 members (Figure 1).

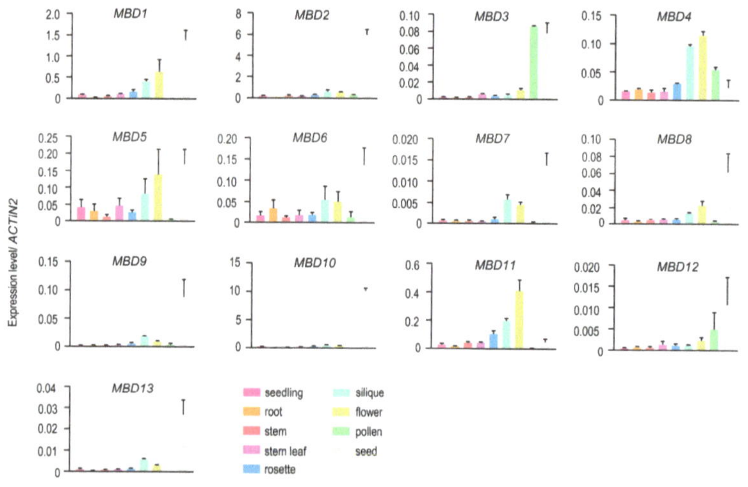

Figure 1. Gene expression of *MBDs* in *Arabidopsis thaliana*. The expression of 13 *MBDs* in 9 tissues of wild-type Col-0 were determined using qPCR. *ACTIN2* was used as an internal control. The error bars represent standard deviations.

2.2. AtMBD3 Preferred to Bind Symmetric Methylated and Unmethylated Sequences

EMSA (electrophoretic mobility shift assays) was used to detect the binding potential of fusion protein GST-MBD3 and DNA sequences in vitro. The result showed that AtMBD3 had a strong binding preference to symmetrically methylated CG and CHG DNA sequences, and to unmethylated DNA sequences, except for methylated CHH (Figure 2). Although AtMBD3 could bind unmethylated DNA sequences, its binding ability was weaker than that of methylated CG and CHG. The EMSA result of fusion protein MBP-SUVH1 was used to test the labeled DNA sequence (Figure S2, Table S2).

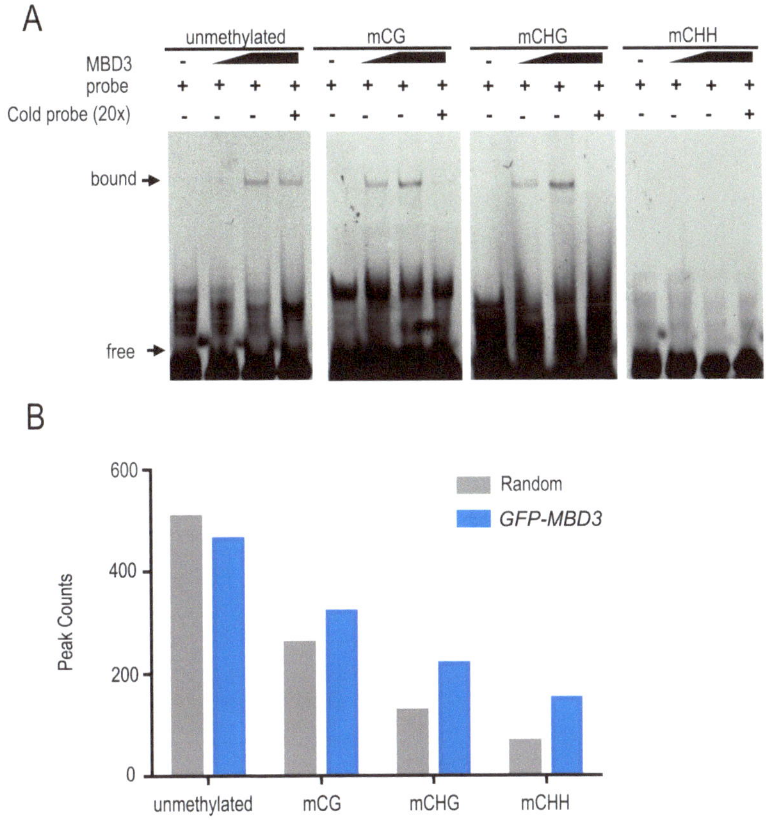

Figure 2. The binding capacity of MBD3 with different mCG, mCHG, mCHH, and unmethylated sequences. (**A**) EMSA results showed that MBD3 protein had the priority to bind mCG, mCHG, and unmethylated sequences. 3′-Fam-labeled substrates (160 nM) were incubated with concentrations of MBD3. The protein concentrations in lanes 1, 2, 3, and 4 were 0, 0.8, 1.6, and 1.6 μM, respectively. For competition analysis, 32 μM of substrates with or without methylation labeling were used as cold probes and incubated with GST-MBD3; and of 3′-Fam-labeled substrates in every fourth lane. For EMSA, at least three independent biological replicates were performed. "Bound" represents the binding band, and "free" represents the free probe. (**B**) ChIP-seq assay was carried out with transgenic *pJIM(Bar)-GFP-linker::MBD3*/Col-0 seedlings, showing the binding potential of AtMBD3 with different methylated regions. The gray column "random" was the control, representing the binding regions in Col-0. The blue columns "GFP-MBD3"represent the binding regions in MBD3.

To further determine the binding potential of MBD3 *in vivo*, ChIP assay was performed by expressing *pJIM(Bar)-GFP:MBD3/Col-0* in seedlings. Then, a total of 711 peaks were obtained (Table S3), among which 687 had a higher DNA methylation in transgenic plants than in Col-0. A total of 320 peaks had an overlap with mCG methylation regions, 218 pea overlapped with mCHG regions, and 149 peaks overlapped with mCHH regions. This difference also confirmed the binding preference of the MBD3 protein.

2.3. *MBD3 Is Important for Plant Embryo Development*

We designed two sgRNAs that target the MBD domain of *MBD3* by using CRISPR/Cas9, and obtained two homozygous transgenic plants, named *mbd3-1* and *mbd3-2* (Figure S3A). Both mutants were found to affect embryo development. In *mbd3-1* (n = 1068) and *mbd3-2* (n = 1144) siliques, 5.71% and 5.77% of seeds underwent early abortion in plants, respectively. A total of 10.86% and 18.88% of seeds in the *mbd3-1* and *mbd3-2* plants were wrinkled, respectively (Figure 3).

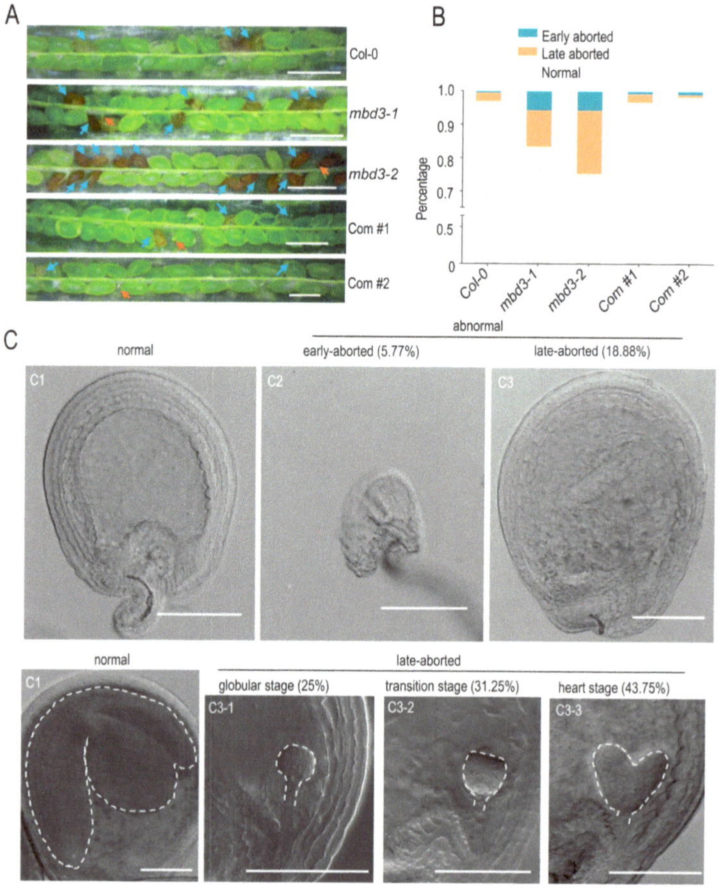

Figure 3. Phenotype of *mbd3* mutant 9 days after pollination. (**A**) Seeds set of Col-0, *mbd3-1*, *mbd3-2*, and complementation lines Com #1 and Com #2. Com #1 and Com #2 represent two individual *proMBD3::MBD3/mbd3-2* transgenic plants. Ovules with development problems are indicated by arrows. "Blue " and "yellow " arrows represent early- and late-aborted seeds, respectively. Bars = 1 mm. (**B**) Seed abortion rates of Col-0 (n = 701), *mbd3-1* (n = 1068), *mbd3-2* (n = 1144), Com #1 (n = 1213), and Com #2 (n = 1156) plants. (**C**) Differential interference contrast (DIC) microscopy of whole-mount-cleared ovules showed abnormal cell division. (**C1–C3**) Embryo patterns and percentages of normal and aborted embryos in *mbd3-2* siliques (n = 96) are shown. (**C3-1–C3-3**) Three stages of late-aborted embryos in *mbd3-2*. Bars = 100 µm.

When the normal embryos in *mbd3-2* (n = 96) were at the cotyledon stage, 25% of abnormal seeds were at the globular stages, 31.25% at the transition stage, and 43.75% at the heart stage. This means that the embryogenesis defects in *mbd3* plants lead to a delayed embryo development, and that these abnormal seed developments could be divided into three stages.

2.4. MBD3 Mutation Impairs Pollen Development

To understand why these seeds underwent abortion, we first examined the anther development in Col-0 and *mbd3* mutant plants by using cryo-SEM. The anther and pollen morphologies were not aberrant obviously (Figure S3B). The pollen viability was evaluated by Alexander's staining, and it was found to produce no effects (Figure S3C). The in vitro germination of pollen was also examined. The germination time, pollen length or quantity of *mbd3* mutant were all found to have no significant difference from Col-0 (Figure S3D).

To further evaluate the effect of *mbd3* mutation on pollen development. Mature pollens from Col-0 and *mbd3* plants were stained with DAPI, and observed with fluorescence microscopy. About 16~21% of pollen from *mbd3-1* and *mbd3-2* plants all suffered a collapse in reproductive transmission. Compared with 92.17% of mature pollen production in Col-0, significantly lower percentages of 84.31% and 78.91% were found in *mbd3-1* and *mbd3-2* plants, respectively. Of them, about 1.61% and 3.45% contained one vegetative nucleus, lacking condensed sperm cells. Approximately 9.82% and 12.42% contained one vegetative nucleus and one condensed sperm cells. Approximately 4.25% and 5.17% did not contain any nuclei. The percentage of two complementary lines, Com #1 and Com #2, approached that of Col-0 (Figure 4).

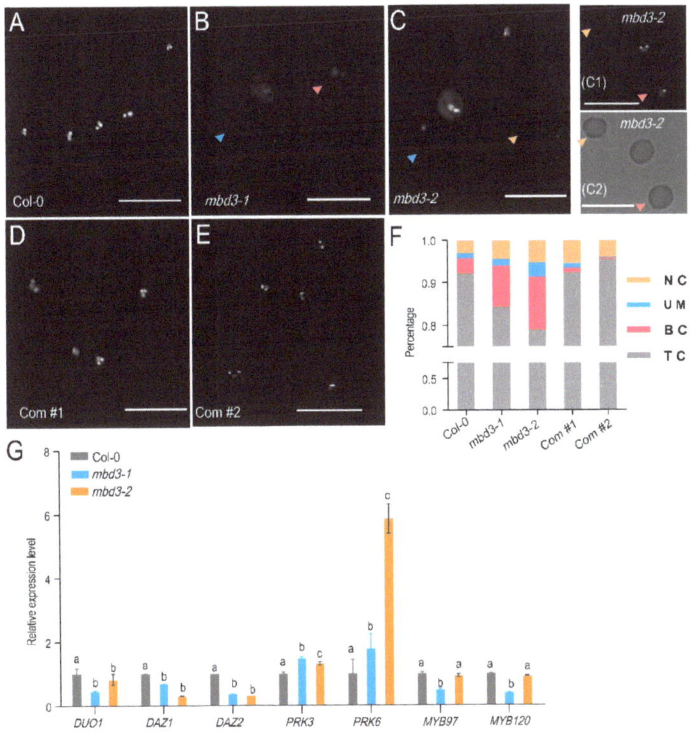

Figure 4. Mutation of *MBD3* impairs pollen development. (**A–E**) DAPI staining of mature pollen from Col-0, *mbd3-1*, *mbd3-2*, Com #1, and Com #2 plants. Different arrows represent abnormal pollen. Red arrows represent binuclear pollen; blue arrows represent mononuclear pollen; yellow arrows

represent nonnuclear pollen. Bright area of (**C1**) was zoomed in (**C2**). Bars = 50 μm. (**F**) Phenotypical percentages in uninucleate microspores (UM), binucleate (BC), trinucleate (TC), and nonnucleate (NC) pollen of Col-0, *mbd3-1*, *mbd3-2*, Com #1, and Com #2 plants ($n > 300$, for each genotype). (**G**) Relative gene expression levels with significant differences in the mature pollen of Col-0 and *mbd3* plants. *ACTIN2* was used as an internal control. Two independent experiments were conducted with similar results. Data from one experiment with four technical replicates. Error bars indicate the mean ± SD; a, b, and c are significant differences ($p < 0.05$, one-way ANOVA).

A quantitative RT-PCR was performed to dissect the gene expression involved in the male gametophyte development of *mbd3*. The expression of seven genes, including male germline-specific R2R3 MYB transcription factor *DUO1*; two *DUO1* target genes *DUO1-ACTIVATED ZINC FINGER1/2 (DAZ1/2)*; *POLLEN RECEPTOR-LIKE KINASE3 (PRK3)*; *PRK6*; and MYB transcription factors MYB97 and MYB120 showed a distinct pattern in *mbd3-1* and *mbd3-2* (Figure 4G). Meanwhile, *duo1* mutant and *daz1/daz2* double-mutant were found to block generative cell division, producing high percentages of binuclear pollen.

2.5. *mbd3 Embryo Development Delay Caused by Asymmetric Divisions*

The RNA-seq results analysis of embryos between Col-0 and *mbd3-2* mutant was performed, in which 950 upregulated genes and 1128 downregulated genes were found (Figure 5; Table S4). Some transcriptional factors, which were selected, were reported to be involved in embryonic development (Table S5), such as *CUC1 (CUP-SHAPED COTYLEDON1)* [24], *WOX2 (WUSCHEL-related homeobox2)*, *WOX8/STIMPY-LIKE/STPL* [25], *PID1 (PINOID)* [26], *ABI4* [27], *RGE1/ZOU (RETARDED GROWTH OF EMBRYO1/ZHOUPI)* [28], and *AGL67* [29]. The expression pattern of these involved genes indicated that *MBD3* may play an important role in embryo development.

2.6. *Interaction Protein Screening of AtMBD3*

Y2H and IP-MS were used to screen the interaction proteins of MBD3. BD-MBD3 plasmid was used as bait, while the prey library was the *A. thalinana* cDNA library. The resulting progenies were selected on SD/-Leu/-Trp/-His/-Ade plates by adding β-galactosidase to eliminate false positives. A total of 69 positive progenies were obtained (Table S6). PCR was adopted to verify these proteins. Through sequencing and blasting the 69 verified proteins, 12 repeatedly identified positive proteins were selected and listed in Table 1. Finally, only the PBL6 (PBS1-like 6, At2G28590) was found to interact with MBD3 (Figures 6 and S4). PBL6 was specifically expressed in pollen, which was confirmed with the IP-MS of GFP-tagged PBL6 transgenic plants (Table 2).

Table 1. Positive proteins selected to screen the interaction with MBD3.

Frequency	ID	Name	Description
13	AT4G25110	ATMC2	Encodes a type I metacaspase.
11	AT2G21620	RD2	Encodes a gene that is induced in response to desiccation
4	AT5G20740		Plant invertase/pectin methylesterase inhibitor superfamily protein
4	AT3G51780	BAG4	Plant homologs of mammalian regulators of apoptosis
3	AT2G28590	PBL6	Protein kinase superfamily protein
2	AT3G18810	PERK6	Encodes a member of the PERK family
2	AT4G39960	DJA5	Molecular chaperone Hsp40/DnaJ family protein
2	AT1G75690	LQY1	DnaJ/Hsp40 cysteine-rich domain superfamily protein
2	AT1G02410	COX11	Encodes a member of the cytochrome c oxidase 11 protein family
2	AT4G17510	UCH3	Ubiquitin C-terminal hydrolase 3
2	AT5G20890	CCT2	TCP-1/cpn60 chaperonin family protein
2	AT4G11260	EDM1	Functions in plant-disease-resistance signaling

'Frequency' indicates the number of positive reactions that the protein was identified through RT-PCR.

Table 2. Identification of co-purified proteins of GFP-tagged PBL6.

ID	Gene	Description	Unique Peptides
AT2G28590.1	PBL6	Protein kinase superfamily protein	40
AT4G00416.1	MBD3	Methyl-CPG-binding domain 3	7
AT5G59380.1	MBD6	Methyl-CPG-binding domain 6	5
AT3G46580.1	MBD5	Methyl-CPG-binding domain 5	2
AT5G20890.1	CCT2	TCP-1/cpn60 chaperonin family protein	1

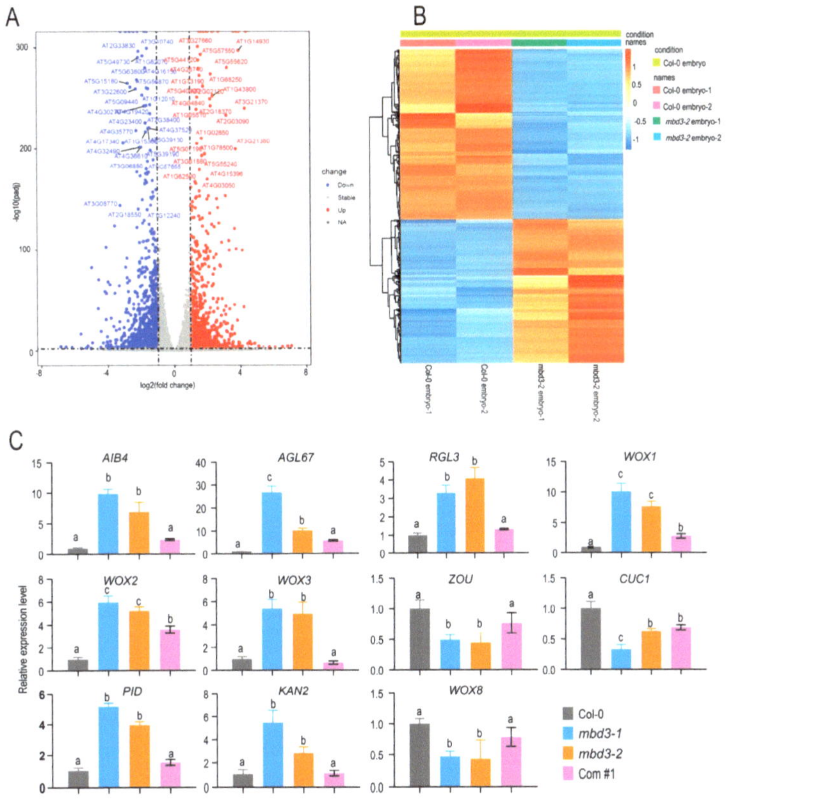

Figure 5. *MBD3* is important for embryogenesis in *A. thaliana*. (**A**) The upregulated and downregulated genes in seeds of Col-0 and *mbd3-2* mutant plants. (**B**) Heat map showing the different expression patterns between Col-0 and *mbd3-2* mutant plants. (**C**) The relative expression level of different transcription factors regulated embryo development by using RT-qPCR analysis. *Tublin8* was used as internal control, given that the expression of *Actins* changed in *mbd3-2*. Two independent experiments were conducted and obtained similar results. Data from one experiment with four technical replicates. Error bars indicate the mean ± SD; a, b, and c are significant differences ($p < 0.05$, one-way ANOVA).

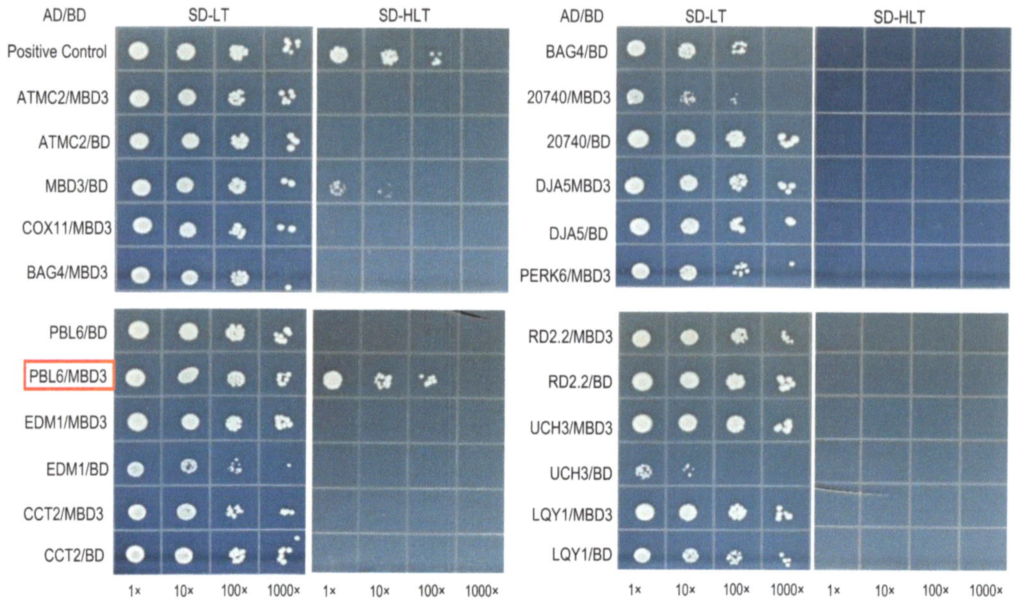

Figure 6. Positive proteins selected to screen interaction with MBD3 by using Y2H. Detailed information on proteins listed in Table 1. Proteins that interacted with AD or BD were negative controls to detect self-activated. The red rectangle is the positive result. The empty wells on the images signify no direct interaction.

3. Discussion

The MBD protein family in plants has been known to act as both "readers" and "erasers" of DNA methylation [30], but the detailed functions of its members are poorly understood. Only a few of them have been reported, and most remain to be characterized. A previous study failed to amplify the RT-PCR products of *AtMBD3* due to its low expression level [22]. In this study, we successfully obtained *mbd3*-mutant plants and found that, despite a low expression level of *AtMBD3,* it played important roles in regulating the plant's reproductive development. Similarly, *mbd3* knockout in mice proved to be embryonically lethal [23]. Furthermore, we verified the binding capability of methylated and unmethylated DNA sequences of MBD3 protein both in vitro and in vivo.

3.1. MBD3 Have the Ability to Bind Sequences More than Methylated DNA

Zemach and Grafi first demonstrated that AtMBD5, AtMBD6, and AtMBD7 can bind the methylated sequences of CG dinucleotides in vitro [14]. Despite its high homology, AtMBD5 can bind both symmetrically methylated CG and asymmetrically methylated CHH sequences, whereas AtMBD6 can only bind methylated CG complexes [31]. Mammalian MBD3 was found not to be able to bind mCG [32]. Our results of AtMBD3, which has a high level of amino acid homology with mammalian MBDs, showed a capability to bind both symmetrically methylated and unmethylated DNA sequences. This result was firstly confirmed both in vitro and in vivo. The capability is more prominent in symmetrically methylated sequences. The ability of MBD5 to bind to mCG may be due to the interaction between conserved 5mCpG residues and residue pairs of R-D and R-E. However, in AtMBD3, residue pairs were changed into R-V and K-E pairs [33]. These substitutions may allow AtMBD3 to not only bind symmetrically methylated DNA sequences, but also guide AtMBD3 to target-specific DNA sequences.

3.2. MBD3 Participates in Embryo Development with Other Genes

DNA methylation affecting plant development has been found in many species. The mutation of *OsMET1b* leads to abnormal seed phenotypes in rice, which is associated with either viviparous germination or early embryonic lethality [3,34]. In this study, *MBD3* was also found to be highly expressed in seeds. At the same time, some transcriptional factor genes involved in embryo development were expressed after transcriptome analysis, including *WOX* and *PID* (Figure 5). *WOXs* have been found to make cell-fate decisions during early embryogenesis. *WOX1*, *WOX2*, and *WOX3* were highly expressed in *mbd3* mutants, whereas *WOX8* was not. Both *WOX2* and *WOX8* are initially co-expressed in egg cells and zygotes, and then specifically expressed in apical and basal cell lineages, respectively, after zygotic division [35]. *PID* regulates *PIN* [36], which is responsible for establishing auxin gradients in early embryogenesis [37]. Except for *WOXs* and *PID*, the other expressed genes also showed clear differences between Col-0 and *mbd3* mutants, such as *CUC1, AIB4, AGL67,* and *ZOU*. Although *AtMBD3* and these genes are significantly regulated by *MBD3* in *mbd3*-mutant plants, but their relationship has not been confirmed in this paper, it is clear that they coordinately participate in embryo development in some specific manner.

3.3. MBD3 Regulates Pollen and Embryo Development with Other Genes

DNA methylation is also important for the fertility of male plants. Many differentially methylated regions have been identified during tomato fruit development [38]. In *Capsella rubella, NRPD1* knockout leads to the interruption of pollen development at the microspore stage [39]. Pollen formation in *A. thaliana* is associated with the reprogramming of CHH methylation in pollen vegetative cells and a locus-specific restoration in sperm [40]. Our investigation found that *AtMBD3* was expressed in pollen, and it obstructed the development of male germ cell division in the pollen of two *mbd3* mutants. The selected genes in this paper were confirmed to be involved in germ cell development, such as *DUO1, DAZ1,* and *DAZ2,* and also found to be downregulated in the pollen of *mbd3* mutants. Moreover, *mbd3*, *duo1*, and *daz1*/*daz2* double mutants all have problems undergoing fertilization. Therefore, we suspect that *MBD3* and *DUO1* may coordinately play roles in the same pathway. Although *DUO1* and *DAZ2* have no DNA methylation site in genes or promoter regions [41], MBD3 may have the ability to bind them.

PBL6, a specific expressed gene in pollen, was found to directly interact with MBD3 to form a complex. Both PBL6 and MBD3 were also found to interact with MBD5 and MBD6 (Table 2). It has been reported that MBD5 and MBD6 are important in pollen vegetative cell development [42], which indicates that the interaction of MBD3 and PBL6 is important for pollen development, and may interact with MBD5 and MBD6. We hypothesize that through the methylation DNA sequences' binding ability, MBD3 and PBL6 complex may be recruited into MBD5 and MBD6 binding sites and then impair pollen development. Our study provides direct clues for further research on AtMBD3 and a theoretical basis for a better understanding of AtMBDs family.

4. Materials and Methods

4.1. Plant Materials and Growth Conditions

Arabidopsis thaliana of Columbia-0 (Col-0) ecotype was used as the wild type reference, and all mutant seeds were Col-0 in this study. Seeds were sterilized with 75% ethanol and 5% sodium hypochlorite and then washed three times with 100% ddH_2O. Sterilized seeds were sown on a 1/2 Murashige and Skoog (1/2 MS) solid medium containing 1% sucrose and 0.7% plant agar powder (*w*/*v*). After being wrapped with aluminum foil at 4 °C for 2 d, the plates were grown at 20–22 °C under cycles of 16 h of light and 8 h of dark in growth chambers. The 2-week-old seedlings were harvested for further experiments or transferred into soil.

4.2. Phylogenetic Analysis of MBD Genes

The search for an HMM profile of the MBD domain (PS50982) in 40 representative species, which are in 11 phyla, including *Rhodophyta, Glaucophyta, Chlorophyta, Streptophyta, Phragmoplastophyta, Marchantiophyta, Bryophyta, Lycopodiophyta, Polypodiophyta, Gymnospermae,* and *Angiospermae* (Table S5), was performed by using the InterProScan software. Phylogenetic trees were constructed in IQTREE software by using maximum likelihood and WAG+R5 with 1000 replicates.

4.3. Generation of the CRISPR Allele of mbd3 Mutants

Given the absence of available T-DNA alleles of *mbd3*, the CRISPR/Cas9 system was used to generate the *mbd3* (AT4G00416) mutant, named as *mbd3-1* and *mbd3-2*. The guide for a 20 bp targeting sequence of sgRNAs was selected from the CRISPR-PLANT platform (https://www.genome.arizona.edu/crispr/CRISPRsearch.html) (accessed on 11 March 2018), and cloned into a YAO promoter-driven CRISPR/Cas9 system in *A. thaliana* [43]. The CRISPR/Cas9 construction was transformed into the *Agrobacterium tumefaciens* strain GV3101, and the floral dip method was used to transform *A. thaliana* Col-0 plants [44]. The T1 transformants were selected on hygromycin plates. After sequencing the AT4G00416 gene from these T1 plants, three of them revealed that they had homozygous deletion, which occurred at the tail of the MBD domain.

4.4. ChIP-seq

ChIP-seq experiments were performed by using 2 g of 11-day-old seedlings with biological duplicates on Col-0 and *pJIM(Bar)-GFP-linker::MBD3/Col-0* transgenic lines. The samples were fixed in 1% formaldehyde. Then, chromatin was extracted from fixed tissue and fragmented using a Bioruptor® Pico (Diagenode, Belgium) of 200–500 bp fragments. The sheared chromatin was immunoprecipitated overnight by using the following antibodies: anti-GFP (ab290, abcam, Cambridge, dilution 1:100) and antibody IgG (ab6730). Protein A Sepharose beads CL-4B (GE Healthcare, California) were used to capture immunocomplexes. Protein-A beads were washed before use. Chromatin was eluted and de-crosslinked at 65 °C overnight. DNA from immunoprecipitated chromatin was treated with RNase and proteinase K, and then purified by phenol–chloroform extraction and ethanol precipitation. For ChIP-seq, two independent biological replicates of immunoprecipitations were treated to prepare the next-generation sequencing libraries. The Ovation® Ultralow Library Systems (NuGEN, Shanghai) was used for end repair, A-tailing, and the ligation of Illumina-compatible adapters. The data analysis was performed as previously described [45]. The clean reads were mapped to TAIR10 with Bowtie2 and default parameters after using fastqc, multiqc, and trimmomatic to obtain clean reads. Bigwig files were generated by bamCoverage from deeptools, and RPKM was normalized to remove PCR duplication by PICARD. Lastly, the data quality was checked by an integrative genomics viewer (IGV).

4.5. Electrophoretic Mobility Shift Assay (EMSA)

MBD3 protein was cloned behind GST (glutathione S-transferase tag). MBD3 was amplified and cloned into bacterial expression vector pGEX-4T-1. The control protein in Figure S2 was cloned behind MBP tag. The construction for expression was transformed into *Escherichia coli* strain BL21 and purified with glutathione Sepharose 4B. A 98 °C water bath was performed on single-stranded nucleotide primers labeled with FAM at the 3′ end, which were used for EMSA, for 10 min and then cooled to room temperature away from light to form labeled double-stranded nucleotides primers. The labeled double-stranded nucleotides primers (160 nM) were incubated with purified AtMBD3 protein or with cold competitors (32 μM) in a total volume of 20 μL reactive system, including 4 μL binding buffer (100 mM Tris-HCl (pH 7.6), 50 mM $MgCl_2$, 1% NP40); 5 mM DTT was added before each reaction. Binding reactions were carried out away from light for 20 min, in room temperature, and 1 μL of loading buffer (Beyotime, GS006, Shanghai) was added to each reaction. The protein–substrate complexes were resolved on 6% nondenaturing polyacrylamide gels at 80 V for at least 90 min on ice in a pre-cold 0.5× TBE buffer (44.6 mM Tris, 44.5 mM boric acid, and 1 mM disodium EDTA). After electrophoresis, the gels were detected by using a Typhoon FLA 9500.

4.6. Affinity Purification and Mass Spectrometry

The GST-MBD3 fusion protein was used for the binding assays. A total of 2.5 g WT seedlings was ground into fine powder, suspended in 10 mL lysis buffer (50 mM Tris-HCl (pH = 8.0), 230 mM NaCl, 5 mM $MgCl_2$, 10% glycerin, 0.2% NP-40, 0.5 mM dithiothreitol (DTT), 1 mM phennylmethysulfonyl fluoride (PMSF) and proteinase inhibitor cocktail tablets (Roche, 14696200)) at 4 °C for 30 min with rotation. Then, 150 μL GST-MBD3 fusion protein was added. After incubation at 4 °C for 3 h, 13,500 g centrifugation for 15 min at 4 °C was performed. The supernatant was incubated with the GST fusion protein beads at 4 °C for 3 h. The bound proteins were boiled in SDS loading buffer after elution with lysis buffer three times, and resolved on 15% denaturing poly-acrylamide gels.

Mass spectrometric identification of the affinity-purified proteins was performed as described previously (Table 2) [46]. Protein bands on the SDS-PAGE were de-stained and in-gel digested with sequencing grade trypsin (0.5 ng/μL). Peptides were extracted by HPLC and sprayed into an LTQ Orbitrap Elite System mass spectrometer (Thermo Scientific, Massachusetts). Database search was performed on an in-house Mascot server (Matrix Science Ltd., London, UK) against the IPI (International Protein Index) *A. thaliana* protein database.

4.7. Microscopy Analysis

The morphologies of anther and pollen were observed using cryogenic scanning electron microscopy (cryo-SEM). The dehiscent anthers and pollens were placed on glass slides with double-sided tape, and transferred to the cold stage of chamber under vacuum for sublimation (90 °C, 5 min) after quickly freezing in liquid nitrogen and coated with platinum sputter (10 mA, 30 s) with three repetitions. Finally, the samples were transferred to another cold stage in the scanning electron microscope and imaged. The morphology of anther and pollen was observed under a spinning disc confocal microscope (Zeiss Cell Observer SD; Zeiss, Oberkochen, Germany).

4.8. Alexander Dye and 4′,6-Diamidino-2-Phenylindole Staining

An appropriate amount of Alexander dye or 4′,6-diamidino-2-phenylindole (DAPI) was dropped onto glass slides, and the dehiscent anthers were placed on glass slides with dissecting needles. The pollen were gently and quickly suspended in Alexander dye or DAPI, and then the glass slides were covered. After staining, the pollen were observed under the fluorescence microscope (Olympus BX53, Miyazaki). DAPI staining was carried out in the dark.

4.9. In Vitro Pollen Germination Analysis

The pollen of newly blooming flowers was collected and placed on the surface of a solid pollen germination medium (SPGM) (18% sucrose, 0.01% boric acid, 2 mM $CaCl_2$, 1 mM $Ca(NO_3)_2$, 1 mM KCl, 1 mM $MgSO_4$, adjusted to pH = 7, 1.5% agar). The medium was placed in a humid box at 22 °C for at least 6 h of incubation, and then the pollen germination was observed. The ImageJ software was used to measure and calculate the length of the pollens' tubes and their germination.

4.10. RNA Isolation and Quantitative Real-Time RT-PCR (qPCR)

Total RNA was extracted by using Quick RNA isolation kit (Huayueyang, Beijing) and treated with in-column DNase. About 2 μg mRNA was used to amplify the first-strand cDNA by using the superScriptIII First-Strand Synthesis SuperMix Kit. The RT-qPCR was performed using SYBR Green Mastermix (Bio-Rad). An amount of 1 μL cDNA reaction mixture was used as template after dilution in a 20 μL reaction system. The reaction conditions were as follows: firstly, 95 °C for 10 min; secondly, 95 °C for 15 s and 60 °C for 1 min, this step is repeated 40 times; and thirdly, 95 °C for 15 s and 60 °C for 1 min, 95 °C for dissociation. The calculation of the expression level was used as $2^{-\Delta\Delta Ct}$. Two biological replicates from WT and *mbd3* mutants were performed. All analyses of variance were performed in SPSS software by using t-test of the ANOVA program.

Mature pollen were collected from 2 to 3 bunches of unbloomed floral buds from each plant in the same tray. Nine DAP seeds were collected from about 15~20 mature siliques. The fruit pod was removed by using a dissecting needle, after placing siliques on a clean glass slide. The seeds were collected by using tweezers and put in liquid nitrogen for quick freezing.

4.11. Yeast Two-Hybrid Screening

Both pGADT7 vector containing the GLA4 AD and pGBKT7 vector containing the GLA4 BD for the yeast (*Saccharomyces cerevisiae*) were used for two-hybrid assay. BD-MBD3 was constructed by cloning the full-length sequence of MBD3 into the pGBKT7 vector at the NdeI and EcoR1 sites to be fused in-frame with the sequence encoding the GAL4 DNA-binding domain (BD). The *A. thaliana* cDNA library cloned into the prey vector pGADT7-RecAB was constructed by Clontech. The AtMBD3 interaction proteins were screened by using the yeast two-hybrid system, according to the manufacturer's instructions (Clontech, Matchmaker GAL4 Two-Hybrid System 3 and Libraries User Manual, PT3247-1). The bait plasmid with BD-MBD3 and the prey library DNA were co-transformed into the yeast strain AH109. The resulting progenies were first selected on SD/-Leu/-Trp/-His/-Ade plates, then, β-galactosidase were added for the activity detection to eliminate false positives. Plasmids harboring positive-prey cDNAs were isolated for RT-PCR and sequenced to verify the proteins that interact with AtMBD3.

Supplementary Materials: The following supporting information can be downloaded at: https://www.mdpi.com/article/10.3390/plants12142654/s1, Figure S1: Comparison analysis of MBDs in plantae.; Figure S2: The binding potential of SUVH1 as a positive control.; Figure S3: Deletion of mbd3 has no influence on pollen viability.; Figure S4: Self-activation of MBD3.; Table S1: Primers and sgRNAs used in this study.; Table S2: The oligonucleotide sequences used in this study.; Table S3: ChIP-seq results of MBD3.; Table S4: Genes involved in embryo development. Related to Figure 5; Table S5: Selected genes from RNA-seq results of *mbd3* embryo development. Related to Figure 5; Table S6: Y2H screening results of BD-MBD3.; Table S7: Identification of co-purified proteins of MBD3 by mass spectrometry. GST-MBD3 protein was used for IP/MS.; Table S8: The species involved in phylogenic tree.

Author Contributions: J.S. performed the experiments, analyzed the data, and wrote the manuscript. X.Y. performed the ChIP experiment, Y.L. and Y.M. performed bioinformatics analysis. B.Z. and A.Z. performed some material collections. J.D. and H.G. designed the project and revised the manuscript. All authors have read and agreed to the published version of the manuscript.

Funding: This research was financially supported by the Natural Science Foundation of Shaanxi Province (2022JM-099) and National College Students' Innovation and Entrepreneurship Training Program (202210712227).

Data Availability Statement: RNA-seq and the ChIP-seq data of *mbd3* have been submitted to the NCBI Sequencing Read Archive under accession no. PRJNA936625. The DNA methylome of Col-0 was previously deposited in SRR005412 [47] and PRJNA686693 [41].

Acknowledgments: We thank Qi-Jun Chen for providing the CRISPR/Cas9 system. We thank JiaJun Wang for providing *pJIM(Bar)*: GFP-linker vector.

Conflicts of Interest: The authors declare no conflict of interest.

References

1. Suzuki, M.M.; Bird, A. DNA methylation landscapes: Provocative insights from epigenomics. *Nat. Rev. Genet.* **2008**, *9*, 465–476. [CrossRef]
2. Xiao, W.; Custard, K.D.; Brown, R.C.; Lemmon, B.E.; Harada, J.J.; Goldberg, R.B.; Fischer, R.L. DNA methylation is critical for *Arabidopsis* embryogenesis and seed viability. *Plant Cell* **2006**, *18*, 805–814. [CrossRef] [PubMed]
3. Yamauchi, T.; Johzuka-Hisatomi, Y.; Terada, R.; Nakamura, I.; Iida, S. The MET1b gene encoding a maintenance DNA methyltransferase is indispensable for normal development in rice. *Plant. Mol. Biol.* **2014**, *85*, 219–232. [CrossRef] [PubMed]
4. Cheng, C.; Tarutani, Y.; Miyao, A.; Ito, T.; Yamazaki, M.; Sakai, H.; Fukai, E.; Hirochika, H. Loss of function mutations in the rice chromomethylase OsCMT3a cause a burst of transposition. *Plant J.* **2015**, *83*, 1069–1081. [CrossRef]
5. Li, Q.; Eichten, S.R.; Hermanson, P.J.; Zaunbrecher, V.M.; Song, J.; Wendt, J.; Rosenbaum, H.; Madzima, T.F.; Sloan, A.E.; Huang, J.; et al. Genetic perturbation of the maize methylome. *Plant Cell* **2014**, *26*, 4602–4616. [CrossRef] [PubMed]
6. Zhang, X.; Yazaki, J.; Sundaresan, A.; Cokus, S.; Chan, S.W.L.; Chen, H.; Henderson, I.R.; Shinn, P.; Pellegrini, M.; Jacobsen, S.E.; et al. Genome-wide High-Resolution Mapping and Functional Analysis of DNA Methylation in *Arabidopsis*. *Cell* **2006**, *126*, 1189–1201. [CrossRef]
7. Cokus, S.J.; Feng, S.; Zhang, X.; Chen, Z.; Merriman, B.; Haudenschild, C.D.; Pradhan, S.; Nelson, S.F.; Pellegrini, M.; Jacobsen, S.E. Shotgun bisulphite sequencing of the *Arabidopsis* genome reveals DNA methylation patterning. *Nature* **2008**, *452*, 215–219. [CrossRef] [PubMed]
8. Bogdanović, O.; Veenstra, G.J.C. DNA methylation and methyl-CpG binding proteins: Developmental requirements and function. *Chromosoma* **2009**, *118*, 549–565. [CrossRef]
9. Unoki, M.; Nishidate, T.; Nakamura, Y. ICBP90, an E2F-1 target, recruits HDAC1 and binds to methyl-CpG through its SRA domain. *Oncogene* **2004**, *23*, 7601–7610. [CrossRef]
10. Filion, G.J.P.; Zhenilo, S.; Salozhin, S.; Yamada, D.; Prokhortchouk, E.; Defossez, P.-A. A Family of Human Zinc Finger Proteins That Bind Methylated DNA and Repress Transcription. *Mol. Cell. Biol.* **2006**, *26*, 169–181. [CrossRef]
11. Bird, A.; Nan, X.; Ng, H.-H.; Johnson, C.A.; Laherty, C.D.; Turner, B.M.; Eisenman, R.N. Transcriptional repression by the methyl-CpG-binding protein MeCP2 involves a histone deacetylase complex. *Nature* **1998**, *393*, 386–389. [CrossRef]
12. Wolffe, A.P.; Jones, P.L.; Jan Veenstra, G.C.; Wade, P.A.; Vermaak, D.; Kass, S.U.; Landsberger, N.; Strouboulis, J. Methylated DNA and MeCP2 recruit histone deacetylase to repress transcription. *Nat. Genet.* **1998**, *19*, 187–191. [CrossRef]
13. Springer, N.M.; Kaeppler, S.M. Evolutionary Divergence of Monocot and Dicot Methyl-CpG-Binding Domain Proteins. *Plant Physiol.* **2005**, *138*, 92–104. [CrossRef]
14. Zemach, A.; Gideon, G. Characterization of *Arabidopsis thaliana* methyl-CpG-binding domain (MBD) proteins. *Plant J.* **2003**, *34*, 565–572. [CrossRef] [PubMed]
15. Zemach, A.; Grafi, G. Methyl-CpG-binding domain proteins in plants: Interpreters of DNA methylation. *Trends Plant Sci.* **2007**, *12*, 80–85. [CrossRef]
16. Parida, A.P.; Sharma, A.; Sharma, A.K. *AtMBD4*: A methylated DNA binding protein negatively regulates a subset of phosphate starvation genes. *J. Biosci.* **2019**, *44*, 14. [CrossRef]
17. Yano, A.; Kodama, Y.; Koike, A.; Shinya, T.; Kim, H.-J.; Matsumoto, M.; Ogita, S.; Wada, Y.; Ohad, N.I.R.; Sano, H. Interaction Between Methyl CpG-Binding Protein and Ran GTPase during Cell Division in Tobacco Cultured Cells. *Ann. Bot.* **2006**, *98*, 1179–1187. [CrossRef] [PubMed]
18. Wang, C.; Dong, X.; Jin, D.; Zhao, Y.; Xie, S.; Li, X.; He, X.; Lang, Z.; Lai, J.; Zhu, J.-K.; et al. Methyl-CpG-Binding Domain Protein MBD7 Is Required for Active DNA Demethylation in *Arabidopsis*. *Plant Physiol.* **2015**, *167*, 905–914. [CrossRef] [PubMed]
19. Lang, Z.; Lei, M.; Wang, X.; Tang, K.; Miki, D.; Zhang, H.; Mangrauthia, S.K.; Liu, W.; Nie, W.; Ma, G.; et al. The Methyl-CpG-Binding Protein MBD7 Facilitates Active DNA Demethylation to Limit DNA Hyper-Methylation and Transcriptional Gene Silencing. *Mol. Cell* **2015**, *57*, 971–983. [CrossRef]
20. Stangeland, B.; Rosenhave, E.M.; Winge, P.; Berg, A.; Amundsen, S.S.; Karabeg, M.; Mandal, A.; Bones, A.M.; Grini, P.E.; Aalen, R.B. AtMBD8 is involved in control of flowering time in the C24 ecotype of *Arabidopsis thaliana*. *Physiol. Plant* **2009**, *136*, 110–126. [CrossRef]
21. Peng, M.; Cui, Y.; Bi, Y.M.; Rothstein, S.J. AtMBD9: A protein with a methyl-CpG-binding domain regulates flowering time and shoot branching in *Arabidopsis*. *Plant J.* **2006**, *46*, 282–296. [CrossRef]
22. Berg, A.; Meza, T.J.; Mahic, M.; Thorstensen, T.; Kristiansen, K.; Aalen, R.B. Ten members of the Arabidopsis gene family encoding methyl-CpG-binding domain proteins are transcriptionally active and at least one, *AtMBD11*, is crucial for normal development. *Nucleic Acids Res.* **2003**, *31*, 5291–5304. [CrossRef]
23. Hendrich, B.; Guy, J.; Ramsahoye, B.; Wilson, V.A.; Bird, A. Closely related proteins MBD2 and MBD3 play distinctive but interacting roles in mouse development. *Genes Dev.* **2001**, *15*, 710–723. [CrossRef] [PubMed]
24. Agarwal, P.; Kapoor, S.; Tyagi, A.K. Transcription factors regulating the progression of monocot and dicot seed development. *Bioessays* **2011**, *33*, 189–202. [CrossRef] [PubMed]
25. Abrash, E.B.; Bergmann, D.C. Asymmetric cell divisions: A view from plant development. *Dev. Cell* **2009**, *16*, 783–796. [CrossRef]
26. Benjamins, R.; Quint, A.; Weijers, D.; Hooykaas, P.; Offringa, R. The PINOID protein kinase regulates organ development in *Arabidopsis* by enhancing polar auxin transport. *Development* **2001**, *128*, 4057–4067. [CrossRef]

27. Penfield, S.; Li, Y.; Gilday, A.D.; Graham, S.; Graham, I.A. Arabidopsis ABA INSENSITIVE4 Regulates Lipid Mobilization in the Embryo and Reveals Repression of Seed Germination by the Endosperm. *Plant Cell* **2006**, *18*, 1887–1899. [CrossRef]
28. Yang, S.; Johnston, N.; Talideh, E.; Mitchell, S.; Jeffree, C.; Goodrich, J.; Ingram, G. The endosperm-specific ZHOUPI gene of *Arabidopsis thaliana* regulates endosperm breakdown and embryonic epidermal development. *Development* **2008**, *135*, 3501–3509. [CrossRef]
29. Lehti-Shiu, M.D.; Adamczyk, B.J.; Fernandez, D.E. Expression of MADS-box genes during the embryonic phase in *Arabidopsis*. *Plant Mol. Biol.* **2005**, *58*, 89–107. [CrossRef] [PubMed]
30. Zhang, H.; Lang, Z.; Zhu, J.-K. Dynamics and function of DNA methylation in plants. *Nat. Rev.* **2018**, *19*, 489–506. [CrossRef]
31. Scebba, F.; Bernacchia, G.; De Bastiani, M.; Evangelista, M.; Cantoni, R.M.; Cella, R.; Locci, M.T.; Pitto, L. *Arabidopsis* MBD proteins show different binding specificities and nuclear localization. *Plant Mol. Biol.* **2003**, *53*, 755–771. [CrossRef] [PubMed]
32. Ho, H.L.; McNae, L.W.; Schmiedeberg, L.; Klose, R.J.; Bird, A.; Walkinshaw, M.D. MeCP2 binding to DNA depends upon hydration at methyl-CpG. *Mol. Cell* **2008**, *29*, 525–531. [CrossRef] [PubMed]
33. Ichino, L.; Boone, B.A.; Strauskulage, L.; Harris, C.J.; Kaur, G.; Gladstone, M.A.; Tan, M.; Feng, S.; Jami-Alahmadi, Y.; Duttke, S.H.; et al. MBD5 and MBD6 couple DNA methylation to gene silencing through the J-domain protein SILENZIO. *Science* **2021**, *372*, 1434–1439. [CrossRef]
34. Hu, L.; Li, N.; Xu, C.; Zhong, S.; Lin, X.; Yang, J.; Zhou, T.; Yuliang, A.; Wu, Y.; Chen, Y.-R.; et al. Mutation of a major CG methylase in rice causes genome-wide hypomethylation, dysregulated genome expression, and seedling lethality. *Proc. Natl. Acad. Sci. USA* **2014**, *111*, 10642–10647. [CrossRef] [PubMed]
35. Wu, X.; Chory, J.; Weigel, D. Combinations of WOX activities regulate tissue proliferation during *Arabidopsis* embryonic development. *Dev. Biol.* **2007**, *309*, 306–316. [CrossRef]
36. Christensen, S.K.; Dagenais, N.; Chory, J.; Weigel, D. Regulation of Auxin Response by the Protein Kinase *PINOID*. *Cell* **2000**, *100*, 469–478. [CrossRef]
37. Friml, J.Y.X.; Michniewicz, M.; Weijers, D.; Quint, A.; Tietz, O.; Benjamins, R.; Ouwerkerk, P.B.; Ljung, K.; Sandberg, G.; Hooykaas, P.J.; et al. A PINOID-dependent binary switch in apical-basal PIN polar targeting directs auxin efflux. *Science* **2004**, *306*, 862–865. [CrossRef]
38. Zhong, S.; Fei, Z.; Chen, Y.R.; Zheng, Y.; Huang, M.; Vrebalov, J.; McQuinn, R.; Gapper, N.; Liu, B.; Xiang, J.; et al. Single-base resolution methylomes of tomato fruit development reveal epigenome modifications associated with ripening. *Nat. Biotechnol.* **2013**, *31*, 154–159. [CrossRef]
39. Wang, Z.; Butel, N.; Santos-Gonzalez, J.; Borges, F.; Yi, J.; Martienssen, R.A.; Martinez, G.; Kohler, C. Polymerase IV Plays a Crucial Role in Pollen Development in *Capsella*. *Plant Cell* **2020**, *32*, 950–966. [CrossRef]
40. Walker, J.; Gao, H.; Zhang, J.; Aldridge, B.; Vickers, M.; Higgins, J.D.; Feng, X. Sexual-lineage-specific DNA methylation regulates meiosis in *Arabidopsis*. *Nat. Genet.* **2018**, *50*, 130–137. [CrossRef]
41. Liang, W.; Li, J.; Sun, L.; Liu, Y.; Lan, Z.; Qian, W. Deciphering the synergistic and redundant roles of CG and non-CG DNA methylation in plant development and transposable element silencing. *New Phytol.* **2022**, *233*, 722–737. [CrossRef] [PubMed]
42. Ichino, L.; Picard, C.L.; Yun, J.; Chotai, M.; Wang, S.; Lin, E.K.; Papareddy, R.K.; Xue, Y.; Jacobsen, S.E. Single-nucleus RNA-seq reveals that MBD5, MBD6, and SILENZIO maintain silencing in the vegetative cell of developing pollen. *Cell Rep.* **2022**, *41*, 111699. [CrossRef]
43. Yan, L.; Wei, S.; Wu, Y.; Hu, R.; Li, H.; Yang, W.; Xie, Q. High-efficiency genome editing in *Arabidopsis* using *YAO* promoter-driven CRISPR/Cas9 system. *Mol. Plant* **2015**, *8*, 1820–1823. [CrossRef] [PubMed]
44. Clough, S.J.; Bent, A.F. Floral dip: A simplified method forAgrobacterium-mediated transformation of *Arabidopsis thaliana*: Floral dip transformation of Arabidopsis. *Plant. J. Cell Mol. Biol.* **1998**, *16*, 735–743. [CrossRef] [PubMed]
45. Yang, T.; Wang, D.; Tian, G.; Sun, L.; Yang, M.; Yin, X.; Xiao, J.; Sheng, Y.; Zhu, D.; He, H.; et al. Chromatin remodeling complexes regulate genome architecture in *Arabidopsis*. *Plant Cell* **2022**, *34*, 2638–2651. [CrossRef]
46. Li, Q.; Wang, X.; Sun, H.; Zeng, J.; Cao, Z.; Li, Y.; Qian, W. Regulation of Active DNA Demethylation by a Methyl-CpG-Binding Domain Protein in *Arabidopsis thaliana*. *PLoS Genet.* **2015**, *11*, e1005210. [CrossRef]
47. Stroud, H.; Do, T.; Du, J.; Zhong, X.; Feng, S.; Johnson, L.; Patel, D.J.; Jacobsen, S.E. Non-CG methylation patterns shape the epigenetic landscape in Arabidopsis. *Nat. Struct. Mol. Biol.* **2014**, *21*, 64–72. [CrossRef]

plants

MDPI

Article

Genotypic Variation of Purple Rice in Response to Shading in Yield, Anthocyanin Content, and Gene Expression

Nantapat Danpreedanan [1], Supapohn Yamuangmorn [2], Sansanee Jamjod [1,2], Chanakan Prom-u-thai [1,2] and Tonapha Pusadee [1,2,*]

[1] Department of Plant and Soil Sciences, Faculty of Agriculture, Chiang Mai University, Chiang Mai 50200, Thailand; liuw_nantapat1@hotmail.com (N.D.); sansanee.j@cmu.ac.th (S.J.); chanakan.p@cmu.ac.th (C.P.-u.-t.)

[2] Lanna Rice Research Center, Chiang Mai University, Chiang Mai 50100, Thailand; teaw_848113@hotmail.com

* Correspondence: tonapha.p@cmu.ac.th

Abstract: Purple rice (*Oryza sativa* L.) contains anthocyanin, which acts as an antioxidant and functional food for humans. The levels of anthocyanin growth and production in rice are mainly controlled by the availability of light. However, shade can affect anthocyanin biosynthesis genes. Therefore, the objective of this study was to determine the yield and anthocyanin content among four purple rice varieties, which provide the difference in colors of purple and green leaves. This study also evaluated gene expression affected by shading treatment to understand the relation of grain anthocyanin and expression level. This research was conducted using a split plot design using four levels of shading (levels of shading from anthesis to maturity) with three replications, no shading, 30% shading, 50% shading, and 70% shading, as the main plots and purple rice varieties as subplots, KJ CMU-107, K2, K4, and KDK10, from anthesis to maturity. Shading significantly decreased yield and yield components, but increased grain anthocyanin content. Nonetheless, the response of yield and grain anthocyanin content to shading did not show a significant different between purple and green leaf varieties. In addition, the level of *OsDFR* gene expression was different depending on the shading level in four rice varieties. The *OsDFR* gene presented the highest expression at shading levels of 30% for K4 and 50% for KDK10, while the expression of the *OsDFR* gene was not detected in the purple rice varieties with green leaves (KJ CMU-107 and K2). The response of grain anthocyanin and gene expression of *OsDFR* to light treatment did not show significantly differences between the purple and green leaf varieties, suggesting that the appearance of anthocyanin in leaves might be not related to anthocyanin synthesis in the grain. Taken together, the results suggest that some purple rice varieties were more suitable for planting under low light intensity based on a lower level of grain yield loss, strong shade tolerance, and high anthocyanin content in leaf and grain pericarp. However, it is necessary to explore the effects of light intensity on genes and intermediates in the anthocyanin synthesis pathway for further study.

Keywords: shading stress; purple rice; anthocyanin content; yield; *OsDFR*; low light intensity

Citation: Danpreedanan, N.; Yamuangmorn, S.; Jamjod, S.; Prom-u-thai, C.; Pusadee, T. Genotypic Variation of Purple Rice in Response to Shading in Yield, Anthocyanin Content, and Gene Expression. *Plants* **2023**, *12*, 2582. https://doi.org/10.3390/plants12132582

Academic Editors: Zhiyong Li and Jian Zhang

Received: 7 June 2023
Revised: 4 July 2023
Accepted: 5 July 2023
Published: 7 July 2023

1. Introduction

Purple rice is commonly grown in the northern and northeastern regions of Thailand. Purple color appears in different parts of the purple rice plant, such as the pericarp, leaf sheath, leaf blade, and petals [1]. Today, increasingly more health-conscious consumers and researchers are turning to various rice varieties with black and purple seed coats that contain anthocyanins [2]. These substances are beneficial for health due to their antioxidant properties, which can improve blood flow in small vessels and reduce the risk of cancer and viral infection [3].

Light is essential for plant growth and development, but excess high-energy UV irradiance can cause damage to a cell. Anthocyanin accumulation in plants benefits plants by enhancing resistance to UV stress [4]. Light is a factor involved in plant growth since light is the source of energy used by plants in photosynthesis to produce sugar and starch. In addition, light plays an important role in various physiological processes within plants, such as protein synthesis, transpiration, and growth [5]. Ultimately, proper light intensity improves plant growth, including through the accumulation of antioxidants, which plants can synthesize to protect against the harmful effects of UV rays [6]. Therefore, the synthesis of anthocyanin, which is an antioxidant in the genetic system of plants, increases with increasing light intensity to a certain level, then decreases when light intensity is too great [7].

Shading with a black shading net reduces light intensity on plants and filters light [8]. Low light reduces photosynthesis and reduces plant yield [9,10]. This factor causes plants to adapt to survive and increases photosynthesis sources, stimulating the production of more anthocyanin in plants compared to plants grown under normal light conditions. In addition, anthocyanin is degraded if the light intensity is coupled with too high a temperature [11]. A previous report showed that anthocyanins in shoot and rice grain are sensitive to low light intensity, which suggests that anthocyanins could serve as a target compound for investigating low-light stress conditions [12]. Nevertheless, the anthocyanin biosynthesis gene related to the appearance of anthocyanin in various rice varieties has not been determined.

Anthocyanin pigment combines to form compounds with the ability to dissolve in the water found in plants. Such pigments play beneficial roles in visual activity, cancer, heart disease, and age-related neurodegenerative disorders [3]. Anthocyanins have protective effects during plant development through absorbing excess UV light, preventing lipid peroxidation, and suppressing the activity of ROS. Plants have evolved such that the biosynthetic pathways of anthocyanins can resist various abiotic stresses including UV irradiation, drought, high salinity, and low temperature [13]. The factors affecting the amounts of anthocyanins are either internal or external. The internal factors are related to plant genetics, which can be studied through genes that help regulate anthocyanin biosynthesis. The functions of genes that work together can be divided into two types: structural genes and regulatory genes [14]. External factors are factors in the natural environment or the occurrence of chemical reactions that affect the stability of anthocyanins, e.g., pH, light, temperature, and nutrients [15]. Both internal and external factors are directly related to increases and decreases in anthocyanin [16]. A previous study found that shade or low light intensity can affect anthocyanin levels. For example, although shade was found to reduce the production of upland rice, the anthocyanin content increased [11].

Usually, rice is a green plant; however, the biosynthesis of anthocyanin can impart purple, red, and black colors in leaf blades, leaf sheaths, stigmas, and pericarps [17]. Previous research and a growing body of experimental evidence suggest that anthocyanin provides plants with physiological benefits. However, the relevant mechanism has not been fully determined. The results of both chlorophyll content analysis and transmission electron microscopy revealed that anthocyanin has a negative impact on the photosynthetic machinery of purple-leaf compared to green-leaf wild-type plants [18]. A purple leaf color is an important morphological marker and valued as an important trait for the study of rice domestication and breeding [19]. In addition, previous research found that shade stress in plants leads to an increased amount of chlorophyll (chlorophyll a, chlorophyll b, and total chlorophyll) corresponding to the higher level of shade [11]. However, a previous study found that in purple rice varieties with purple leaves, the relative gene expression of *OsPL6* led to greater anthocyanin accumulation than that in purple rice varieties with green leaves [18]. Thus, we studied the differences in purple rice with both purple-leaf and green-leaf varieties under low-light conditions.

Anthocyanin biosynthesis is regulated by structural genes, yet the expression intensity of those genes is controlled by the interaction of regulatory genes, which are encoded by transcription factors (TFs) known as the MBW complex. Rice tissues are mainly controlled by three factors, C (Chromogen), A (activator), and P (Purple, distributor), where C and

A are essentially color-producing genes, and P is a tissue-specific regulator of both C and A [20]. A combination called *OsDFR* results in a red grain color [21]. Two classes of genes are required for anthocyanin biosynthesis, the structural genes encoding the enzymes that directly participate in the formation of anthocyanins and other flavonoids and the regulatory genes that control the transcription of structural genes [22]. A previous study showed that enzyme activities in the various branch pathways are highly regulated (Figure 1). Genes code for the enzyme dihydroflavonol 4-reductase (DFR), which is involved in the process of changing dihydroflavonols to leucoanthocyanidins, which are colorless compounds, before being synthesized into anthocyanidin as the main structure of anthocyanins in the next step [23]. Anthocyanin biosynthesis was also found to be controlled in response to different developmental and environmental cues [22]. Therefore, the present study aimed to determine the yield and anthocyanin content among four purple rice varieties that provide the difference in color of purple and green leaves. This study also evaluated gene expression affected by shading treatment to understand the relation of grain anthocyanin to the expression level. The variety that responds well to light stress in terms of stability of grain yield and anthocyanin content would be useful for breeders in selecting plant varieties as parents in the assembly of new varieties.

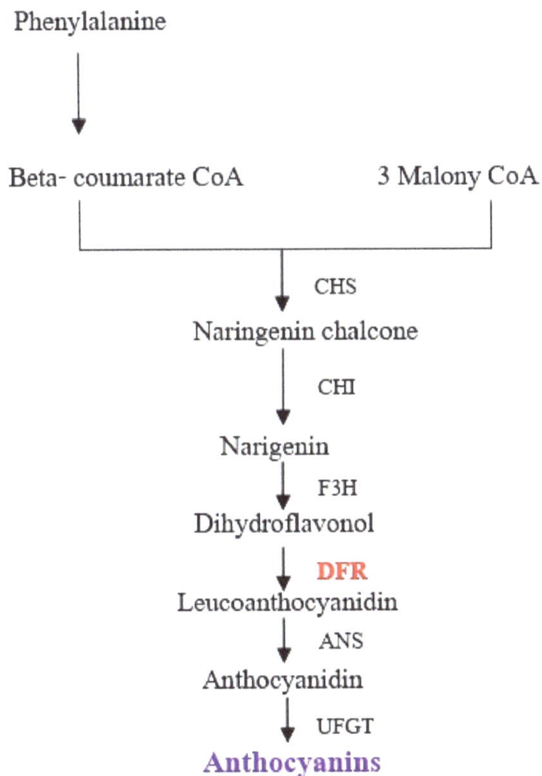

Figure 1. Pathway of anthocyanin biosynthesis. The *DRF* gene in red represents the candidate gene in the present study. Adapted from information in [24].

2. Results

2.1. Yield and Yield Components

Grain yield, straw yield, grain weight, and filled grain amounts were significantly affected by the interactions between shading treatment and rice variety (Figure 2). The shading treatment strongly decreased grain yield by 17% to 81% in all varieties compared to no shading treatment. However, the level of decrease was dependent on rice variety. Shading at 30% resulted in 44% and 31% decreases in KJ CMU-107 and KDK10, respectively, and 19% and 17% decreases in K2 and K4. Similarly, shading treatment significantly decreased the straw yield. The straw yields of KJ CMU-107 and KDK10 grown under shading at 30% were reduced by 8% and 15%, respectively, compared to no shading treatment, whereas shading treatment showed no significant difference in K2 and K4. The result of the 100-grain weight of all varieties steadily reduced by 11% to 36% with an increase in shading level compared to the control treatment. A reduction in filled grain was found in all rice varieties and was obviously reduced when plants were treated with the 50% and 70% shading treatments.

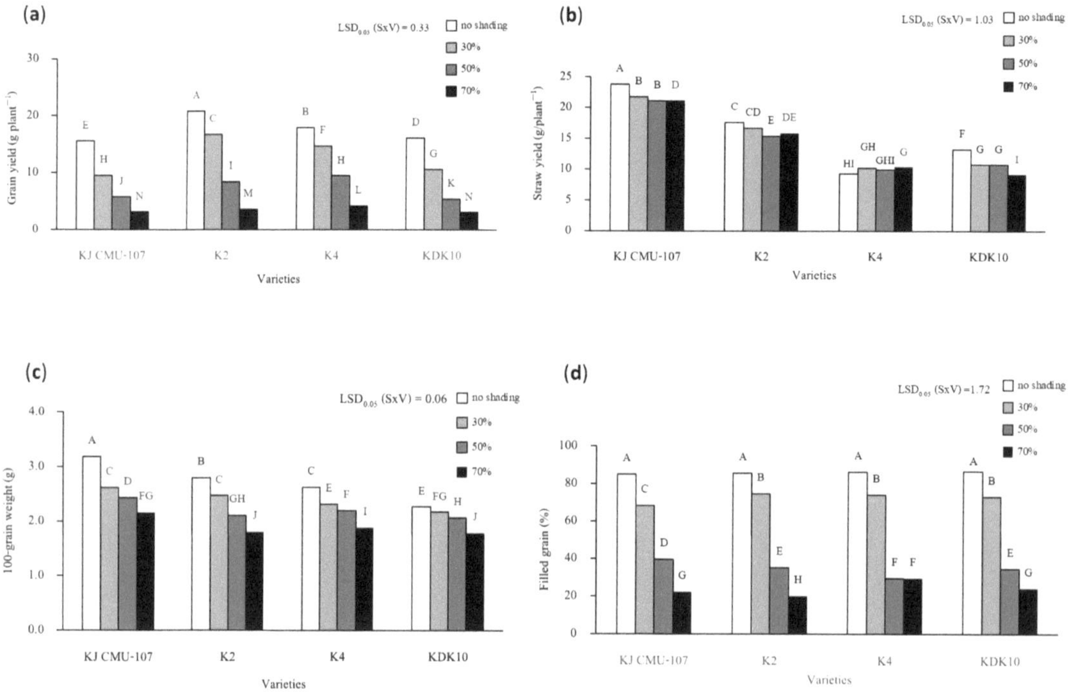

Figure 2. The interaction effects of yield and yield components between shading treatments (S); no shading with 30%, 50%, and 70% and purple rice varieties (V); KJ CMU-107, K2, K4, and KDK10. (**a**) Grain yield; (**b**) straw yield; (**c**) 100-grain weight; (**d**) filled grain. Different letters above bars indicate significant differences based on the least significant difference (LSD) at $p < 0.05$.

Table 1 shows the number of tillers per plant, number of panicles per plant, number of spikelets per panicle, panicle length, and culm length in response to the shading treatment performed on the four varieties. Shading treatment had no effect on number of tillers per plant, number of panicles per plant, number of spikelets per panicle, or panicle length; however, the result was significantly different depending on rice variety. For instance, the highest number of tillers and panicles per plant, number of spikelets per panicle, and panicle length were observed in K2, while the lowest values were found in KDK10. By contrast, culm length differed according to shading treatment and rice variety. Shading at 50% resulted in a 3% reduction in culm length compared to the control treatment. KJ CMU-107 was the tallest followed by KDK10, K2, and K4, in that order.

Table 1. Responses of number of tillers per plant, number of panicles per plant, number of spikelets per panicle, panicle length, and culm length.

	No. of Tillers Plant^{-1}	No. of Panicles Plant^{-1}	No. of Spikelets Panicle^{-1}	Panicle Length (cm)	Culm Length (cm)
Shading treatment					
No shading	6.51 ± 0.27	5.96 ± 0.26	118.63 ± 0.24	24.65 ± 0.35	86.72 ± 1.66 AB
30%	6.42 ± 0.28	5.81 ± 0.28	118.93 ± 0.23	24.29 ± 0.34	87.10 ± 1.65 A
50%	6.28 ± 0.26	5.76 ± 0.29	119.06 ± 0.24	24.71 ± 0.31	84.25 ± 1.66 C
70%	6.25 ± 0.27	5.63 ± 0.28	119.24 ± 0.25	24.83 ± 0.32	85.40 ± 1.64 BC
Variety					
KJ CMU-107	6.53 ± 0.21 B	6.25 ± 0.22 B	98.57 ± 0.23 D	24.14 ± 0.33 B	121.98 ± 1.81 A
K2	8.23 ± 0.19 A	7.58 ± 0.23 A	140.97 ± 0.25 A	25.65 ± 0.34 A	71.07 ± 1.85 D
K4	4.93 ± 0.21 C	4.33 ± 0.22 D	122.08 ± 0.23 B	25.23 ± 0.33 A	74.18 ± 1.82 C
KDK10	6.43 ± 0.22 B	5.72 ± 0.21 C	113.97 ± 0.24 C	23.39 ± 0.32 C	76.23 ± 1.81 B
F-test					
Shading treatment (S)	ns	ns	ns	ns	*
$LSD_{0.05}$ (S)					1.65
Variety (V)	***	***	***	***	***
$LSD_{0.05}$ (V)	0.42	0.45	0.48	0.63	1.81
SxV	ns	ns	ns	ns	ns

The data were analyzed using F-tests (*: $p < 0.05$, ***: $p < 0.001$, ns: not significant $p < 0.05$). Different letters indicate the least significant differences in each parameter within the column at $p < 0.05$. The values are expressed as the mean ± SE.

2.2. *Total Anthocyanin and Total Chlorophyll Content*

The total anthocyanin contents in the leaves of two purple-leaf varieties at days 7, 14, and 21 after shading treatment present a significant interaction between shading treatment and rice variety (Figure 3a–c). Overall, an increase in shading level increased anthocyanin contents in the leaves, with an increase of 30% to 66% in K4 and 30% to 54% in KDK10 compared to the control plants. In addition, the grain anthocyanin contents of the four varieties were found to be significantly different based on the interaction between shading treatment and variety (Figure 3d). Shading at 30% yielded no significant increase in grain anthocyanin content in KJ CMU-107, K4, or KDK10 compared to the control shading but did show a 22% increase in K2. Across all varieties, shading at 50% resulted in the highest grain anthocyanin content, with increases of approximately 48%, 51%, 77%, and 51% in KJ CMU-107, KDK10, K2, and K4, respectively, compared to the control treatment. Nonetheless, grain anthocyanin content decreased when plants were grown under 70% shade, which was observed for all varieties, except KDK10. However, the color of pericarp when put under 30%, 50%, and 70% shade became darker compared to the plants grown with no shading. We also observed that the shapes of the seeds became smaller and abnormal when subjected to shading at 70% (Figure 4).

There was a significant interaction between shading treatment and rice variety in terms of total leaf chlorophyll content at days 7, 14, and 21 after shading treatment (Figure 5). At day 7, the total chlorophyll content steadily increased from 30% to 70% based on the shading level. However, the magnitude of the chlorophyll response was different between the green and purple-leaf varieties (Figure 5a). Total chlorophyll contents in the green leaves of KJ CMU-107 and K2 increased by approximately 13% to 32% compared to that for the plant with no shading treatment. Meanwhile, an increase of 26% to 60% was observed in the purple-leaf varieties, K4 and KDK10. Similarly, at day 14, the response of leaf chlorophyll content to the shading level continued to increase with an increase in shading level (Figure 5b). Shading treatment from 30% to 70% increased the chlorophyll content by 15% to 37% and 22% to 37% in the green-leaf varieties of KJ CMU-107 and K2, respectively, compared to the control plant. However, K4 and KDK10 presented strong increases in chlorophyll content of 37% to 77%. In addition, shading treatment had an effect on leaf chlorophyll content at day 21 for all varieties (Figure 5c). The chlorophyll contents of KJ CMU-107 and K2 increased by 41% to 79% in plants grown under 30% to 70% shade compared to the control plants and increased by 31% to 82% in K4 and KDK10, respectively.

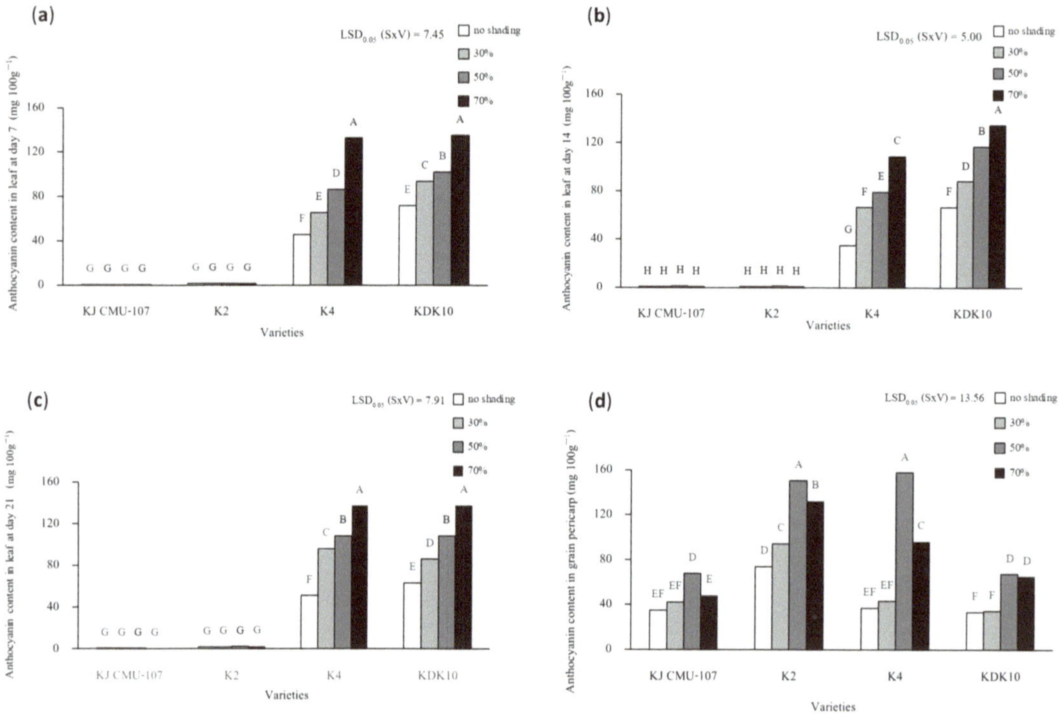

Figure 3. The interaction effect of total anthocyanin content between shading treatment (S); no shading treatment at 30%, 50%, and 70% with purple rice varieties (V); and KJ CMU-107, K2, K4, and KDK10. (**a**) Anthocyanin content in leaves at day 7; (**b**) anthocyanin content in leaves at day 14; (**c**) anthocyanin content in leaves at day 21; (**d**) anthocyanin content in grain pericarp. Different letters above the bars indicate significant differences based on the least significant difference (LSD) at $p < 0.05$.

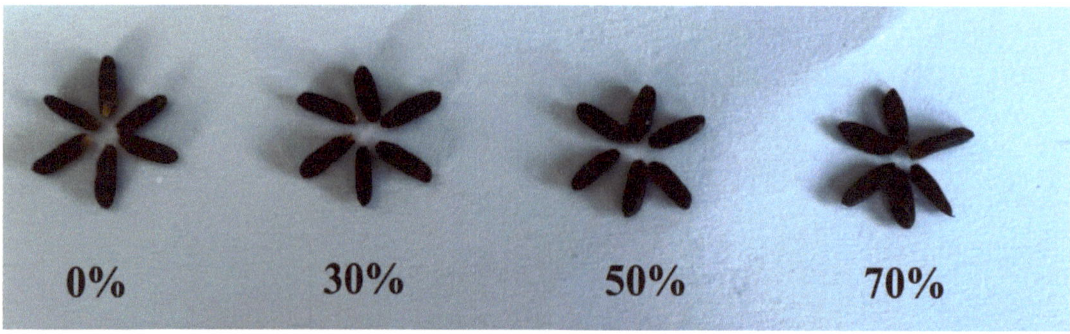

Figure 4. Grains of the K2 variety grown under varied shading: no shading and 30%, 50%, and 70% shading.

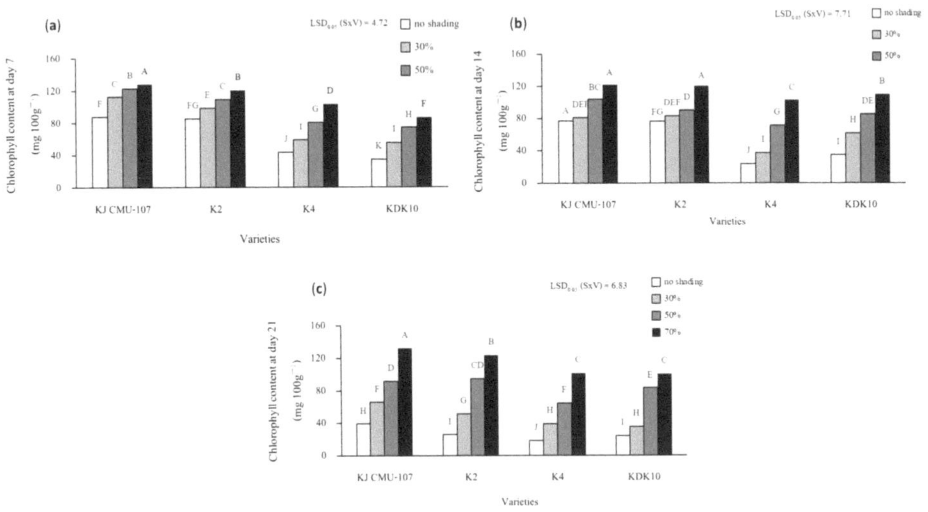

Figure 5. The interaction effect of total chlorophyll content between shading treatment (S); no shading at 30%, 50%, and 70 with purple rice varieties (V); and KJ CMU-107, K2, K4, and KDK10. (**a**) Chlorophyll content in leaves at day 7; (**b**) chlorophyll content in leaves at day 14; (**c**) chlorophyll content in leaves at day 21. Different letters above the bars indicate significant differences based on the least significant difference (LSD) at $p < 0.05$.

2.3. *Expression of OsDFR*

The expression of *OsDFR* in the leaves at day 14 after shading was detected only in purple-leaf varieties K4 and KDK10, but no expression was detected in the green-leaf varieties of KJ CMU-107 and K2 (Figure 5). The expression level of *OsDFR* was significantly affected by the interaction between shading treatment and rice variety (Figure 6). The treatments of shading at 30% and 50% resulted in high expression levels of *OsDFR* in K4, which were 18% and 35% higher, respectively, than those in plants with no shading. However, a 70% shading treatment decreased the expression level by 57%. In contrast, the expression level of *OsDFR* of KDK10 grown under 30% shading treatment was not different compared to the level under no shading treatment, but the expression levels did significantly decrease by 78% and 82% when the plants were grown under shading treatments of 50% and 70%, respectively.

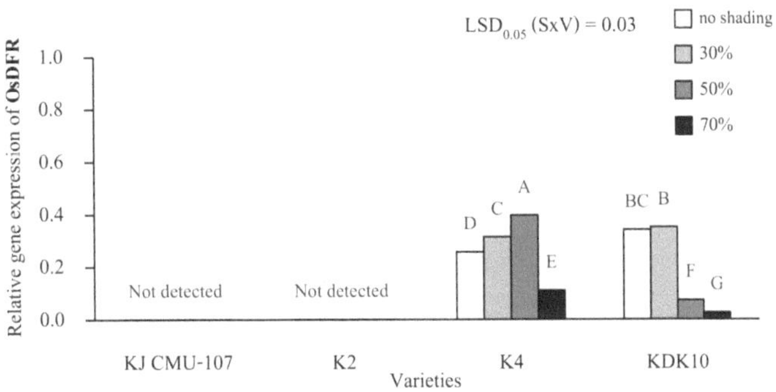

Figure 6. Relative gene expression of the *OsDFR* gene compared with *OsActin* (housekeeping gene) grown under varied shading; no shading, 30%, 50% and 70% with purple rice varieties; KJ CMU-107, K2, K4, and KDK10 in leaf at day 14 after shading. Different letters above the bars indicate significant differences based on the least significant difference (LSD) at $p < 0.05$.

3. Discussion

This study shows that yield and yield components significantly decreased with increased levels of shading. A shading level of 70% produced the lowest yield as the plants received the lowest light intensity. Based on the results, grain filling was strongly affected by low light intensity compared to normal light conditions. Moreover, shading stress decreased the dry matter accumulation of rice plants, as well as the grain weight. Plant productivity was found to be mainly controlled by the photosynthesis rate since reduced light intensity generally reduced the amount of source [11,25]. A previous study reported that shading stress reduces the supply of photosynthetic products, thereby remarkably decreasing starch biosynthesis in grains and postponing caryopsis development [26]. When plants were disrupted by shading at the heading stage, a decrease in sink capacity was observed, leading to a strong reduction in grain-filling rate and grain yield [9,27]. Meanwhile, shading reduced the energy and nutrient supply, which affected grain-filling progress, pollen germination, and tube elongation [28]. Therefore, the grain-filling progress of rice, especially for spikelets at the bottom and middle positions, was significantly decreased by shading after the heading stage.

The magnitude of the decline in rice production due to shade depends on the level of tolerance and growth phases of each variety [29]. The results of this experiment suggest that shading treatment interfered with the source of the plant, which reduced grain filling, but the sink storage remained the same. As a result, the seeds became more withered because the leaves were unable to synthesize enough carbohydrates to produce sufficient amounts of source for all the seeds. This study found an 11–36% reduction in grain weight across all rice varieties in the shading treatment. Nonetheless, the KDK10 variety showed more stable 100-grain weight under the various shade levels compared to other varieties. A rice variety with stable yield parameters would be useful for selecting rice varieties to be planted under natural low-light conditions in order to achieve higher productivity in the future.

Shading treatment significantly increased anthocyanin levels in both the leaf and grain pericarp of purple rice. This result indicates that light could be the main factor in the anthocyanin synthesis of rice plants grown under shade conditions, which is in accordance with previous reports [10,12,30]. This study is the first observation on the responses of anthocyanin synthesis to low light intensity among various rice varieties. Interestingly, this research shows that different varieties yielded different anthocyanin accumulation results. For instance, the level of grain anthocyanin content in K2 and K4 varieties responded less strongly to reduced light intensity compared to other varieties. Meanwhile, the severe condition of low light intensity did not increase anthocyanin in the rice grains. Under the lowest light intensity treatment (70% of shading), the anthocyanin contents of all rice varieties tended to decrease, except in KDK10. However, the mechanism and function of anthocyanins in each variety remain unclear. Even though anthocyanins are not directly involved in plant growth, they can protect against plant damage from abiotic stress. A previous study reported that abiotic stresses inhibits plant growth and reduces crop productivity and that plants produce anthocyanins after ROS signaling via the transcription of anthocyanin biosynthesis genes, enabling increased anthocyanin to alleviate plants under stress conditions [13,31]. Nonetheless, the increased anthocyanin content in leaf and grain samples in the present study was not found to be related to the stability of grain yield. This result agrees with the response of anthocyanin to the shading of green- and purple-leaf varieties. The appearance of leaf anthocyanin in the purple-leaf variety could not maintain stable yield productivity. This result suggests that light is a necessary factor for rice production, especially during the flowering stage because carbohydrate accumulation in grains mainly depends on the photosynthesis rate. Although anthocyanin synthesis in leaf and pericarp was not found to be correlated in this experiment, this phenomenon should be studied in further research on anthocyanin transport from leaf to pericarp in purple rice. The gene expression of *OsDFR* was induced by low light intensity, which is similar to the response of grain anthocyanin. Under increased shading levels, the gene expression and grain anthocyanin of K4 tended to increase, but this result contrasts with that of KDK10. This result suggests that *OsDFR* gene expression could alert anthocyanin synthesis in the grain but might differ by rice variety. Other genes, such as *OsANS*, were reported to have an impact on grain anthocyanin [32]. However, a previous study found that *OsC1* and *OsRb* are tissue-specific regulators that do not influence anthocyanin biosynthesis in the pericarp [33].

Dihydroflavonol 4-reductase (DFR) uses NADPH as a cofactor to catalyze the reduction of dihydroflavonols to their respective colorless, unstable leucoanthocyanidins, which are common precursors for anthocyanin and proanthocyanidin biosynthesis [34]. The results show that the *OsDFR* gene was expressed only in purple leaves in K4 and KDK10 since these genes are important genes in anthocyanin biosynthesis that contain the enzyme generation code dihydroflavonol 4-reductase (DFR) in the process of changing dihydroflavonols to leucoanthocyanidins, which is a colorless compound, before being synthesized into anthocyanidin as the main structure of anthocyanins [35]. The genes that encode DFR and related proteins have been isolated from many plant species and have been well characterized in terms of their functions [36]. The gene expression at a shading level of 50% in K4 featured the highest gene expression of *OsDFR* due to the acceleration of anthocyanin biosynthesis in plants caused by high light shading stress. However, the gene expression of *OsDFR* was decreased under a shading level of 70%, which indicates that the plants were exposed to too little light for anthocyanin biosynthesis to occur. The *OsDFR* gene may not be involved in anthocyanin synthesis in K4 and KDK10, which should be further studied in other genes if the genes involved in anthocyanin content in K4 and KDK10 are to be investigated. The *OsDFR* gene has the ability to synthesize dihydroflavonol into leucoanthocyanidin. These compounds have no color of their own, but in acidic environments and at elevated temperatures, they are converted to the color of anthocyanidins [37]. A previous study found that *OsDFR* and other anthocyanin biosynthesis genes become purple in the apiculi and stigmas [36]. However, in the present research, we studied the *OsDFR* gene with shading treatment in the leaf and grain pericarp. No correlation was found in either rice tissue, so we suggest studying the other genes that result in a direct correlation of shading treatment to explore the total anthocyanin content and relative gene expression. The results of this experiment suggest that the *OsDFR* gene may not directly affect rice leaves' transformation into a purple color caused by anthocyanin accumulation and that there may be another gene that interacts with the *OsDFR* gene. A previous study found that the accumulation of anthocyanin in rice leaves is caused by the interaction of *OsC1*, *OsRb*, and *OsDFR* genes, which are key genes for determining anthocyanin biosynthesis in rice leaves [38]. Experiments on the relative gene expression affected by shading treatment may be of greater interest when the ANS gene is studied in purple rice in the future in order to directly identify the genes affecting the purple color caused by anthocyanin accumulation in purple rice leaves during shading.

4. Materials and Methods

4.1. Expression of OsDFR

Purple rice seeds were obtained from the Division of Agronomy, Faculty of Agriculture, Chiang Mai University. This study used four purple pericarp rice varieties consisting of two purple colors in the shoot and grain varieties, K4 and KDK10, and two non-purple colors in the shoot varieties, K2 and Kum Jao Morchor 107 (KJ CMU-107). The experiment used a split-plot design for three replications. Four rice varieties were used as a subplot, and four shading treatments were used as the main plot, arranged as follows: no shading (control) and 30%, 50%, and 70% light reduction compared to normal light.

The study was conducted in a glasshouse from August to December 2021, at the Agronomy Division, Faculty of Agriculture, Chiang Mai University, Chiang Mai, Thailand. Seeds (paddy rice) were sown after soaking in water overnight. Two-week-old seedlings were transferred into pots (30 cm in diameter and 25 cm in height). The plants were grown with a single seedling per hill and five hills per pot. The shading treatment was set up with black polypropylene netting that reduced the light intensity to 30%, 50%, and 70% of full light from anthesis to the mature stage in planting pots 30 cm in diameter and 25 cm in height, with 5 plants per pot. The light intensity was measured daily using a light meter (AS one LM-332, Osaka, Japan) above the plant canopy. A fertilizer formula of 15-15-15 (N-P-K) was applied at a rate of 3 g pot^{-1}, and fertilizer in the form of urea (46-0-0) was applied at a rate of 5 g pot^{-1} before planting.

4.2. Yield Measurement and Sample Preparation

At maturity, five plants in each pot were harvested, from which the yield components were determined. The paddy rice was threshed manually and dried until the moisture content reached 14% before being weighed and measured for grain yield. The collected data on yield components consisted of tiller number plant^{-1}, panicle number plant^{-1}, percentage of filled grain, 100-grain weight, number of spikelets panicle^{-1}, and culm and panicle lengths. The straw samples were weighed and recorded as straw dry weight after being dried in a hot air oven at 75 °C for 72 h [12]. Two fully expanded leaves from the top position were sub-sampled at days 7, 14, and 21 after shading treatment to determine chlorophyll content and gene expression, and the remaining samples were freeze-dried in a freeze dryer (CHRIST, Beta 2–8 LSCbasic, Harz, Germany) for 24 h and mechanically ground in a hammer mill (Scientific Technical Supplies D–6072 Dreieich, West, Germany) to determine anthocyanin content. The paddy rice was de-husked with a laboratory husking machine (Model P-1 from Ngek Huat Co., Ltd., Bangkok, Thailand) to produce brown rice for the determination of anthocyanin content.

4.3. Determination of Total Anthocyanin Content

The total anthocyanin contents in leaves at days 7, 14, and 21 after shading treatment and grains were determined using the modified pH-differential method of [39]. About 2.5 g of freeze-dried sample was added into a 25 mL tube. Then, we pipetted 24 mL of acidified methanol (70% methanol and 30% of 1.5 mol L^{-1} HCl) into the tube and shook the tube for 60 min. Exactly 2 mL of supernatant was added to two buffer solutions. The potassium chloride buffer (0.025 mol^{-1}, pH 1) was measured for absorbance at a wavelength of 520 nm, and a sodium acetate buffer (0.400 mol L^{-1}, pH 4.5) was measured for absorbance at a wavelength of 700 nm using a UV–VIS spectrophotometer (Biochrom Libra S22, Cambridge, UK). The absorbance was calculated in the diluted sample (A) as A = (A520 − A700) − (A520′ − A700′), where A520 and A700 are the absorbance values at 520 and 700 nm under pH 1.0, respectively, while A520′ and A700′ are the absorbance values at 520 and 700 nm under pH 4.5, respectively. The total anthocyanin content was calculated as follows:

$$\text{Total anthocyanin content} = (A \times MW \times DF \times 1000)/\varepsilon \times L$$

where MW is the molecular weight of cyanidin-3-glucoside (449.2 g mol^{-1}); DF is the dilution factor; ε is 26,900 M absorbance, and L is the cell path length (1 cm). Total anthocyanin content was expressed as milligrams of cyanidin-3-glucoside per kg of dry weight (mg kg^{-1}) [12].

4.4. Determination of Total Chlorophyll Content

The total chlorophyll content in leaves at days 7, 14, and 21 after shading was determined using the method of Lichtenthaler and Wellburn, 1983. About 0.2 g of fresh sample was extracted with methanol under cold conditions in the darkness. The extraction process was performed with four replicates. The supernatant was warmed to room temperature before measurement in a spectrophotometer at 665 and 652 nm to assess chlorophyll a and chlorophyll b, respectively. The chlorophyll a (Chl a), b (Chl b), and total chlorophyll (Total Chl) content (mg L^{-1}) were calculated as follows:

$$\text{Chl a} = (16.29 \times A665) - (8.54 \times A652)$$

$$\text{Chl b} = (30.66 \times A652) - (13.58 \times A665)$$

$$\text{Total Chl} = (22.12 \times A652) - (2.71 \times A665)$$

where A665 and A652 are the absorbance of Chl a and Chl b, respectively. Chlorophyll contents were expressed as milligrams per gram of fresh weight [12].

4.5. Gene Expression of OsDFR

4.5.1. RNA Extraction

Total RNA was extracted from the leaves at day 14 after shading using a PureLinkTM RNA Mini Kit (Invitrogen, Thermo Fisher Scientific, MA, USA). The fresh tissues were ground in liquid nitrogen. The extracted RNA samples were verified for quantity and quality using a nanodrop spectrophotometer and 1.5% agarose gel electrophoresis. The genomic DNA was removed from the RNA preparations in the following conditions: 1 μg of total RNA, 2 μL of 10× reaction buffer, 1 μL of DNase1, and 9–15 μL of DEPC-treated water (total volume 20 μL); next, the reaction was incubated at 37 °C for 30 min, 1 μL of 50 mM EDTA was added, and the mixture was incubated at 65 °C for 10 min. The total RNA was diluted to a 100 ng/μL concentration and used for the qRT-PCR experiments.

4.5.2. cDNA Synthesis

The cDNA was synthesized from 1 μg of total RNA using a RevertAid first-strand cDNA synthesis Kit (Thermo scientific). PCR was carried out using 1 μg of total RNA (DNase I-treated), 1 μL of Oligo (dT)18, 2 μL of 10 mM dNTP mix, 6 μL of 5× RT buffer, 1 μL of RiboLock Rnase Inhibitor (20 U/μL), 1 μL of RevertAid RT (200 U/μL), and 1–5 μL of DEPC-treated water (total volume 30 μL) and a RevertAid first strand cDNA synthesis Kit (Thermo Scientific). Next, we incubated the reaction at 42 °C for 60 min and terminated the reaction by incubating it at 70 °C for 5 min. Then, we stored the mixture at −20 °C. The cDNA was verified for quantity and quality using a nanodrop spectrophotometer and 1.5% agarose gel electrophoresis.

4.5.3. Gene Expression via Semi-Quantitative RT-PCR Analysis

The gene expression levels of *OsDFR* were analyzed via semi-quantitative RT-PCR using gene-specific primers of *OsDFR* and *OsActin1* (housekeeping gene) (Table 2). The PCR was performed in triplicate to amplify cDNA templates with *OsDFR* and OsActin using the following reaction: 2 μL of 1:20 diluted cDNA from 1 μg total RNA, 14 μL of water (ddH2O), 4 μL of 5× MyTaq Reaction Buffer, 0.2 μL of forward and reverse primer, 0.1 μL of 5 unit MyTaqTM HS DNA Polymerase (Bioline, London, UK), and 0.6 μL of DMSO (total volume 20 μL). The PCR was performed by denaturing the solution at 95 °C for 2 min, followed by 35 cycles at 95 °C for 30 s, primer annealing at 53 °C for 30 s, extension at 72 °C for 30 s, and a final extension at 72 °C for 5 min. For the semi-quantitative RT-PCR assays, the total amount of cDNA in the samples was standardized after the amount of actin mRNA was evaluated with an *OsActin* primer pair.

Table 2. Primer used for studying the gene expression of *OsDFR* and *OsActin* genes.

Gene	Primer Names and Sequences (5′→3′)	References
OsDFR	*OsDFRF*: CGGGTTCAGGTTCAGGTACA	[40]
	OsDFRR: TGAAACCGGAGGGAGTAAC	[40]
OsActin	*OsActinF*: GACTCTGGTGATGGTGTCAGC	[41]
	OsActinR: GGCTGGAAGAGGACCTCAGG	[41]

F: forward primer; R: reverse primer.

4.5.4. Statistical Analysis

Statistical analyses were carried out using analysis of variance (ANOVA) (Statistic version 8.0 for Windows) for a spit plot design, and shading treatment and rice variety were used as the main plot and sub plot, respectively. The least significant difference (LSD) at $p < 0.05$ was used to compare the means for significant differences among parameters. The significance of the correlation coefficients was analyzed using Pearson correlations at $p < 0.05$.

The gene expression levels were analyzed by relative intensity compared to the reference gene (Actin) using the ImageJ software version 1.50i (Wayne Rasband National Institutes of Health, MD, USA). The relative intensity of gene expression was subjected to statistical analysis using the Statistica 8 software (analytical software SX, version 8, Tallahassee, FL, USA).

5. Conclusions

This study demonstrates that shading at the reproductive stage corresponds to a significant decrease in the yield productivity of all rice varieties. However, some varieties presented stable yield parameters such as grain filling, suggesting that there are potential useful characteristics for selecting varieties to be grown under natural low-light conditions. Shading increased the biosynthesis of anthocyanin and chlorophyll contents in the leaves of all rice varieties. Similarly, an increase in anthocyanin was found in the grain pericarp when the level of shading was increased, whereas the response was different between rice varieties. In addition, the responses of grain yield and increasing anthocyanin content to shading were not different between green- and purple-leaf varieties, suggesting that light intensity plays an important role in rice productivity and anthocyanin synthesis, especially during the flowering stage. Expression of the *OsDFR* gene, which aids in the biosynthesis of total anthocyanin content, was found only in K4 and KDK-10 plants with purple leaves. Nonetheless, the *OsDFR* gene expression levels of K4 and KDK-10 varieties showed different responses to shading treatments. Meanwhile, a correlation between expression and grain anthocyanin was not found in this study. This result suggests that under low-light conditions, other genes may be related to anthocyanin biosynthesis in rice grains. These results provide useful data for future studies to understand the biosynthesis of anthocyanin in purple rice grown under biotic stress conditions to maintain grain anthocyanin content and grain yield. This research suggests some varieties of purple rice can adapt and grow under low-light conditions, which may be useful for commercial cultivation of purple rice using an ideal shading technique to increase antioxidants from purple rice grains.

Author Contributions: Conceptualization, T.P., S.J. and C.P.-u.-t.; methodology, N.D., S.Y., S.J., C.P.-u.-t. and T.P.; writing—original draft preparation, N.D., S.Y. and T.P.; writing—review and editing, N.D., S.Y. and T.P.; supervision, T.P.; project administration, T.P.; funding acquisition, T.P. All authors have read and agreed to the published version of the manuscript.

Funding: This research project was supported by the Agricultural Research Development Agency (Public Organization) or "ARDA" (GSCMU(HRD65050085)/02/2565) and the Fundamental Fund 2022 (FF65) of Chiang Mai University.

Data Availability Statement: All data, tables, and figures in this manuscript are original.

Acknowledgments: We thank the members of CMUPN lab, Division of Agronomy, Faculty of Agriculture, Chiang Mai University for advice throughout this study.

Conflicts of Interest: The authors declare no conflict of interest.

References

1. Fongfon, S.; Pusadee, T.; Prom-u-thai, C.; Rerkasem, B.; Jamjod, S. Diversity of Purple Rice (*Oryza sativa* L.) Landraces in Northern Thailand. *Agronomy* **2021**, *11*, 2029. [CrossRef]
2. Yamuangmorn, S.; Prom-U-Thai, C. The Potential of High-Anthocyanin Purple Rice as a Functional Ingredient in Human Health. *Antioxidants* **2021**, *10*, 833. [CrossRef] [PubMed]
3. Khoo, E.H.; Azlan, A.; Tang, T.S.; Lim, M.S. Anthocyanidins and anthocyanins: Colored pigments as food, pharmaceutical ingredients, and the potential health benefits. *Food Nutr. Res.* **2017**, *61*, 1361779. [CrossRef] [PubMed]
4. Landi, M.; Tattini, M.; Gould, K.S. Multiple functional roles of anthocyanins in plant-environment interactions. *Environ. Exp. Bot.* **2015**, *119*, 4–17. [CrossRef]
5. Formisano, L.; Moreno, M.B.; Ciriello, M.; Zhang, L.; Pascale, D.S.; Lucini, L.; Rouphael, Y. Between Light and Shading: Morphological, Biochemical, and Metabolomics Insights into the Influence of Blue Photoselective Shading on Vegetable Seedlings. *Front. Plant Sci.* **2022**, *13*, 830. [CrossRef]
6. Chen, Y.; Li, T.; Yang, Q.; Zhang, Y.; Zou, J.; Bian, Z.; Wen, X. UVA Radiation Is Beneficial for Yield and Quality of Indoor Cultivated Lettuce. *Front. Plant Sci.* **2019**, *10*, 1563. [CrossRef]
7. Zoratti, L.; Sarala, M.; Carvalho, E.; Karppinen, K.; Martens, S.; Giongo, L.; Häggman, H.; Jaakola, L. Monochromatic light increases anthocyanin content during fruit development in bilberry. *BMC Plant Biol.* **2014**, *14*, 377. [CrossRef]
8. Petrella, D.; Sessoms, B.F.; Watkins, E. Layering contrasting photo selective filters improves the simulation of foliar shade. *Plant Methods* **2022**, *18*, 16. [CrossRef]
9. Li, Q.; Deng, F.; Chen, H.; Zeng, Y.; Li, B.; Zhong, X.; Wang, L.; Zhou, W.; Chen, Y.; Ren, W. Shading decreases rice yield by impeding grain-filling progress after heading. *Agron. J.* **2020**, *112*, 4018–4030. [CrossRef]
10. Zhu, H.; Li, X.; Zhai, W.; Liu, Y.; Gao, Q.; Liu, J.; Ren, L.; Chen, H.; Zhu, Y. Effects of low light on photosynthetic properties, antioxidant enzyme activity, and anthocyanin accumulation in purple pak-choi (*Brassica campestris* ssp. Chinensis Makino). *PLoS ONE* **2017**, *12*, e0179305. [CrossRef]
11. Syam'un, M.; Musa, Y.K.; Sadimantara, R.G.; Leomo, S.U.; Rakian, C.T. Shading effect on generative characters of upland red rice of Southeast Sulawesi, Indonesia. *IOP Conf. Ser. Earth Environ. Sci.* **2018**, *157*, 012017.
12. Yamuangmorn, S.; Jumrus, S.; Jamjod, S.; Sringarm, K.; Arjin, C.; Prom-u-thai, C. Responses of purple rice variety to light intensities and soil zinc application on plant growth, yield and bioactive compounds synthesis. *J. Cereal Sci.* **2022**, *106*, 103495. [CrossRef]
13. Naing, A.H.; Kim, C.K. Abiotic stress-induced anthocyanins in plants: Their role in tolerance to abiotic stresses. *Physiol. Plant.* **2021**, *172*, 1711–1723. [CrossRef]
14. Enaru, B.; Dretcanu, G.; Pop, D.T.; Stănilă, A.; Diaconeasa, Z. Anthocyanins: Factors Affecting Their Stability and Degradation. *Antioxidants* **2021**, *10*, 1967. [CrossRef]
15. Diaconeasa, Z.; Stirbu, I.; Xiao, J.; Leopold, N.; Ayvaz, Z.; Danciu, C.; Ayvaz, H.; Stănilă, A.; Nistor, M.; Socaciu, C. Anthocyanins, Vibrant Color Pigments, and Their Role in Skin Cancer Prevention. *Biomedicines* **2020**, *8*, 336. [CrossRef]
16. Liu, H.; Liu, Z.; Wu, Y.; Zheng, L.; Zhang, G. Regulatory Mechanisms of Anthocyanin Biosynthesis in Apple and Pear. *Int. J. Mol. Sci.* **2021**, *22*, 8441. [CrossRef]
17. Kim, M.-K.; Kim, H.-A.; Koh, K.; Kim, H.-S.; Lee, Y.S.; Kim, Y.H. Identification and quantification of anthocyanin pigments in colored rice. *Nutr. Res. Pr.* **2008**, *2*, 46–49. [CrossRef]
18. Khan, A.; Jalil, S.; Cao, H.; Tsago, Y.; Sunusi, M.; Chen, Z.; Shi, C.; Jin, X. The Purple Leaf (*pl6*) Mutation Regulates Leaf Color by Altering the Anthocyanin and Chlorophyll Contents in Rice. *Plants* **2020**, *9*, 1477. [CrossRef]
19. Choudhury, B.I.; Khan, M.L.; Dayanandan, S. Patterns of nucleotide diversity and phenotypes of two domestication related genes (OsC1 and Wx) in indigenous rice varieties in Northeast India. *BMC Genet.* **2014**, *15*, 71. [CrossRef]
20. Meng, L.; Qi, C.; Wang, C.; Wang, S.; Zhou, C.; Ren, Y.; Cheng, Z.; Zhang, X.; Guo, X.; Zhao, Z.; et al. Determinant Factors and Regulatory Systems for Anthocyanin Biosynthesis in Rice Apiculi and Stigmas. *Rice* **2021**, *14*, 37. [CrossRef]
21. Furukawa, T.; Maekawa, M.; Oki, T.; Suda, I.; Iida, S.; Shimada, H.; Takamure, I.; Kadowaki, K.-I. The Rc and Rd genes are involved in proanthocyanidin synthesis in rice pericarp. *Plant J.* **2006**, *49*, 91–102. [CrossRef] [PubMed]
22. Jaakola, L.; Määttä, K.; Pirttilä, A.M.; Törrönen, R.; Kärenlampi, S.; Hohtola, A. Expression of genes involved in anthocyanin biosynthesis in relation to anthocyanin, proanthocyanidin, and flavonol levels during bilberry fruit development. *Plant Physiol.* **2002**, *130*, 729–739. [CrossRef] [PubMed]
23. Ithal, N.; Reddy, A.R. Rice flavonoid pathway genes, OsDfr and OsAns, are induced by dehydration, high salt and ABA, and contain stress responsive promoter elements that interact with the transcription activator, OsC1-MYB. *Plant Sci.* **2004**, *166*, 1505–1513. [CrossRef]
24. Tian, J.; Chen, M.-C.; Zhang, J.; Li, K.-T.; Song, T.-T.; Zhang, X.; Yao, Y.-C. Characteristics of dihydroflavonol 4-reductase gene promoters from different leaf colored Malus crabapple cultivars. *Hortic. Res.* **2017**, *4*, 17070. [CrossRef] [PubMed]
25. Yang, M.; Liu, M.; Lu, J.; Yang, H. Effects of shading on the growth and leaf photosynthetic characteristics of three forages in an apple orchard on the Loess Plateau of eastern Gansu, China. *PeerJ* **2019**, *7*, e7594. [CrossRef]
26. Deng, F.; Wang, L.; Pu, L.S.; Mei, F.X.; Li, X.S.; Li, P.Q.; Ren, J.W. Shading stress increases chalkiness by postponing caryopsis development and disturbing starch characteristics of rice grains. *Agric. For. Meteorol.* **2018**, *263*, 49–58. [CrossRef]

27. Cheng, F.; Bin, S.; Iqbal, A.; He, L.; Wei, S.; Zheng, H.; Yuan, P.; Liang, H.; Ali, I.; Xie, D.; et al. High Sink Capacity Improves Rice Grain Yield by Promoting Nitrogen and Dry Matter Accumulation. *Agronomy* **2022**, *12*, 1688. [CrossRef]
28. Cantagallo, J.; Medan, D.; Hall, A. Grain number in sunflower as affected by shading during floret growth, anthesis and grain setting. *Field Crop. Res.* **2004**, *85*, 191–202. [CrossRef]
29. Cruz, P.; Sierra, J.; Wilson, R.J.; Dulormne, M.; Tournebize, R. Effects of Shade on the Growth and Mineral Nutrition of Tropical Grasses in Silvopastoral Systems. *Ann. Arid Zone* **1999**, *38*, 335–361.
30. Shoeva, Y.O.; Gordeeva, I.E.; Klhestkina, E.K. The Regulation of Anthocyanin Synthesis in the Wheat Pericarp. *Molecules* **2014**, *19*, 20266–20279. [CrossRef]
31. Fukuoka, S.; Yamamoto, S.-I.; Mizobuchi, R.; Yamanouchi, U.; Ono, K.; Kitazawa, N.; Yasuda, N.; Fujita, Y.; Nguyen, T.T.T.; Koizumi, S.; et al. Multiple functional polymorphisms in a single disease resistance gene in rice enhance durable resistance to blast. *Sci. Rep.* **2014**, *4*, 4550. [CrossRef]
32. Xia, D.; Zhou, H.; Wang, Y.; Li, P.; Fu, P.; Wu, B.; He, Y. How rice organs are colored: The genetic basis of anthocyanin biosynthesis in rice. *Crop J.* **2021**, *9*, 598–608. [CrossRef]
33. Zheng, J.; Wu, H.; Zhu, H.; Huang, C.; Liu, C.; Chang, Y.; Kong, Z.; Zhou, Z.; Wang, G.; Lin, Y.; et al. Determining factors, regulation system, and domestication of anthocyanin biosynthesis in rice leaves. *New Phytol.* **2019**, *223*, 705–721. [CrossRef]
34. Pourcel, L.; Irani, N.G.; Koo, A.J.K.; Bohorquez-Restrepo, A.; Howe, G.A.; Grotewold, E. A chemical complementation approach reveals genes and interactions of flavonoids with other pathways. *Plant J.* **2013**, *74*, 383–397. [CrossRef]
35. Wang, H.; Fan, W.; Li, H.; Yang, J.; Huang, J.; Zhang, P. Functional characterization of Dihydroflavonol-4-reductase in anthocyanin biosynthesis of purple sweet potato underlies the direct evidence of anthocyanins function against abiotic stresses. *PLoS ONE* **2013**, *8*, e78484. [CrossRef]
36. Katsuhiro, M.; Suzuki, R.; Tsuchiya, W.; Inagaki, N.; Yamazaki, T.; Hisano, T.; Yasui, Y.; Komori, T.; Koshio, M.; Kubota, S.; et al. A new buckwheat dihydroflavonol 4-reductase (DFR), with a unique substrate binding structure, has altered substrate specificity. *BMC Plant Biol.* **2017**, *17*, 239. [CrossRef]
37. Lina, M.B.; Robert, V.; Peter, G.L.; Klinkhamer; Choi, H.Y. Thin-Layer Chromatography Metabolomics. In *Encyclopedia of Analytical Science*; Elsevier: Amsterdam, The Netherlands, 2019; pp. 59–75.
38. Zheng, T.; Tan, W.; Yang, H.; Zhang, L.; Li, T.; Liu, B.; Zhang, D. Regulation of anthocyanin accumulation via MYB75/HAT1/TPL-mediated transcriptional repression. *PLoS Genet.* **2019**, *15*, e1007993. [CrossRef]
39. Abdel, A.; Hucl, P. A rapid method for quantifying total anthocyanins in blue aleurone and purple pericarp wheats. *Cereal Chem.* **1999**, *76*, 350–354. [CrossRef]
40. Sakulsingharoj, C.; Inta, P.; Sukkasem, R.; Pongjaroenkit, S.; Chowpongpang, S.; Sangtong, V. Overexpression of OSB2 gene in transgenic rice up-regulated expression of structural genes in anthocyanin biosynthesis pathway. *Thai J. Genet.* **2014**, *7*, 173–182.
41. Yamaji, N.; Ma, J.F. A Transporter at the Node Responsible for Intervascular Transfer of Silicon in Rice. *Plant Cell* **2009**, *21*, 2878–2883. [CrossRef]

plants

Article

Transcriptome Analysis Reveals the Genes Related to Water-Melon Fruit Expansion under Low-Light Stress

Wenrui Gao [1,2], Fuchun She [1,2,3], Yanjun Sun [1,2], Bing Han [1,2], Xiansheng Wang [4,5,*] and Gang Xu [1,2,*]

1 Institute of Vegetable Crop, Jiangsu Province Academy of Agricultural Sciences, Nanjing 210014, China
2 Jiangsu Key Laboratory for Horticultural Crop Genetic Improvement, Nanjing 210014, China
3 School of Horticulture, Anhui Agricultural University, Hefei 230036, China
4 Institute of Germplasm Resources and Biotechnology, Jiangsu Academy of Agricultural Sciences, Nanjing 210014, China
5 Nanjing Station for DUS Testing Center of New Varieties of Plants of MARA, Nanjing 210014, China
* Correspondence: wangxiansheng80@126.com (X.W.); xugang90@163.com (G.X.)

Abstract: Watermelon is one of people's favorite fruits globally. Fruit size is one of the important characteristics of fruit quality. Low light can seriously affect fruit development, but there have been no reports concerning molecular mechanism analysis in watermelons involved in fruit expansion under low-light stress. To understand this mechanism, the comparative transcriptomic file of watermelon fruit flesh at four different developmental stages under different light levels was studied. The results showed that the fruit size and content of soluble sugar and amino acids at low-light stress significantly decreased compared to the control. In addition, 0–15 DAP was the rapid expansion period of watermelon fruit affected by shading. In total, 8837 differentially expressed genes (DEGs) were identified and 55 DEGs were found to play a role in the four different early fruit development stages. We also found that genes related to oxidation-reduction, secondary metabolites, carbohydrate and amino acid metabolism and transcriptional regulation played a key role in watermelon fruit expansion under low-light stress. This study provides a foundation to investigate the functions of low-light stress-responsive genes and the molecular mechanism of the effects of low-light stress on watermelon fruit expansion.

Keywords: watermelon; low-light stress; fruit expansion; RNA-seq; differentially expressed genes; transcription factors

Citation: Gao, W.; She, F.; Sun, Y.; Han, B.; Wang, X.; Xu, G. Transcriptome Analysis Reveals the Genes Related to Water-Melon Fruit Expansion under Low-Light Stress. *Plants* **2023**, *12*, 935. https://doi.org/10.3390/plants12040935

Academic Editors: Zhiyong Li and Jian Zhang

Received: 31 December 2022
Revised: 12 February 2023
Accepted: 13 February 2023
Published: 18 February 2023

1. Introduction

Fruit setting, cell division and cell expansion also play a crucial role in fruit development. These processes of growth largely determine the initiation and final form of fruit growth. Cell cycle-related genes play an important role in cell division and nuclear replication. Some enzyme-coding genes involved in cell wall extension play an important role in cell enlargement and maturation [1,2].

Light is essential for plant growth, development and productivity. Low light unbalances carbon assimilation, resulting in the inhibition of physiological attributes, morphology parameters and yields [3]. Many studies have shown that low light can seriously affect fruit development [4,5]. Shading treatments have been shown to significantly decrease photosynthesis, total nitrogen (T-N) of the stems and roots and the fruits and yields of strawberry plants. At the same time, the lower yield was due to the decreased photosynthesis of plants under low light [6]. The reduced growth of apple fruit in shade treatment was because of reduction in cell production and expansion. Moreover, shading results in coordinated changes in the expression of carbohydrate metabolism-related genes, key transcription factors related to fruit growth and genes associated with cell production and expansion [7]. During the early peach developmental stage, the formation of starch grains was inhibited and fewer photoassimilates were translocated from source leaves to fruit sinks [8]. After shading, the imported assimilates of the melon were reduced in number and galactosyl-sucrose oligosaccharide unloading was inhibited. In addition, the activity of key enzymes related to sucrose synthesis changed after shading, which affected sucrose content in melon fruit [9]. Crosstalk between the flavonoid biosynthetic genes and the involvement of key transcription factors such as McMYB4, McMYB7, McMYB10 and McMYB16 in the regulation of the ratio of anthocyanins and flavanols accounted for the different coloration of the fruit peel and softening of the flesh under shade conditions [10].

Watermelon (*Citrulluslanatus*) is a highly popular and economic horticultural crop. Approximately 3.48 million hectares were planted with watermelon in 2014 all over the world, making it among the top five most consumed fresh fruits, and China is the largest watermelon producer in the world. The watermelon plant produces large, edible fruit, the flesh of which is about 91% water by weight, and is a rich source of bioavailable compounds including lycopene and other carotenoids, vitamins A and C, and the non-essential amino acid L-citrulline, and is about 6% sugar by weight [11]. The size and weight of watermelon fruit are important aspects in agriculture and important directions of breeding research. In addition, fruit size is not only the important factor of yield, but also strongly associated with the watermelon's commercial value. The size of fruit is regulated by internal heredity and environmental factors. Watermelon is a photophilic plant and often suffers from low-light stress in protective cultivation in early spring and autumn [12]. Research has shown that watermelon fruit size is significantly decreased as the planting density is increased, whereas the soluble solids content of the fruits is affected little. The fruit size of the watermelon is closely related to both the total solar radiation and the photosynthetic production per plant [13]. Our previous studies also show that the overcast and rainy climate in early spring and autumn leads to weak light in the facility, which seriously affects the growth of watermelon plants and the expansion of fruit. However, there have been no reports concerning transcriptome analysis in watermelon involved in fruit expansion under low-light stress.

Therefore, in this study, natural full light (CK) was recruited as a control to study the mechanism of watermelon fruit enlargement in response to low light (LL). We first analyzed the dynamic changes of fruit development under different light conditions. Then, the transcriptome of the watermelon fruit at four stages of development under low light and CK were analyzed using RNA-seq technology. We study and define the differential expression of genes of watermelon fruit at four different early fruit development stages. Meanwhile, gene annotation, GO classification, KEGG metabolic pathway, etc., were studied. Our hypothesis is that (1) the number and type of genes would differ in two treatments at different stages, and (2) that genes might play important roles during the process of fruit expansion under low light. Our aim is to find key genes and metabolic pathways that control fruit enlargement in the early fruit development stage and the molecular mechanism of low light regulation of the fruit expansion of watermelon. The outcome provides an important theoretical basis for the regulation of watermelon fruit size and fruit development.

2. Results and Discussion

2.1. Response of Watermelon Fruits Morphological and Physiological Dynamic Changes to Low-Light Stress

The early stages of fruit development, including fruit set and exponential growth, are clearly essential for all fruit. Early fruit development is typified by phases of cell division and expansion, which are critical determinants of size and yield. So far, most of studies, including transcriptomic studies, have focused on late development, or a broad range of developmental stages, with only a few studies focused on early fruit development stages [14].

In order to clarify the key period of watermelon fruit enlargement affected by low light, watermelon fruits' vertical and transverse diameters at different time points after pollination were studied (Figure 1A,B). The results showed that the fruits' vertical and transverse diameter of CK and LL increased sharply in 0–15 DAP, and then changed smoothly. The fruits' vertical and transverse diameter of CK in 0–30 days were significantly larger than LL. This also indicated that 0–15 DAP was the rapid expansion period of watermelon fruit enlargement affected by low light intensity. Based on this result, we compared and analyzed the transcriptome of watermelon fruit at 0–15 DAP under LL and CK.

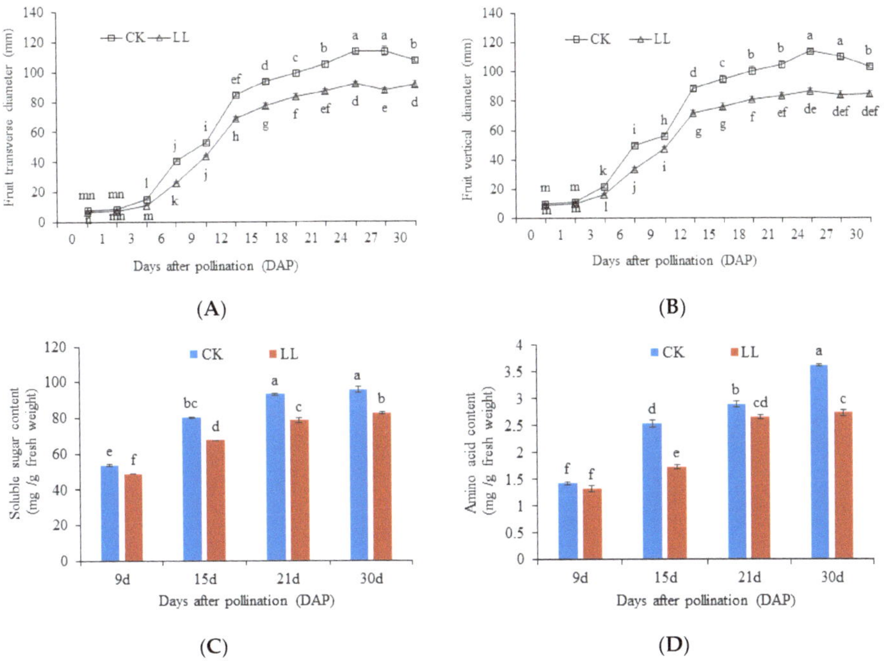

Figure 1. Watermelon fruit dynamic development under low-light stress. (**A**) Fruit vertical diameter. (**B**) Fruit transverse diameter. (**C**) Fruit soluble sugar content. (**D**) Fruit amino acid content. Error bars represent standard errors. Different letters (a, b, c, d, e, f) indicate significant statistical differences ($p < 0.05$).

Sugar can provide metabolic energy for plants and act as substrates and signaling molecules in various metabolic pathways, and could provide carbon skeletons for the synthesis of amino acids and nucleotides [15]. Amino acids are mainly nitrogen-based metabolites that affect watermelon quality. During the increase in pollination days, the soluble sugar content and amino acid content of the LL group and the CK group showed a continuous upward trend, and the soluble sugar content and amino acid content of the LL group were significantly lower than the control group ($p < 0.05$), except that the amino acid at 9 DAP between the LL group and the control group was not significant (Figure 1C,D). Low-light stress reduces the photosynthetic performances of the plants, and the transportation of photosynthetic products from leaves to fruits are also affected [6,9], which ultimately leads to the reduction in the sugar content in fruits. Similar research results have been reported on melon [16], strawberry [6] and jujube [17]. Nitrogen metabolism was disturbed by shading, which induced the decrease in dry matter accumulation, ultimately resulting in the failure of watermelon fruits to expand normally [18]. This means that low-light stress affects the carbon and nitrogen metabolism of watermelon fruit, hindering the formation and transport of substances when watermelon fruit expands under low-light stress, thus making it difficult for watermelon fruit to expand normally.

2.2. Overview of Sequencing and Transcriptome Assembly

In order to explore the transcriptome of watermelon fruit expansion under low-light stress more comprehensively, we established 24 cDNA libraries and carried out RNA-seq, and 545.71 million raw reads were generated (Supplementary Table S1). In total, 527.59 million (96.68% of the raw reads) clean reads (approximately 158.28 Gb clean data) were obtained. On average, 21.98 million clean reads (6.59 Gb clean data) were obtained from each sample. The Q30 percentages ranged from 90.89% to 94.56%, and the average GC percentage was 44.32%. Among the 24 samples, 95.76–96.55% of the clean reads were mapped to the reference genome, and 89.77–91.36% of clean reads were uniquely mapped (Supplementary Table S1).

Analysis of saturation curves of 24 cDNA libraries (genes with FPKM ≥ 0.01) reveals that the gene coverage began to become saturated when clean reads exceeded 10 million (Supplementary Figure S1a). The average clean reads were 21.98 million, which exceeded the saturation threshold, indicating that the sequencing depth was enough for transcriptome studies. The sequencing reads of 24 samples were uniformly distributed from 5′ to 3′ of genes (Figure S1b).

In all, 27,063 unigenes were generated through cufflinks program, including 23,800 genes aligned to the watermelon reference genome and 3623 new genes. Among them, 82.70% (21,116 mapped unigenes and 1263 new genes) of genes were successfully annotated in at least one of seven databases, and only 5.73% (1550) of genes annotated in all databases (Table 1).

Table 1. Summary of the gene annotations in seven databases.

Annotation Database [1]	Annotated_Number	Percentage
COG_Annotation	8060	36.02%
GO_Annotation	9037	40.38%
KEGG_Annotation	5156	23.04%
KOG_Annotation	11,715	52.35%
Pfam_Annotation	17,006	75.99%
Swissprot_Annotation	16,206	72.42%
nr_Annotation	22,343	99.84%
Annotated in all seven Database	1550	5.73%
Annotated in at least one Database	22,379	82.70%
All_Annotated	27,063	100.00%

[1] Clusters of orthologous groups of proteins (COG) database, gene ontology (GO), KEGG ontology (KO), eukaryotic orthologous groups (KOG), protein family (Pfam), reviewed protein sequence database (Swiss-Prot), NCBI non-redundant protein database (NR).

2.3. Identification of Differentially Expressed Genes (DEGs) in Watermelon Fruit

Compared with control, 8837 genes were differentially expressed in watermelon fruits under low-light stress using the criteria of FDR ≤ 0.01 and $|\log_2 \text{fold change}| \geq 1$. There were 466, 1576, 752 and 3380 up-regulated genes and 689, 2346, 1007 and 2112 down-regulated genes after 0, 3, 9 and 15 DAP of low-light stress, respectively (Figure 2a). The numbers of up- and down-regulated genes were similar. At 0 DAP under low-light stress, alanine, aspartate and glutamate metabolism (ko00250) were the most significantly enriched KEGG entries in the down-regulated DEGs, while phenylpropanoid biosynthesis (ko00940) were the most in the up-regulated DEGs. At 3 DAP under low-light stress, genes involved in DNA replication (ko03030) were most enriched in down-regulated DEGs, while genes involved in plant hormone signal transduction (ko04075) were the most enriched in up-regulated DEGs. At 9 DAP of low-light stress, phenylpropanoid biosynthesis (ko00940) were the most significantly entries in down-regulated DEGs, while plant hormone signal transduction (ko04075) were the most entries in up-regulated DEGs. At 15 DAP, genes involved in the biosynthesis of secondary metabolites (ko01110) were most enriched in down-regulated DEGs, while genes involved in phenylpropanoid biosynthesis (ko00940) were enriched in up-regulated DEGs.

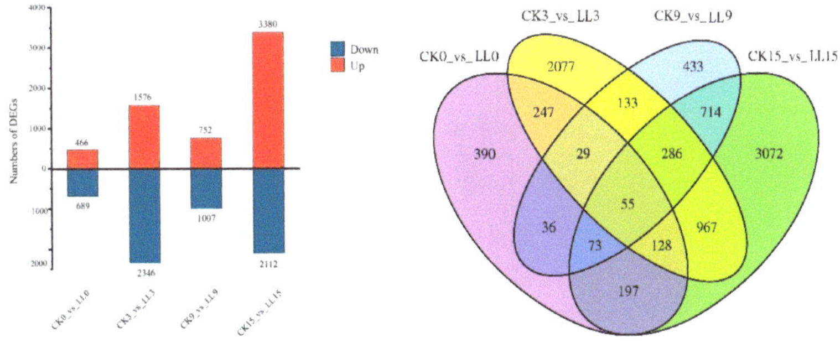

(a) (b)

Figure 2. The numbers of up- and down-regulated differentially expressed genes (DEGs) between low light and control under different developments (**a**), Venn diagram of the differentially expressed genes (DEGs) between low light and control (**b**).

The results of Venn diagram showed that 390, 2077, 433 and 3072 DEGs were specifically at 0, 3, 9 and 15 DAP, respectively. 84 DEGs were all at 0, 3 and 9 DAP, 55 DEGs were shared in four time points (Figure 2b), suggesting their importance in the watermelon fruit expansion response to low-light stress. Among these 55 genes, three genes were up-regulated at four time points, suggesting their importance in watermelon fruit expansion response to low-light stress, while there was no gene down-regulated at four time points. The three genes included iron-sulfur binding oxidoreductase (*Cla008418*), two-component response regulator ARR12-like (*Cla016659*) and AT-hook motif nuclear-localized protein (*Cla006543*).

These 55 genes could be classified into five different categories according to KEGG pathway classification: metabolism, environmental information procession, genetic information processing, organismal systems and stress-related genes (genes with no KEGG pathway classification were assigned to this item). Most of the genes were classified into stress-related genes (Table 2).

Table 2. 55 DEGs that were shared at all four time points.

Gene ID	Gene Description	CK0_vs_LL0	CK3_vs_LL3	CK9_vs_LL9	CK15_vs_LL15
	Metabolism				
	Carbohydrate metabolism				
Cla005337	glucan endo-1,3-beta-glucosidase 13	Up	Up	Down	Up
Cla018009	pyruvate decarboxylase 1	Down	Down	Up	Down
Cla004587	pyrophosphate—fructose 6-phosphate 1-phosphotransferase subunit beta-like	Up	Down	Down	Down
Cla017722	fructose-bisphosphate aldolase cytoplasmic isozyme isoform X2	Down	Down	Up	Down
Cla003920	probable ribose-5-phosphate isomerase 1	Up	Up	Down	Down
Cla015028	alcohol dehydrogenase 1	Down	Down	Up	Down
	Amino acid metabolism				
Cla010147	LOW QUALITY PROTEIN: protein NRT1/PTR FAMILY 2.13-like	Up	Up	Down	Down
Cla010146	protein NRT1/PTR FAMILY 2.13-like	Up	Up	Down	Down
Cla019134	protein NRT1/PTR FAMILY 2.11-like	Up	Up	Down	Down
Cla013371	asparagine synthetase [glutamine-hydrolyzing]	Down	Up	Up	Up
	Nucleotide metabolism				
Cla022500	deoxyuridine 5'-triphosphate nucleotidohydrolase	Up	Down	Down	Down
	Lipid metabolism				
Cla013862	acyl-[acyl-carrier-protein] desaturase 6, chloroplastic	Down	Down	Up	Down

Table 2. *Cont.*

Gene ID	Gene Description	CK0_vs _LL0	CK3_vs _LL3	CK9_vs _LL9	CK15_vs _LL15
Cla012276	cytochrome P450 734A1-like	Up	Up	Down	Up
Cla001235	cytochrome P450 86A8-like	Up	Up	Down	Up
	Biosynthesis of other secondary metabolites				
Cla001815	LOW QUALITY PROTEIN: salicylic acid-binding protein 2-like	Up	Up	Down	Up
Cla016032	scopoletin glucosyltransferase-like	Up	Down	Down	Down
Cla017208	cinnamoyl-CoA reductase 2-like	Up	Down	Up	Up
Cla017207	cinnamoyl-CoA reductase 2-like	Up	Down	Up	Up
	Environmental Information Procession				
	Signal transduction				
Cla002572	BTB/POZ and TAZ domain-containing protein 1	Down	Down	Up	Up
Cla017611	uncharacterized protein LOC103486691	Down	Down	Up	Up
Cla016659	two-component response regulator ARR12-like	Up	Up	Up	Up
Cla008326	putative cell division cycle ATPase	Up	Up	Down	Up
Cla008169	uncharacterized protein LOC101203772	Down	Down	Up	Up
MSTRG.1055	defensin-like protein 6	Up	Up	Up	Down
Cla006543	AT-hook motif nuclear-localized protein 15	Up	Up	Up	Up
	genetic information processing				
	Folding, sorting and degradation				
Cla023118	U-box domain-containing protein 26	Down	Down	Up	Up
	Mitochondrial biogenesis				
Cla009892	mitochondrial translocator assembly and maintenance protein 41 homolog isoform X1	Down	Down	Up	Down
	Organismal Systems				
	Environmental adaptation				
Cla007307	WRKY70	Down	Down	Down	Up
Cla007656	probable WRKY transcription factor 31	Up	Up	Down	Down
Cla018486	probable receptor-like protein kinase At5g39030	Up	Down	Down	Down
Cla007979	probable transcription factor GLK2	Down	Down	Up	Up
	stress-related genes				
Cla023376	Early nodulin-like protein 2 [Glycine soja]	Down	Down	Down	Up
MSTRG.17433	hypothetical protein Csa_6G059230	Up	Up	Down	Down
Cla010283	hypothetical protein Csa_2G237730	Down	Down	Up	Up
Cla002904	hypothetical protein Csa_3G822280	Down	Down	Up	Up
Cla015065	hypothetical protein Csa_7G397010	Down	Down	Down	Up
Cla008418	Iron-sulfur binding oxidoreductase	Up	Up	Up	Up
Cla010875	BAG family molecular chaperone regulator 6-like	Up	Down	Down	Down
Cla012888	classical arabinogalactan protein 10-like	Up	Up	Down	Up
Cla013835	formin-like protein 20-like	Up	Up	Down	Down
Cla013527	seed biotin-containing protein SBP65	Down	Down	Up	Up
Cla004456	uncharacterized protein LOC103491021	Down	Down	Up	Down
Cla007700	uncharacterized protein LOC103495564	Down	Down	Up	Up
Cla011574	14 kDaproline-rich protein DC2.15-like	Up	Up	Down	Down
Cla018055	probably inactive leucine-rich repeat receptor-like protein kinase At3g28040	Down	Down	Up	Up
Cla019959	universal stress protein A-like protein isoform X2	Down	Down	Up	Up
Cla014570	dehydrin DHN1-like	Down	Up	Up	Up
Cla017570	F-box/kelch-repeat protein At2g44130-like	Down	Down	Up	Up
Cla009891	haloacid dehalogenase-like hydrolase domain-containing protein 3	Down	Down	Up	Down
Cla021693	sugar transporter ERD6-like 16 isoform X2	Up	Down	Down	Down
Cla011361	bidirectional sugar transporter SWEET4-like	Down	Up	Down	Up
MSTRG.15291	no hit found	Down	Down	Up	Up
MSTRG.20101	no hit found	Down	Up	Up	Up
MSTRG.967	no hit found	Up	Up	Down	Down
MSTRG.968	no hit found	Up	Up	Down	Down

Eighteen genes were divided into the metabolism category. Among these, plant glucan endo-1,3-β-glucosidases have been involved in diverse physiological and developmental processes including microsporogenesis, pollen germination, fertilization, response to wounding and cell division [16]. In our study, glucan endo-1,3-β-glucosidases was up-regulated by low-light stress, except for 9 DAP. Plant pyruvate decarboxylases (PDC) catalyze the decarboxylation of pyruvate to form acetaldehyde and CO_2, and PDC derives the fermentation pathway involved in energy and material metabolism, responding to development, biotic and abiotic stresses [19]. PDC was down-regulated by low-light stress (except for 9 DAP) of watermelon fruit in our study, it may be a way to reduce the energy supply to fruit development under low-light stress. Through generating various intermediate products (such as reductant and pyruvate), glycolysis is the main pathway that supports respiration in plants. During the phosphorylation of Fru-6-P to Fru-1, 6-P2, PFP utilizes pyrophosphate (PPi) as an alternative phosphoryl donor in place of ATP, thereby providing an energy advantage to plants [20]. In our study, gene expression of *Cla004587* (except for 0 DAP) and *Cla017722* (except for 9 DAP) was down-regulated by low-light stress. This showed that low light can inhibit the energy supply of watermelon fruit development. Alcohol dehydrogenase (ADHs) gene expression produces enzymes that are not only active when plants are subjected to various stresses, but also during all developmental stages of plants under suitable growing conditions [21]. During the development of plant organs, ADH participates in glycolysis to provide energy for plants and regulate the metabolism of substances [22]. ADHs showed down-regulated patterns by low-light stress in this study (except for 9 DAP). This means that low light can affect the energy supply by inhibiting the expression of ADH during watermelon fruit development.

Three NRT1/PTR family (NPF) proteins can be detected in amino acid metabolism and were originally identified as nitrate or di/tri-peptide transporters. Recent studies revealed that this transporter family also transports the plant hormones and secondary metabolites (glucosinolates) [23]. The gene expression patterns of three NPF genes were up-regulated at 0 and 3 DAP and down-regulated at 9 and 15 DAP by low-light stress; this may lead to the changes in amino acid composition, hormone content and secondary metabolites during watermelon fruit development, which affected the size of watermelon fruit. Asparagine synthetase is able to assimilate ammonium in plants and it might be involved in N remobilization [24]. Asparagine synthetase was up-regulated at 0 DAP and down-regulated at 3, 9 and 15 DAP in our study. The reason may be that low-light stress can affect N metabolism. *Cla013862* encode acyl-[acyl-carrier-protein] desaturase (AAD) 6 (chloroplastic), and *Cla012276* and *Cla01235* encode cytochrome P450, which were classified into lipid metabolism. Interestingly, the expression patterns of AAD (chloroplastic) and cytochrome P450 are just opposite. Four genes (*Cla001815*, *Cla016032*, *Cla017208*, *Cla017207*) were classified into biosynthesis of other secondary metabolites. Except for copoletinglucosyltransferase-like, the expression patterns of the other three genes were first increased, then decreased and then increased again.

Seven genes involved in signal transduction, such as two-component response regulator ARR12-like (ARR12), ARR12 mediate the cytokinin regulated gene expression as myeloblastosis (MYB)-like transcription factors (TFs) and have both receiver and output domains. Type B Arabidopsis Response Regulators (ARRs) of Arabidopsis thaliana are transcription factors that act as positive regulators in the two-component cytokinin signaling pathway [25]. In our study, the expression level of ARR12 under low-light stress was higher than that of CK at four time points. The results showed that high level ARR12 affects watermelon fruit expansion and development by promoting the cytokinin signaling pathway under low-light stress. The AT-hook motif nuclear-localized protein (AHL) gene family, which encodes embryophyte-specific nuclear proteins with DNA binding activity, regulate gene expression and affect various developmental processes in plants, such as the modulation of GA and auxin biosynthesis, ABA-mediated stress growth regulation [26], etc. The AHL gene family regulates plant growth and development through forming DNA-protein and protein-protein homo−/hetero-trimericcomplex [27]. *Cla006543* (AT-hook motif nuclear-localized protein (AHL) 15) was also shown to exist at a higher level under low-light stress than CK at four time points. This means the AHL genes may play important roles through regulation of GA, ABA or auxin biosynthesis in watermelon expansion under low-light stress. *Cla023118* and *Cla009892* encode U-box domain-containing protein and mitochondrial translocator assembly and maintenance protein 41 homolog isoform X1, respectively, which are involved in genetic information processing. Two genes showed higher gene expression level at 0 and 3 DAP under low-light stress than CK, which mean that low light can affect watermelon fruit expansion through inhibiting genetic information processing.

Four genes are involved in environmental adaptation, which belong to organismal systems. Among them, two genes (*Cla007307*, *Cla007656*) encode WRKY transcription factor, one gene (*Cla018486*) encodes receptor-like protein kinase, and one gene (*Cla007979*) encodes transcription factor GLK2. Gene expression level of *Cla007307* (WRKY70) was down-regulated by low-light stress (except for 15 DAP), while *Cla007656* was up-regulated by low-light stress at 0 and 3 DAP and down-regulated by low-light stress at 9 and 15 DAP. Receptor-like protein kinase was down-regulated by low-light stress (except for 0 DAP) and transcription factor GLK2 was down-regulated at 0 and 3 DAP and up-regulated at 9 and 15 DAP by low-light stress. These results suggested that there is a complicated network for watermelon expansion and development under low-light stress.

A total of 24 genes were classified as stress-related genes. Among them, *Cla023376*, similar to early nodulin-like protein, was important for the transport of nutrients, solutes, amino acids or hormones and for major aspects of plant development [28]. In our study, nodulin-like protein 2 was down-regulated by low-light stress (except for 15 DAP), which showed that low light can inhibit the watermelon expansion and development by affecting the transportation of nutrients, amino acid, etc., to cause the fruit to be difficult to expand. Classical arabinogalactan protein 10-like (*Cla012888*) can play important roles in abiotic stress resistant, such as cold stress [29], and was up-regulated by low-light stress (except for 9 DAP). *Cla014570*, encoding dehydrins, belongs to the late embryogenesis abundant (LEA) protein family and was involved in responses to multiple abiotic stresses, such as cold and drought stress [30]. In our study, dehydrins were up-regulated by low-light stress (except for 0 DAP). Two genes (*Cla021693*, *Cla011361*) were both sugar transporters, which play important roles in plant growth and development, as well as biotic and abiotic stresses [31]. In addition, *Cla021693* (sugar transporter ERD6-like 16 isoform X2) was down-regulated by low-light stress (except for 0 DAP), while *Cla011361* (bidirectional sugar transporter SWEET4-like) was down-regulated at 0 and 9 DAP and up-regulated at 3 and 15 DAP by low-light stress. This indicated that low light affected the sugar transport, which led to the difficulty of watermelon fruit expansion under low light. It was noteworthy that *Cla008418* (Iron-sulfur binding oxidoreductase) was up-regulated by low-light stress at four time points.

2.4. Hierachical Clustering Analysis

Hierarchical clustering was performed for the 8837 DEGs based on similarity of gene expression (Figure 3). These DEGs were classified into nine gene clusters, and the genes' expression pattern of CK and LL at different stages was different (Figure 3). Cluster 1 showed a relatively high expression level of CK compared with LL. The genes in cluster 1 mainly included genes encoding proteins related to replication, recombination and repair (*Cla003671*), posttranslational modification, protein turnover, chaperones (*Cla004677*, *Cla005777*), transcription (*Cla000971*, *Cla015165*) and carbohydrate transport and metabolism (*Cla011246*, *Cla020836*). The expression of the above genes was inhibited under low-light stress. The results showed that these genes may be related to watermelon fruit expansion under low-light stress. Meanwhile, cluster 3 and cluster 5 showed a relatively high transcript level in LL compared with CK. For example, these two clusters mainly included genes encoding proteins related to carbohydrate transport and metabolism (*Cla020563*, *Cla016546* and *Cla015950*), signal transduction mechanism (*Cla016899* and Cla016659), replication, recombination and repair (*Cla015651*, *Cla013640*), amino acid transport and metabolism (*Cla019133*, *Cla003285*) and transcription (*Cla007586*, *Cla016837*). Gene expression levels in cluster 7 and cluster 8 decreased with the development of fruit, but the gene expression pattern in cluster 2 and cluster 9 was contrary. The rest of the DEGs were clustered into cluster 4 and cluster 6.

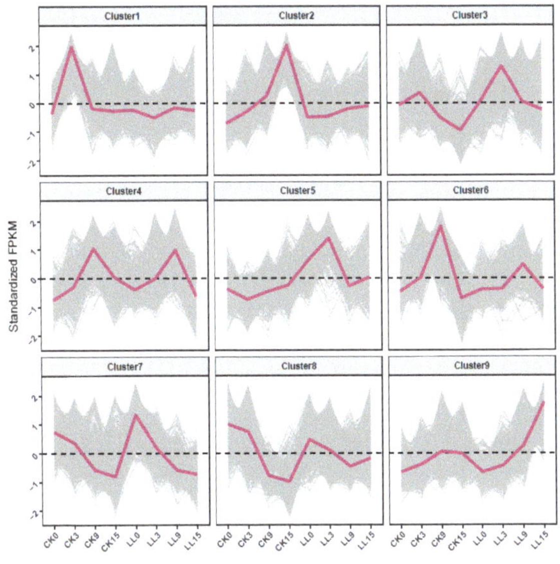

(**a**)

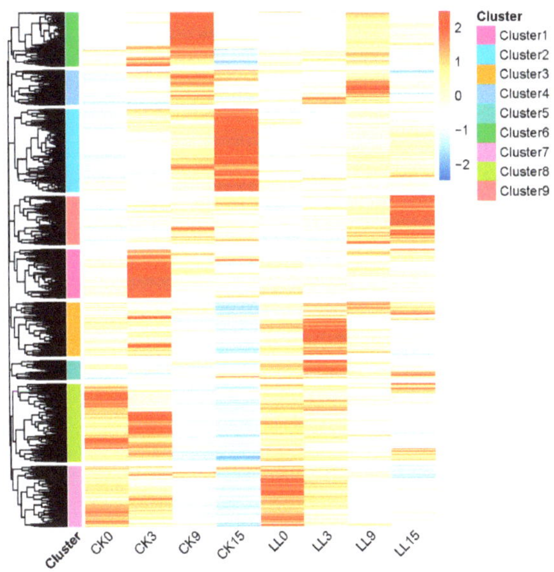

(**b**)

Figure 3. Clustering analysis of 8837 DEGs between CK and LL. (**a**). Hierarchical clustering of 8837 DEGs. (**b**). Expression patterns of the 8837 DEGs in the nine clusters.

GO enrichment analysis of the three clusters may be relevant to watermelon expansion under low-light stress. The results showed that the GO terms related to replication, recombination and repair, posttranslational modification, protein turnover, chaperones and transcription were highly enriched in cluster 1, such as GO:0006261 (DNA-dependent DNA replication), GO:0042138 (meiotic DNA double-strand break formation), GO:0048589 (developmental growth), GO:0003713 (transcription coactivator activity). In addition, we found that cluster 3 and 5 showed that GO:0007155 (cell adhesion) and GO:0045010 (actin nucleation) were most highly enriched.

2.5. Weighted Correlation Network Analysis of Low-Light Stress Responsive Genes Related to Watermelon Fruit Expansion

Weighted gene coexpression network analysis (WGCNA) is used frequently to explore effectively the relationships between genes and phenotypes. WGCNA can also analyze the target genes at a network-level [32] in order to identify the low light responsive genes correlated with watermelon fruit expansion. In this study, 10 modules were identified from the RNA-seq data (Figure S2a).

The module eigengenes for 10 modules were correlated with different test indicators. Analysis of the module-trait relationship showed that the module 'MElightpink4' and 'MElavenderblush1' were highly correlated with watermelon fruit expansion under low-light stress. 'MElightpink4' module was positively correlated with FVD (cor = 0.88, $P = 1 \times 10^{-8}$), FTD (cor = 0.87, $P = 3 \times 10^{-8}$), SFW (cor = 0.81, $P = 2 \times 10^{-8}$), AA (cor = 0.88, $P = 2 \times 10^{-8}$) and SS (cor = 0.85, $P = 2 \times 10^{-7}$). 'MElavenderblush1' module was negatively correlated with FVD (cor = −0.95, $P = 4 \times 10^{-13}$), FTD (cor = −0.95, $P = 3 \times 10^{-12}$), SFW (cor = −0.78, $P = 7 \times 10^{-6}$), AA (cor = −0.89, $P = 4 \times 10^{-9}$) and SS (cor = −0.88, $P = 1 \times 10^{-8}$) (Figure S2b). The module 'MElightpink4' included 4059 genes mainly containing genes related to "posttranslational modification, protein turnover, chaperones", "carbohydrate and amino acid transport and metabolism" and transcription. 5643 genes, mainly including genes encoding transcription, replication, recombination and repair and signal transduction mechanisms were identified in the module 'MElavenderblush1'. The results were almost consistent with earlier hierarchical clustering analysis.

To further understand the mechanism of watermelon fruit expansion under low-light stress, the GO enrichment and KEGG pathway of genes in the three modules were analyzed. The unigenes in the "MElightpink4' module were mainly enriched in aromatic amino acid family biosynthetic process (GO:0009073), proteolysis (GO:0006508), isopentenyl diphosphate biosynthetic process (GO:0019288), photorespiration (GO:0009853), ubiquitin-dependent protein catabolic process (GO:0006511) and glycolytic process (GO: 0006096). While the highly enriched terms of the 'MElavenderblush1' were associated with translation (GO:0006412) and embryo sac egg cell differentiation (GO:0009560). The most significantly entries in KEEG analysis of the 'MElightpink4' and 'MElavenderblush1' module were biosynthesis of secondary metabolites (Ko:map01110) and ribosome (Ko:map03010), respectively (Tables S2 and S3).

2.6. Functional Classification of DEGs

Through the WEGO database, the DEGs were classified into 14 categories for cellular components, 11 categories for molecular functions and 22 categories for biological processes. 9037 unigenes and 3217 DEGs were assigned GO terms and a hierarchical relationship picture is drawn in Figure S3. The DEGs involved in cell part (1187), cells (1179) and organelles (867) were the most abundant entries in cellular components. The two most common categories for molecular functions were catalytic activity (1794) and binding (1559). For biological processes, metabolic processes (2018) and cellular processes (1517) were the most two abundant catalogs.

GO enrichment analyses of DEGs were also carried out by topGO. Oxidation-reduction process (GO: 0055114), salicylic acid-mediated signaling pathway (GO: 0009863) and cell wall organization (GO: 0071555), were the top three enriched in biological process (Table S4). Integral component of membrane (GO: 0016021), extracellular region (GO: 0005576), apoplast (GO: 0048046) and plant-type cell wall (GO: 0009505) were the top four enriched terms in cellular processes (Table S5). Cellulose synthase (UDP-forming) activity (GO: 0016760), oxidoreductase activity, acting on paired donors, with incorporation or reduction in molecular oxygen (GO: 0016705) and oxidoreductase activity (GO: 0016491) were the top three enriched terms in molecular functions (Table S6). It is not surprising that many of the DEGs identified in this study are involved in or associated with cell wall metabolism. Plant cell walls are dynamic structures that determine and maintain the size and the shape of the cells and act as a protective barrier, and they both resist the stresses generated by the hydrostatic pressure of the cells and have the ability to undergo the irreversible deformation associated with cell expansion [33]. In our study, gene expression patterns of the DEGs (with KOG classification was cell wall/membrane/envelop biogenesis and cell cycle control, cell division, chromosome partitioning) were analyzed (Figure 4). The results showed that the number of down-regulated genes was higher than that of the up-regulated genes at 0 DAP and 3 DAP, especially at 3 DAP, which means that genes related to cell wall and cell division were inhibited by low-light stress and low light affects cell division and replication at the initial stage of watermelon fruit pollination. The number of up-regulated genes was significantly more than that of down-regulated genes at 15 DAP, indicating that these genes can resist low-light stress in watermelon fruit development.

Pathway enrichment analysis can identify the major biochemical and signal transduction pathways in which the DEGs were involved. As shown in Figure 5, biosynthesis of secondary metabolites (ko01110), phenylpropanoid biosynthesis (ko00940), phenylalanine metabolism (ko00360), carotenoid biosynthesis (ko00906), alanine, aspartate and glutamate metabolism (ko00250), starch and sucrose metabolism (ko00500), glycolysis/gluconeogenesis (ko00010), carbon fixation in photosynthetic organisms (ko00710), flavonoid biosynthesis (ko00941), amino sugar and nucleotide sugar metabolism (ko00520) were the ten most significantly entries.

As a substrate of protein biosynthesis, amino acid phenylalanine is necessary for the survival of all cells. Thus, the routing of newly synthesized phenylalanine into protein is a primary metabolic pathway. In plants, phenylalanine is also a substrate of phenylpropanoid metabolism [34]. Phenylpropanoids contribute to plant responses to biotic and abiotic stress. They are not only indicators of plant stress responses upon variation of light or mineral treatment, but also key mediators of the plant's resistance towards pests. The general phenylpropanoid metabolism generates an enormous array of secondary metabolites based on the few intermediates of the shikimate pathway as the core unit [35]. Secondary metabolites play an important role in the adaptation of plants to the changing environment and in overcoming stress constraints. Carotenoids, including lycopene, and flavonoids are secondary metabolites in watermelon fruit, as well as in other fruits such as tomato, suggesting carotenoids and flavonoids are important to watermelon fruit expansion or development under low-light stress [36]. These results showed that these multiple pathways might contribute to fruit development and expansion under low-light stress within a complicated pathway network. In addition, the roles of oxidation-reduction, secondary metabolites, carbohydrate and amino acid metabolism are particularly critical.

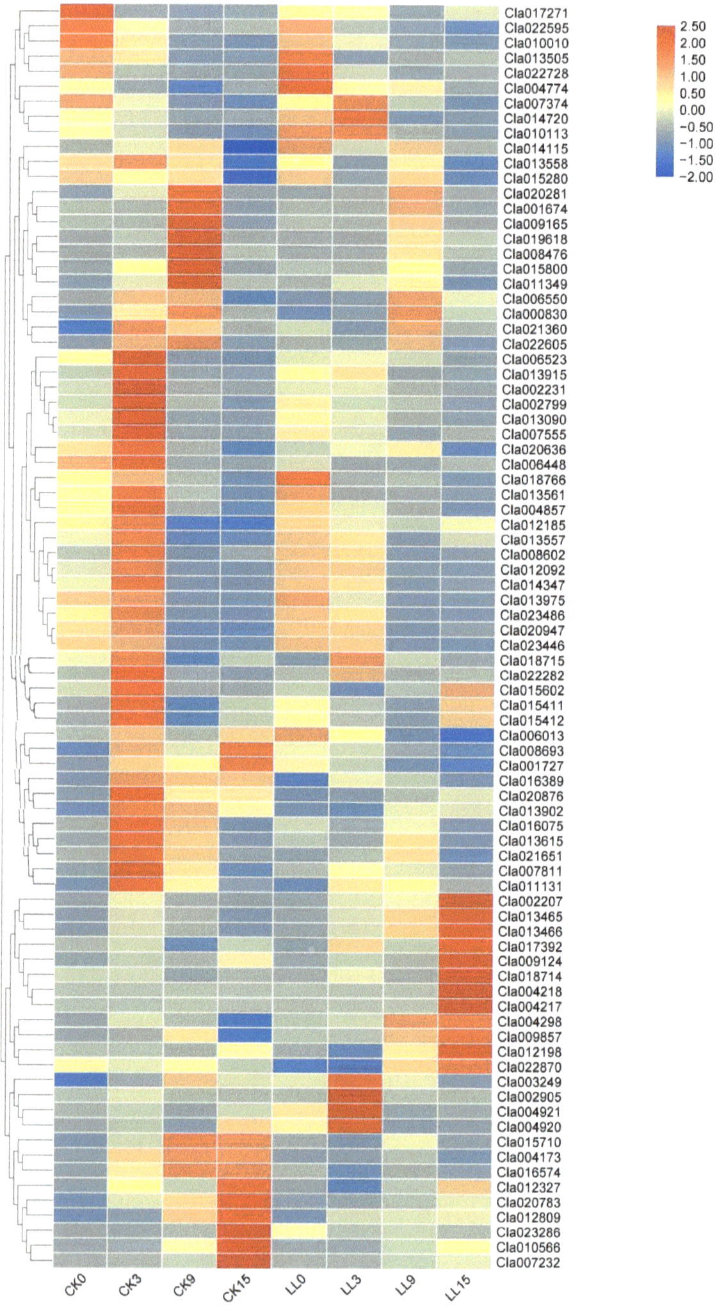

Figure 4. The raw FPKM values of the DEGs (with KOG classification was cell wall/membrane/envelop biogenesis and cell cycle control, cell division, chromosome partitioning) were first normalized by logarithmic method and a heat map was constructed to show the different expression profiles using the Heml software (version 1.0, http://hemi.biocuckoo.org/. accessed on 9 July 2020). *X*-axis, samples; *Y*-axis, differentially expressed gene names.

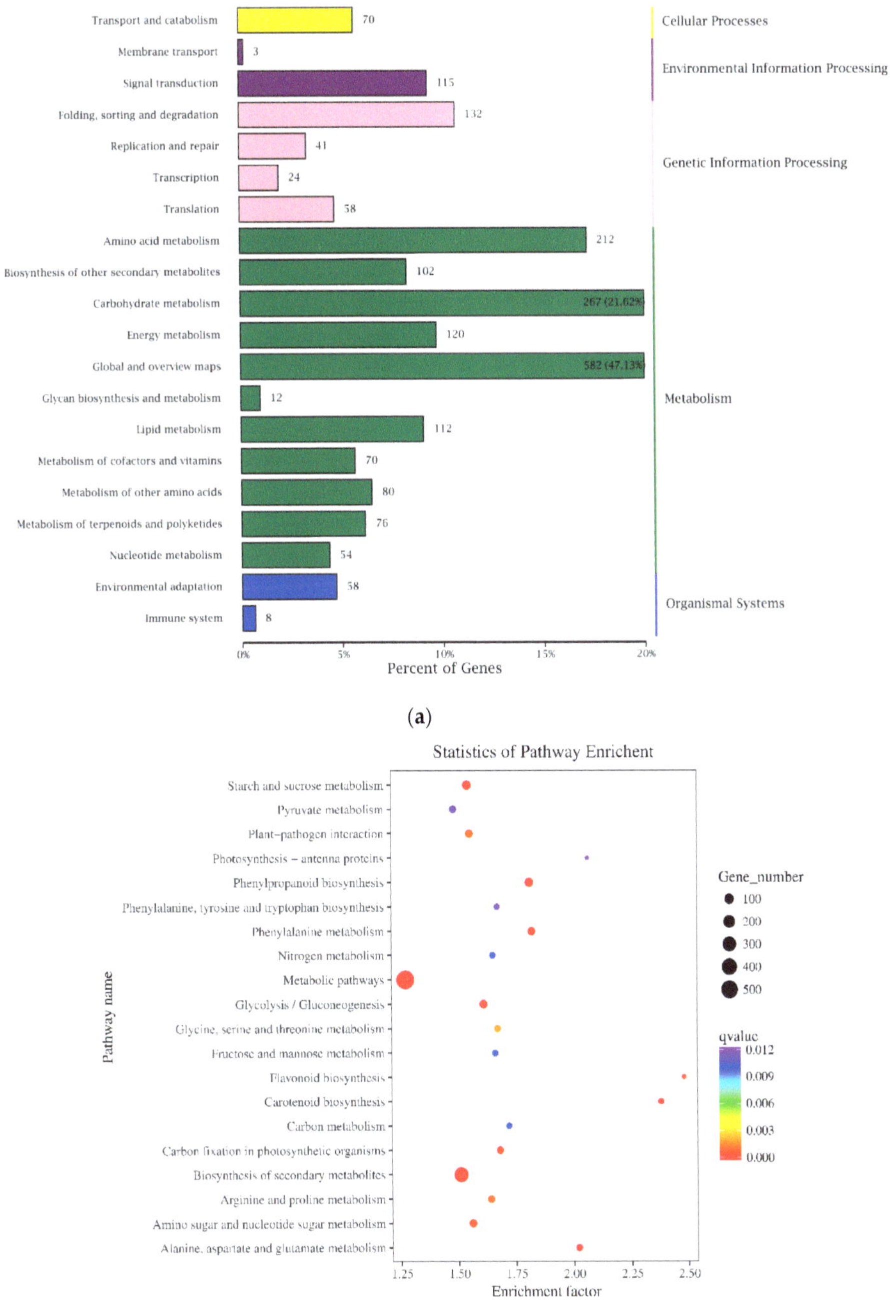

(a)

(b)

Figure 5. KEGG pathway enrichment analysis. (**a**) Statistic analysis of annotated genes in KEEG pathways, (**b**) scatterplot of KEEG pathway enrichment for DEGs.

2.7. Analysis of Transcription Factors (TFs) Involved in Watermelon Fruit Development and Response to Low-Light Stress

Transcription factors are essential for the regulation of gene expression through binding to specific *cis*-acting elements in their regulated genes. We found 644 TFs in DEGs, which can be classified into 53 TF families that were present in the plantTFcat database including ERF (12.27%), bHLH (11.65%), MYB (9.32%), NAC (6.21%), C2H2 (5.90%) and WRKY (4.66%) (Figure 6). Of these differentially expressed TFs, bHLH, ERF, NAC and WRKY are all the most abundant in the 0, 3, 9 and 15 DAP (Figure S4). This means that these four TFs were important for watermelon fruit expansion under low-light stress. ERF have been shown to be intimately connected to plant development, defense responses and stress signaling pathways. The development of a plant organ to a specific size and shape is controlled by cell proliferation and cell expansion. Differentiation of xylem elements involves cell expansion, secondary cell wall (SCW) deposition, etc. The study shows ERF139 as a transcriptional regulator of xylem cell expansion and secondary cell wall formation, and possibly in response to the osmotic changes of the cells [37]. Many studies showed that bHLH plays critical roles in plant growth and development, metabolic regulation, and response to environmental changes, such as fruit development [38] and drought and salt stress [39]. MYB plays a vital role in organ development by directly affecting cell wall structure and/or cytoplasmic growth or indirectly regulating through the ethylene and/or ABA signaling pathways [40]. As a major component of plant cell walls, lignin plays important roles in mechanical support, water transport, and stress responses. The study shows NAC was associated with fruit lignification by activating genes involved in lignin biosynthesis [41]. WRKY can also play significant roles in stress responses and cellular growth [42]. Some TFs from the same family were up-regulated while others were down-regulated (Figure S5). The results demonstrated that members of the same TF family may play different roles in fruit development under low-light stress, and the transcriptional network for watermelon fruit expansion under low-light stress is intricate and complex.

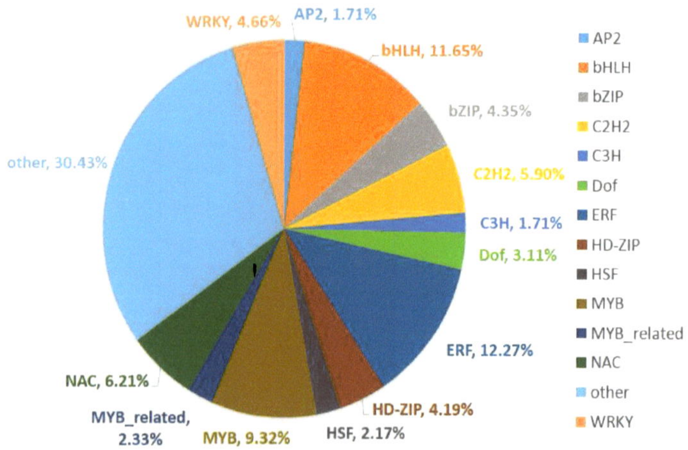

Figure 6. Distribution of differentially expressed TFs.

2.8. Validation of RNA-Seq Data by qRT-PCR Analysis

To verify the RNA-seq data, 14 DEGs were selected randomly for qRT-PCR analysis (Figure 7a). In order to facilitate the comparison of the expression data between RNA-seq and qRT-PCR, the relative expression level was converted to $\log_2$ fold change. The qRT-PCR results showed a high consistency (linear regression equation $y = 0.7687x - 0.0047$, $R^2 = 0.7475$) with RNA-seq data (Figure 7b), indicating the high reliability of RNA-seq expression profile in this study.

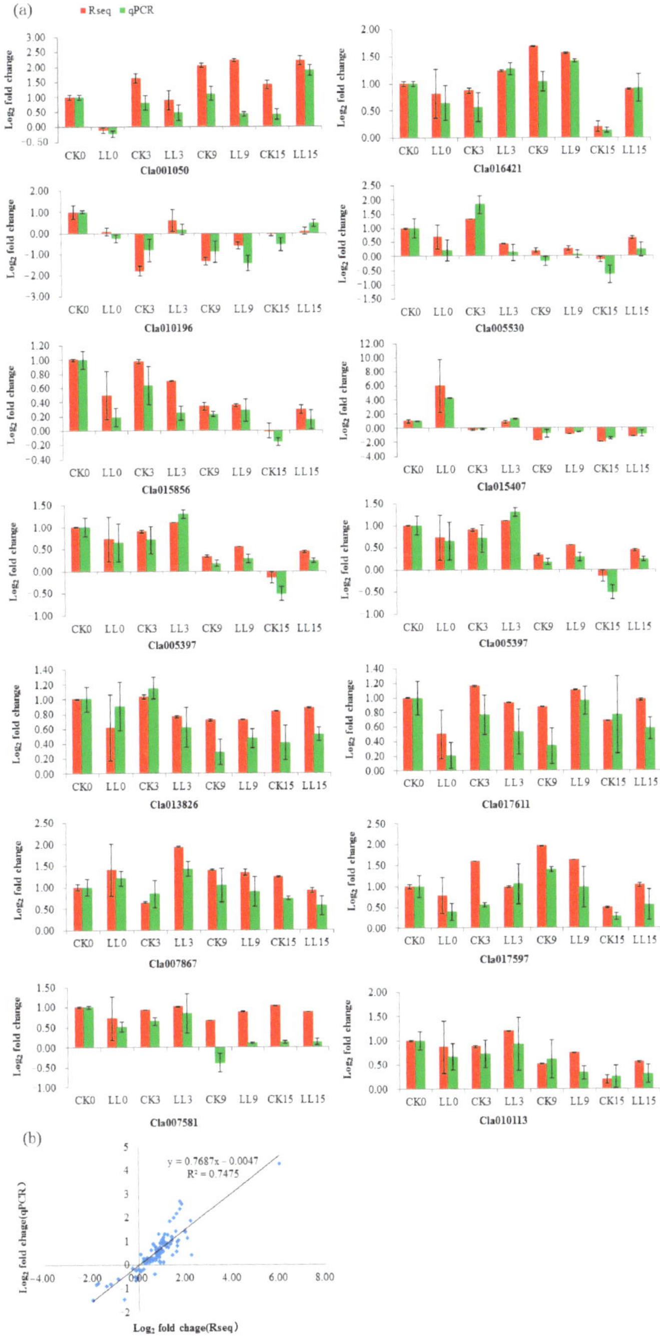

Figure 7. Verification of the expression of selected DEGs by qRT-PCR. (**a**) Log_2 fold change of 14 genes under control and low-light stress at four developments of watermelon fruit. The bars represent the mean of three biological replicates ± standard errors. (**b**) Log_2 fold change of RNA-seq (*X*-axis) and qRT-PCR (*Y*-axis).

3. Materials and Methods

3.1. Plant Materials

This research was performed in a steel frame plastic greenhouse with internal circulation fan, which can ensure the same temperature in the plastic greenhouse, at Luhe production base in Jiangsu Academy of Agricultural Sciences. Watermelon variety 'Sumi No.8' (Jiangsu Academy of Agricultural Sciences) was used as plant materials. 'Sumi No.8' is a new watermelon hybrid with early maturity and small fruit (an average single fruit weight is around 1.5 kg). The fruit flesh is yellow, and the fruit development period is about 30 d after pollination. Seeds were sown in plug tray (54 cm length, 27 cm width, 50 holes) filled with vegetable seedling commercial organic substrate (HengAoda Fertilizer Technology Co., Ltd., Lianyungang, China). The healthy and uniform size seedlings at the four-leaf stage were transplanted into the plastic pots (height, 26.5 cm; diameter, 31 cm) containing the organic culture substrate (cassava residue: peat: vermiculite = 2:1:1 in volume). Fertilizer, control of diseases and insects and other agronomy control was performed according to standard management practices.

Treatments were carried out 7 days before pollination. The low light (LL) groups of watermelon were grown under one layer of black sunshade net (Xinqiao agricultural screen Co., Ltd., Taizhou, China) and the light intensity of LL was 150–650 μmol quanta m^2 s^{-1}, while the control group (CK) was grown under natural full light (310–1300 μmol quanta m^2 s^{-1}). All flowers for each experiment were hand pollinated in the 9–10 node order, and only one fruit remained per plant. The flesh of fruit center was harvested at 0, 3, 9 and 15 days after flowering, and immediately frozen in liquid nitrogen and stored in a −80 °C freezer until use. At each sampling point, three fruits' flesh were collected in one biological replicate, and three separate, biological replications of the fruits' flesh were performed for RNA-seq.

3.2. Morphological and Physiological Index Measurements and Statistical Analysis

Prior to the harvests, fruit vertical and transverse diameter were measured at 0, 1, 3, 6, 9, 12, 15, 18, 21, 24, 27 and 30 DAP. Soluble sugar content and amino acid (AA) content were measured using a Plant Soluble Sugar Content Assay Kit (Suzhou Keming Biotechnology Co., Ltd. Suzhou, China) with 0.1 g samples of watermelon flesh at 9, 15, 21, 30 DAP.

3.3. cDNA Preparation and RNA Sequencing

The RNA of 24 samples was extracted using RNAprep Pure Plant Kit (TIAGEN, Beiing, China). After DNAse I treatment, mRNA was purified from the total RNA using the Oligotex mRNA Midi Kit (QIAGEN, Dusseldorf, Germany). The first cDNA strand was synthesized from purified mRNA fragments by reverse transcription with random hexamer adaptors (Invitrogen, California, MD, USA). Double strand cDNA was then synthesized using the SMART cDNA Library Construction kit (Clontech, Mountain View, CA, USA) according to the manufacturer's instructions. The cDNA fragments with suitable lengths and insert sizes were selected by AMPure XP beads to construct cDNA libraries. The cDNA libraries' quality was checked by Qubit2.0 and Agilent 2100 before they were sequenced using the IlluminaHiSeq 2500 platform, and the 150 bp paired-end reads were generated by Genepioneer Biotechnologies Co, Ltd. (Nanjing, China).

3.4. Transcriptomic Analysis

The raw sequence data have been submitted to the National Center for Biotechnology Information (NCBI) databases, for which the study accession was SRP243255. Clean reads were obtained by removing duplication sequences, adaptor sequences and low-quality reads (ambiguous sequences with 'N' percentage values ≤ 3 and the percentage of low-quality bases less than 3 is ≥50%) from raw reads. The cleaned reads were subsequently mapped to the watermelon reference genome (ftp://cucurbitgenomics.org/pub/cucurbit/genome/watermelon/97103/v1/watermelon_v1.genome.gz. accessed on 29 September 2018) by HISAT2 software (v2.1.0). Subsequently, the sequencing saturation, coverage analysis and the genomic distributions in CDS (exons), introns, and intergenic regions were analyzed.

3.5. Digital Gene Expression Tag Profiling and Identification of Differentially Expressed Genes (DEGs)

The expression of all genes was determined with FPKM (Fragments PerKilobase of exon per million fragments Mapped) values using the software Cufflinks software (v2.2.1) (http://cole-trapnell-lab.github.io/cufflinks, accessed on 29 September 2018). R package DESeq was used to identify DEGs by comparing the low-light stress and control at the same time point, with the false discovery rate (FDR) < 0.05 and $|\log_2$ fold change$| \geq 1$.

3.6. Gene Annotation and Functional Annotation and Analysis

All genes were compared against NR, GO, COG, KOG, Swiss-Prot, KO using BLASTn (v2.2.26) with an E-value cut off at 10^{-5} and searched against the Protein family (Pfam) database by hmmscan (v3.0) (https://www.ebi.ac.uk/Tools/hmmer/search/hmmscan/, accessed on 29 September 2018). Based on GO classification, all genes were analyzed for their functional characteristics and gene expression profiles under low-light stress and controlled conditions. GO enrichment analyses of DEGs were performed using singular enrichment analysis (SEA) method with $p < 0.01$ and FDR < 0.05 by topGO, and the watermelon genome was set as background. KOBAS (v2.0) (https://www.biostars.org/p/200126/) was employed in validation of gene expression (FDR < 0.05) for pathway enrichment analysis. All genes were BLAST against the Plant Transcription Factors Database (v3.0) (http://plntfdb.bio.uni-potsdam.de/v3.0/, accessed on 29 September 2018) with an E-value cut off at 10^{-5}.

3.7. Weighted Correlation Network Analysis (WGCNA)

We built a correlation network using R package WGCNA. The adjacency matrix was created through calculating the pearson's correlations. The pickSoftThreshold was used to calculate the soft thresholding power β. The topological overlap measure (TOM) was calculated by the adjacency matrix and TOM dissimilarity was used to build the dendrogram. The modules were detected as branches of the dendrogram using the dynamic tree-cut with a cut off height of 0.30 to merge the branches to final modules. The module eigengene (ME) value was calculated and used to estimate the association of modules with low-light stress responsive genes related to physiological index.

3.8. qRT-PCR Analysis

We chose 14 DEGs with different expression patterns for validation RNA-seq data by quantitative reverse transcription-polymerase chain reaction (qRT-PCR). *Actin* (*Cla007792*) was used as reference gene. The primers were designed using Primer 5.0. Primer sequences and gene annotations are listed in Table S7. QRT-PCR assays were conducted with three biological and three technical replicates. RNA for each sample was reverse-transcribed into first-strand cDNA using M-MLV reverse transcriptase (TaKaRa, Tokyo, Japan). QRT-PCR was conducted on an ABI 7300 real-time PCR system (Applied Biosystems, Foster City, CA, USA) using the SYBR Premix Ex Taq II reagent (Takara, Tokyo, Japan). QPCR conditions were set at 95 °C pre-denaturation for 3 min, 40 cycles of 95 °C for 15 s, 60 °C for 30 s, and 72 °C for 15 s. Gene expression levels were calculated by previous $2^{-\Delta\Delta Ct}$ method [43]. The correlation of RNA-seq with qRT-PCR analysis was calculated using SAS (v9.3) proccorr based on $\log_2$ fold changes.

4. Conclusions

In conclusion, this study confirmed that 0–15 DAP was the rapid expansion period of watermelon fruit under low-light stress and CK. Low light significantly inhibited fruit expansion, and the contents of soluble sugar and amino acids were also decreased. As far as we know, this work is the first study to provide comprehensive sequencing and

DEG profiling data for a dynamic view of the transcriptomic variation in watermelon fruit development under low-light stress. In total, total 8837 DEGs were obtained in watermelon fruit flesh under low-light stress compared with the control, and 55 DEGs were shared among four stages of fruit development. At 0 DAP under low-light stress, alanine, aspartate and glutamate metabolism (ko00250) were the most significantly enriched KEGG entries in the down-regulated DEGs, while phenylpropanoid biosynthesis (ko00940) were the most enriched in the up-regulated DEGs. GO and KEGG analysis of the DEGs showed that these DEGs were mainly involved in oxidation-reduction, secondary metabolites, carbohydrate and amino acid metabolism, which indicated that the genes and pathways may be related to fruit expansion under low-light stress. In addition, the function and metabolic pathway of up-regulated and down-regulated DEGs at different time points were also different. At 3 DAP under low-light stress, genes involved in DNA replication (ko03030) were the most enriched in down-regulated DEGs, while genes involved in plant hormone signal transduction (ko04075) were the most enriched up-regulated DEGs. At 9 DAP of low-light stress, phenylpropanoid biosynthesis (ko00940) were the most significantly entries in down-regulated DEGs, while plant hormone signal transduction (ko04075) had the most entries in up-regulated DEGs. At 15 DAP, genes involved in the biosynthesis of secondary metabolites (ko01110) were most enriched in down-regulated DEGs, while genes involved in phenylpropanoid biosynthesis (ko00940) were enriched in up-regulated DEGs. Through WGCNA analysis, the modules of 'MElightpink4' and 'MElavenderblush1' were highly correlated with watermelon fruit expansion under low-light stress. The unigenes in the 'MElightpink4' module were mainly enriched in ubiquitin-dependent protein aromatic amino acid family biosynthetic process (GO:0009073), proteolysis (GO:0006508), isopentenyl diphosphate biosynthetic process (GO:0019288), photorespiration (GO:0009853), ubiquitin-dependent protein catabolic process (GO:0006511) and glycolytic process (GO: 0006096), while the highly enriched terms of the 'MElavenderblush1' were associated with translation (GO:0006412) and embryo sac egg cell differentiation (GO:0009560). The results were almost consistent with earlier hierarchical clustering analysis and physiological index analysis. Of differentially expressed TFs, bHLH, ERF, NAC and WRKY are all the most abundant at 0, 3, 9 and 15 DAP, which means that these four TFs were important for watermelon fruit expansion under low-light stress. Members from the same TF family may play different roles in fruit development under low-light stress, and the transcriptional network for watermelon fruit expansion under low-light stress is intricate and complex. Therefore, our study elucidated the key genes and pathways related to watermelon fruit expansion under low-light stress, which provided a theoretical basis and foundation for further research on fruit expansion under low-light stress.

Supplementary Materials: The following supporting information can be downloaded at: https://www.mdpi.com/article/10.3390/plants12040935/s1, Table S1: The integrated function of 22379 annotated genes. Table S2: The genes and annotation in module 'lightpink4'. Table S3: The genes and annotation in module 'lavenderblush1'. Table S4: Top 20 enriched in biological process by GO enrichment analyses of DEGs. Table S5: Top 20 enriched in cellular processes by GO enrichment analyses of DEGs. Table S6: Top 20 enriched in molecular functions by GO enrichment analyses of DEGs. Table S7: Summary of the RNA-Seq data. Figure S1: Sequencing saturation curves (a) and sequencing gene coverage analysis (b) of the 24 RNA-Seq samples. Figure S2: Weighted correlation network analysis of related genes response to lowlight stress of watermelon fruit expansion. (a) Heriarchical clustering tree showing co-expression modules. Each leaf in the tree is one gene. The major tree branches constitute 10 modules labeled by different colors. (b) Module-trait relationship. The left lane indicates 10 module eigengenes. The right lane indicates the module-trait correlation from −1 to 1. FVD: fruit vertical diameter; FTD: fruit trans-verse diameter; SFW: single fruit weight; SS: soluble sugar; AA: amino acid. Figure S3: Functional classifications of DEGs in watermelon fruit flesh using WEGO. The *X*-axis shows the GO functional categories of cellular components, molecular functions and biological processes. The left *Y*-axis shows the percentage of each category; the right *Y*-axis shows the number of DEGs in each category. Figure S4: TF families of the DEGs at 0 DAP (a), 3 DAP (b), 9 DAP (c), and 15 DAP (d) of watermelon fruit expansion under low light stress. A total of 97, 277, 143 and 419 TFs were differentially expressed at 0 DAP, 3 DAP, 9 DAP and 15 DAP after low light stress, respectively, and the percentages of TF families in DEGs and all genes in watermelon genome were shown as blue bars and red bars ($P < 0.01$ and FDR < 0.05), respectively. Figure S5: The raw FPKM values of TF DEGs were first normalized by logarithmic method and a heat map was constructed to show the different expression profiles by the Heml software (version 1.0, http://hemi.biocuckoo.org/. accessed on 9 July 2020). *X*-axis, samples; *Y*-axis, differentially expressed gene names. (a) BHLH, (b) ERF, (c) NAC, (d) WRKY.

Author Contributions: Conceptualization, W.G. and X.W.; methodology, W.G.; software, W.G.; validation, W.G., F.S. and Y.S.; formal analysis, W.G.; investigation, W.G., F.S., B.H. and Y.S.; resources, W.G.; data curation, W.G. and X.W.; writing—original draft preparation, W.G.; writing—review and editing, X.W.; visualization, W.G. and X.W.; supervision, G.X.; project administration, W.G. and G.X.; funding acquisition, W.G. All authors have read and agreed to the published version of the manuscript.

Funding: This research was funded by "Jiangsu Agricultural Science And Technology Innovation Fund, CX(20)3106", and "National Natural Science Foundation of China, 31601795".

Data Availability Statement: The raw sequence data (in FastQ format) have been submitted to the NCBI SRA database (https://www.ncbi.nlm.nih.gov/sra/?term=. accessed on 19 January 2020) with the study accession (SRP243255). The plant materials are available from the corresponding author.

Acknowledgments: We also thank Genepioneer Biotechnologies (Nanjing, China) for the assistance in analysis of RNA-seq date.

Conflicts of Interest: The authors declare no conflict of interest. The funders had no role in the design of the study; in the collection, analyses, or interpretation of data; in the writing of the manuscript; or in the decision to publish the results.

References

1. Musseau, C.; Just, D.; Jorly, J.; Gevaudant, F.; Moing, A.; Chevalier, C.; Lemaire-Chamley, M.; Rothan, C.; Fernandez, L. Identification of two new mechanisms that regulate fruit growth by cell expansion in tomato. *Front. Plant Sci.* **2017**, *8*, 988. [CrossRef] [PubMed]
2. Guo, S.G.; Sun, H.H.; Zhang, H.Y.; Liu, J.A.; Ren, Y.; Gong, G.Y.; Jiao, C.; Zheng, Y.; Yang, W.C.; Fei, Z.J.; et al. Comparative transcriptome analysis of cultivated and wild watermelon during fruit development. *PLoS ONE* **2015**, *10*, e0130267. [CrossRef] [PubMed]
3. Li, Y.F.; Fan, Y.; Ma, Y.; Zhang, Z.; Yue, H.B.; Wang, L.J.; Li, J.; Jiao, Y. Effects of Exogenous gamma-Aminobutyric Acid (GABA) on photosynthesis and antioxidant system in pepper (*Capsicum annuum* L.) seedlings under low light stress. *J. Plant Growth Regul.* **2017**, *36*, 436–449. [CrossRef]
4. Ilic, Z.S.; Milenkovic, L.; Stanojevic, L.; Cvetkovic, D.; Fallik, E. Effects of the modification of light intensity by color shade nets on yield and quality of tomato fruits. *Sci. Hortic.* **2012**, *139*, 90–95. [CrossRef]
5. Tinyane, P.P.; Soundy, P.; Sivakumar, D. Growing 'Hass' avocado fruit under different coloured shade netting improves the marketable yield and affects fruit ripening. *Sci. Hortic.* **2018**, *230*, 43–49. [CrossRef]
6. Choi, H.G.; Moon, B.Y.; Kang, N.J.; Kwon, J.K.; Bekhzod, K.; Park, K.S.; Lee, S.Y. Yield loss and quality degradation of strawberry fruits cultivated under the deficient insolation conditions by shading. *Hortic. Environ. Biotechnol.* **2014**, *55*, 263–270. [CrossRef]
7. Dash, M.; Johnson, L.K.; Malladi, A. Severe shading reduces early fruit growth in apple by decreasing cell production and expansion. *J. Am. Soc. Hortic. Sci.* **2012**, *137*, 275–282. [CrossRef]
8. Wang, X.Q.; Huang, W.D.; Zhan, J.C. Effects of low light on phloem ultrastructure and subcellular localization of sucrose synthase in *Prunus persica* L. var. nectarina Ait. fruit. *Russ. J. Plant Physiol.* **2009**, *56*, 462–469. [CrossRef]
9. Yang, L.Y.; Chen, J.J.; Sun, X.M.; Li, J.X.; Chen, N.L. Inhibition of sucrose and galactosyl-sucrose oligosaccharide metabolism in leaves and fruits of melon (*Cucumis melo* L.) under low light stress. *Sci. Hortic.* **2019**, *244*, 343–351. [CrossRef]
10. Lu, Y.F.; Bu, Y.F.; Hao, S.X.; Wang, Y.R.; Zhang, J.; Tian, J.; Yao, Y.C. MYBs affect the variation in the ratio of anthocyanin and flavanol in fruit peel and flesh in response to shade. *J. Photochem. Photobiol. B Biol.* **2017**, *168*, 40–49. [CrossRef]
11. Shanely, R.A.; Nieman, D.C.; Perkins-Veazie, P.; Henson, D.A.; Meaney, M.P.; Knab, A.M.; Cialdell-Kam, L. Comparison of watermelon and carbohydrate beverage on exercise-induced alterations in systemic inflammation, immune dysfunction, and plasma antioxidant capacity. *Nutrients* **2016**, *8*, 518. [CrossRef] [PubMed]
12. Sun, Y.P.; Zhang, Z.P.; Wang, L.J. Promotion of 5-aminolevulinic acid treatment on leaf photosynthesis is related with increase of antioxidant enzyme activity in watermelon seedlings grown under shade condition. *Photosynthetica* **2009**, *47*, 347–354. [CrossRef]
13. Watanabe, S.; Nakano, Y.; Okano, K. Effect of planting density on fruit size, light-interception and photosynthetic activity of vertically trained watermelon (*Citrullus lanatus* (Thunb.) Matsum. et Nakai) plants. *J. Jpn. Soc. Hortic. Sci.* **2003**, *72*, 497–503. [CrossRef]

14. Guo, S.G.; Liu, J.A.; Zheng, Y.; Huang, M.Y.; Zhang, H.Y.; Gong, G.Y.; He, H.J.; Ren, Y.; Zhong, S.L.; Fei, Z.J.; et al. Characterization of transcriptome dynamics during watermelon fruit development: Sequencing, assembly, annotation and gene expression profiles. *BMC Genom.* **2011**, *12*, 454. [CrossRef] [PubMed]
15. Ruan, Y.L. Sucrose metabolism: Gateway to diverse carbon use and sugar signaling. *Annu. Rev. Plant Biol.* **2014**, *65*, 33–67. [CrossRef]
16. Beffa, R.S.; Neuhaus, J.M.; Meins, F. Physiological compensation in antisense transformants- Specific induction of an eresatz Gluan Endo-1,3-Beta-Glucosidase in plants infected with necrotizing viruses. *Proc. Natl. Acad. Sci. USA* **1993**, *90*, 8792–8796. [CrossRef]
17. Wang, Y.; Ren, S.; Li, X.L.; Luo, X.; Deng, Q.X. Shading inhibits sugar accumulation in leaf and fruit of jujube (*Ziziphus jujuba* Mill.). *Horticulturae* **2022**, *8*, 592. [CrossRef]
18. Wang, J.; Shi, K.; Lu, W.P.; Lu, D.L. Post-silking shading stress affects leaf nitrogen metabolism of spring maize in Southern China. *Plants* **2020**, *9*, 210. [CrossRef]
19. Long, X.Y.; He, B.; Wang, C.; Fang, Y.J.; Qi, J.Y.; Tang, C.R. Molecular identification and characterization of the pyruvate decarboxylase gene family associated with latex regeneration and stress response in rubber tree. *Plant Physiol. Biochem.* **2015**, *87*, 35–44. [CrossRef]
20. Duan, E.; Wang, Y.; Liu, L.; Zhu, J.; Zhong, M.; Zhang, H.; Li, S.; Ding, B.; Zhang, X.; Guo, X.; et al. Pyrophosphate: Fructose-6-phosphate 1-phosphotransferase (PFP) regulates carbon metabolism during grain filling in rice. *Plant Cell Rep.* **2016**, *35*, 1321–1331. [CrossRef]
21. Yi, S.Y.; Ku, S.S.; Sim, H.J.; Kim, S.K.; Park, J.H.; Lyu, J.I.; So, E.J.; Choi, S.Y.; Kim, J.; Ahn, M.S.; et al. An alcohol dehydrogenase gene from *Synechocystis* sp Confers salt tolerance in transgenic tobacco. *Front. Plant Sci.* **2017**, *8*, 1965. [CrossRef] [PubMed]
22. Strommer, J. The plant ADH gene family. *Plant J.* **2011**, *66*, 128–142. [CrossRef] [PubMed]
23. Chiba, Y.; Shimizu, T.; Miyakawa, S.; Kanno, Y.; Koshiba, T.; Kamiya, Y.; Seo, M. Identification of *Arabidopsis thaliana* NRT1/PTR FAMILY (NPF) proteins capable of transporting plant hormones. *J. Plant Res.* **2015**, *128*, 679–686. [CrossRef] [PubMed]
24. Avila-Ospina, L.; Marmagne, A.; Talbotec, J.; Krupinska, K.; Masclaux-Daubresse, C. The identification of new cytosolic glutamine synthetase and asparagine synthetase genes in barley (*Hordeum vulgare* L.), and their expression during leaf senescence. *J. Exp. Bot.* **2015**, *66*, 2013–2026. [CrossRef] [PubMed]
25. Bacete, L.; Melida, H.; Lopez, G.; Dabos, P.; Tremousaygue, D.; Denance, N.; Miedes, E.; Bulone, V.; Goffner, D.; Molina, A. *Arabidopsis* response regulator 6 (ARR6) modulates plant cell-wall composition and disease resistance. *Mol. Plant-Microbe Interact.* **2020**, *33*, 767–780. [CrossRef]
26. Wong, M.M.; Bhaskara, G.B.; Wen, T.N.; Lin, W.D.; Nguyen, T.T.; Chong, G.L.; Verslues, P.E. Phosphoproteomics of Arabidopsis Highly ABA-Induced1 identifies AT-Hook-Like10 phosphorylation required for stress growth regulation. *Proc. Natl. Acad. Sci. USA* **2019**, *116*, 2354–2363. [CrossRef]
27. Zhao, L.J.; Lu, Y.J.; Chen, W.; Yao, J.B.; Li, Y.; Li, Q.L.; Pan, J.W.; Fang, S.T.; Sun, J.; Zhang, Y.S. Genome-wide identification and analyses of the AHL gene family in cotton (*Gossypium*). *BMC Genom.* **2020**, *21*, 69. [CrossRef]
28. Denance, N.; Szurek, B.; Noel, L.D. Emerging functions of nodulin-like proteins in non-nodulating plant species. *Plant Cell Physiol.* **2014**, *55*, 469–474. [CrossRef]
29. Gong, S.Y.; Huang, G.Q.; Sun, X.; Li, P.; Zhao, L.L.; Zhang, D.J.; Li, X.B. GhAGP31, a cotton non-classical arabinogalactan protein, is involved in response to cold stress during early seedling development. *Plant Biol.* **2012**, *14*, 447–457. [CrossRef]
30. Bao, F.; Du, D.L.; An, Y.; Yang, W.R.; Wang, J.; Cheng, T.R.; Zhang, Q.X. Overexpression of *Prunus mume* dehydrin genes in tobacco enhances tolerance to cold and drought. *Front. Plant Sci.* **2017**, *8*, 151. [CrossRef]
31. Lu, J.; Sun, M.-H.; Ma, Q.-J.; Kang, H.; Liu, Y.-J.; Hao, Y.-J.; You, C.-X. MdSWEET17, a sugar transporter in apple, enhances drought tolerance in tomato. *J. Integr. Agric.* **2019**, *18*, 2041–2051. [CrossRef]
32. Liang, W.W.; Sun, F.F.; Zhao, Y.N.; Shan, L.Z.; Lou, H.Y. Identification of susceptibility modules and genes for cardiovascular disease in diabetic patients using WGCNA analysis. *J. Diabetes Res.* **2020**, *2020*, 4178639. [CrossRef] [PubMed]
33. Voxeur, A.; Hofte, H. Cell wall integrity signaling in plants: "To grow or not to grow that's the question". *Glycobiology* **2016**, *26*, 950–960. [CrossRef] [PubMed]
34. Adams, Z.P.; Ehlting, J.; Edwards, R. The regulatory role of shikimate in plant phenylalanine metabolism. *J. Theor. Biol.* **2019**, *462*, 158–170. [CrossRef]
35. Vogt, T. Phenylpropanoid biosynthesis. *Mol. Plant* **2010**, *3*, 2–20. [CrossRef]
36. Wechter, W.P.; Levi, A.; Harris, K.R.; Davis, A.R.; Fei, Z.J.; Katzir, N.; Giovannoni, J.J.; Salman-Minkov, A.; Hernandez, A.; Thimmapuram, J.; et al. Gene expression in developing watermelon fruit. *BMC Genom.* **2008**, *9*, 275. [CrossRef]
37. Wessels, B.; Seyfferth, C.; Escamez, S.; Vain, T.; Antos, K.; Vahala, J.; Delhomme, N.; Kangasjarvi, J.; Eder, M.; Felten, J.; et al. An AP2/ERF transcription factor ERF139 coordinates xylem cell expansion and secondary cell wall deposition. *New Phytol.* **2019**, *224*, 1585–1599. [CrossRef]
38. Zhao, M.R.; Li, J.; Zhu, L.; Chang, P.; Li, L.L.; Zhang, L.Y. Identification and characterization of MYB-bHLH-WD40 regulatory complex members controlling anthocyanidin biosynthesis in blueberry fruits development. *Genes* **2019**, *10*, 496. [CrossRef]
39. Mao, K.; Dong, Q.L.; Li, C.; Liu, C.H.; Ma, F.W. Genome wide identification and characterization of apple bHLH transcription factors and expression analysis in response to drought and salt stress. *Front. Plant Sci.* **2017**, *8*, 480. [CrossRef]

40. Wu, P.P.; Peng, M.S.; Li, Z.G.; Yuan, N.; Hu, Q.; Foster, C.E.; Saski, C.; Wu, G.H.; Sun, D.F.; Luo, H. DRMY1, a Myb-like protein, regulates cell expansion and seed production in *Arabidopsis thaliana*. *Plant Cell Physiol.* **2019**, *60*, 285–302. [CrossRef]
41. Xu, Q.; Wang, W.Q.; Zeng, J.K.; Zhang, J.; Grierson, D.; Li, X.; Yin, X.R.; Chen, K.S. A NAC transcription factor, EjNAC1, affects lignification of loquat fruit by regulating lignin. *Postharvest Biol. Technol.* **2015**, *102*, 25–31. [CrossRef]
42. Ayadi, M.; Mzid, R.; Ben Ayed, R.; Kharrat, N.; Merchaoui, H.; Rebai, A.; Hanana, M. Genome-wide identification and molecular characterization of *Citrus unshiu* WRKY transcription factors in *Satsuma mandarin*: Clues for putative involvement in cell growth, fruit ripening, and stress response. *Turk. J. Agric. For.* **2019**, *43*, 209–231. [CrossRef]
43. Livak, K.J.; Schmittgen, T.D. Analysis of relative gene expression data using real-time quantitative PCR and the 2(T)(-Delta Delta C) method. *Methods* **2001**, *25*, 402–408. [CrossRef] [PubMed]

MDPI AG
Grosspeteranlage 5
4052 Basel
Switzerland
Tel.: +41 61 683 77 34

Plants Editorial Office
E-mail: plants@mdpi.com
www.mdpi.com/journal/plants

www.ingramcontent.com/pod-product-compliance
Lightning Source LLC
Chambersburg PA
CBHW041412210926
51850CB00013B/3246
9783725830442